Marinas: a working guide to their development and design

Second edition

First published in Great Britain by The Architectural Press Ltd: London 1975
Second edition 1977

ISBN: 0 85139 381 0 (British edition)

First published in the United States by Cahners Books/ A Division of Cahners Publishing
Company Inc: Boston, Mass. 1975
Second edition published in the United States by Nichols Publishing Company: New York
1977

Library of Congress Cataloging in Publication Data

Adie, Donald W
  Marinas: a working guide to their development and design.
  Bibliography: p.
  Includes index.
  1. Marinas. I. Title.
VK369.A34   1977   627'.38   77-136
ISBN 0-89397-018-2 (American edition)

Filmset and printed in Great Britain by BAS Printers Limited, Over Wallop, Hampshire

MARINAS

a working guide to their development and design

Second edition

Donald W. Adie
Dip Arch, RIBA, Dip TP, MRTPI

THE ARCHITECTURAL PRESS LTD: LONDON

NICHOLS PUBLISHING COMPANY: NEW YORK

Acknowledgements

Contents

This book is based on information collected during a travelling fellowship to the United States and the Bahamas awarded by the Winston Churchill Memorial Trust. I am most grateful to the Trust and my employer The Greater London Council for their generosity and encouragement. I am also indebted to many people and organizations throughout the world whose experience and ideas have contributed so much and whose kindness made my travels so pleasant. I would also like to thank my publishers and in particular, Alexandra Artley for her encouragement and expertise, and Cluny Gillies for such a painstaking layout and for dealing with scores of drawings, tables and charts.

I am particularly grateful to the following individuals and organisations: Mr Ian Bailey and Mr Chester Smith of the Bailey, Piper Stevens Partnership; Mr D. P. Bertlin of Bertlin and Partners; Mr Peter Bloomfield; Mr G. E. Broadhead of the Hydraulics Division of Wimpey Laboratories; Mr P. A. Bond of Raglan Squire and Partners; Mr A. S. Cork of the National Yacht Harbours Association; Mrs Daphen Dunster (née Collins); Mr William Farnham of Portsmouth Polytechnic; The French Government Tourist Office; Mr B. S. Folley of Harleyford Marina (Thames) Ltd.; Mrs Linda Irvine of the Natural History Department of the British Museum; Mr R. H. Mims of Shell-Mex and B.P. Ltd.; Mr J. A. Maudsley, City Architect of Birmingham; The staff of the Planning, Research and Education Libraries of the Greater London Council; Miss Nancy Smith, Mr I. W. Strickland of British Hovercraft International, Mr H. Pokorny of Resabo Inter GmbH, Hamburg, Mr A. J. White of the Countryside Commission, Mr Peter Mackintosh of the I.L.E.A., Mr Peter Drew of the Taylor Woodrow Property Company, Professor H. Lundgren of the Technical University of Denmark, Mr Christopher Scarlett of Wallace Evans and Partners, Mr and Mrs Earl Smith, Sanford, Florida, the staff of the R.I.B.A. Library, Mr N. B. Webber and Mr R. J. S. Harris of Southampton University, Mrs M. Hadfield of Economic Associates Ltd., Mr D. H. Sessions of Marina Management Ltd., Mr J. Burrows of Ove Arup & Partners, Mr D. R. Morrison of Riverside, Florida; Westminster Dredging Group Ltd.

To Jo

Illustration credits

Unless stated otherwise below illustrations are by the author. Thanks are due to Tielman and Roselyne Nicolopoulos for their work in re-drawing the majority of diagrams throughout this book and to Greg Edwards who also contributed to the re-drawing.

Chapter 1. 1.A–G Lionel Mendoza of Donaldson & Sons; 1.1 British Travel Association and the University of Keele; 1.6 Gloucestershire County Council.

Chapter 2. 2.3–6, 14–15 Eileen Ramsay; 2.7 Thanet District Council; 2.8–12 Greater London Council; 2.13 Bailey Piper Stevens Partnership; 2.18 *The Daily Telegraph*.

Chapter 3. 3.3, 8–11, 13 French Government Tourist Office; 3.4 *The Daily Telegraph* Colour Library; 3.5 Sidney Kaye/Kaye Firmin & Partners; 3.6 Bernard Beaujard; 3.7 G. Petit/La Mission Interministérielle Languedoc-Roussillon; 3.12 John Watney/French Government Tourist Office; 3.17, 19 Greater London Council; 3.18 Frank Martin/*The Guardian*; 3.28 Chichester Yacht Marina; 3.29, 46–49 Bertlin & Partners, London; 3.30–31 Fairey Marine Ltd.; 3.32 Brighton Marina Co.; 3.33 Nationaal Foto Persbureau, Amsterdam; 3.37–39 Peter Ogden/World Trade Centre, St. Katharine Haven Ltd.; 3.40, 42, 44–45 *Yachting World*; 3.43 Bob Fisher.

Chapter 4. 4.1, 12 Harleyford Marina (Thames) Ltd.; 4.2–3 Greater London Council; 4.4–7 Birmingham City Architect's Department; 4.8–9, 14–15 The Architectural Press Ltd.; 4.13, 16–17 Resabo Inter GmbH, Hamburg; 4.B–C *The Architects' Journal*.

Chapter 5. 5.12, 18–19 Gordon Cullen/The Architectural Press Ltd.; 5.D–E, G–H D. P. Bertlin/Bertlin & Partners, London; 5.17, 41–42, 65, 67 Daphen Dunster; 5.25–26 Ivor de Wolfe/The Architectural

Press Ltd.; 5.28 Jack Howe/The Architectural Press Ltd.; 5.29 Sir James Richards/The Architectural Press Ltd.; 5.30, 64 The Architectural Press Ltd.; 5.32 Hugh de Burgh-Galwey/The Architectural Press Ltd.; 5.33 Ian McCallum/The Architectural Press Ltd.; 5.34–36 Kenneth Browne/The Architectural Press Ltd.; 5.39 Mono Concrete; 5.45–47, 59 Bailey Piper Stevens Partnership; 5.51–52 Kaye Firmin & Partners; 5.66 Eileen Ramsay/ Raglan Squire & Partners in association with Leslie Bewes A.R.I.N.A.; 5.F–G International Commission for Sport and Pleasure Navigation of the Permanent International Association of Navigation Congresses (PIANC); 5.J National Yacht Harbour Association, London; 5.71–72 Outboard Boating Club of America, Chicago; 5.84 Bercleve Uni Float, Florida; 5.85–87 Walter Bower & Co.; 5.88 Resabo Inter GmbH, Hamburg; 5.95 Bob Johnson.

Chapter 6. 6.3 I. W. Stickland; 6.4 Bailey Piper Stevens Partnership; 6.5 Charles Chaney/National Association of Engine and Boat Manufacturers Inc., New York; 6.8–13 Baltimore Dock and Harbour Board; 6.15–21, 23–25 Wimpey Laboratories Ltd.; 6.22, 57 River and Sea Gabions (London) Ltd.; 6.28 John Harland/ Taylor Woodrow; 6.29–31, 41–42 Professor H. Lundgren/Danish Hydraulic Institute; 6.33 Archibald Shaw & Partners; 6.34 Sea Services Agency; 6.35–39 N. B. Webber and R. J. S. Harris; 6.40 Beken of Cowes, Marine Photographers; 6.45–46 U.S. Navy; 6.47, 59–61 The City Commission of Sanford, Florida; 6.50–51 L. R. Wootton, M. H. Warner, R. N. Sainsbury, D. H. Cooper for the Construction Industry Research and Information Association (CIRIA); 6.52 reproduced by gracious permission of H.M. The Queen; 6.58, 64, 68 S.P.A. Officine Maccaferri, Milan; 6.65 Ivor de Wolfe/The Architectural Press Ltd.; 6.71 Rendel Palmer & Tritton for the Greater London

Council; 6.73–74, 76–78 Henry Cornick/ Charles Griffin & Co. Ltd.; 6.75 Chichester Photographic Service Ltd.

Chapter 7. 7.7 The Architectural Press Ltd.; 7.14 Linda Irvine/The British Museum.

Chapter 8. 8.2–3 *The Architects' Journal*; 8.14–15 National Fire Protection Association, Boston, Massachusetts; 8.D The Architectural Press Ltd.

Chapter 9. 9.2 Bob Johnson; 9.17 Rotork (Controls) Ltd., London; 9.11, 18 Outboard Boating Club of America, Chicago; 9.21 Sidney Kaye/Kaye Firmin & Partners; 9.25 Daphen Dunster.

Chapter 10. 10.A, 4 Shell International Petroleum Co. Ltd.; 10.5 Greater London Council.

Chapter 11. 11.1 Daphen Dunster; 11.7 RFD Ltd., London; 11.8–9 Ambler Engineering, Walsall; 11.10–11 Her Majesty's Stationery Office.

Chapter 12. 12.A–B, D–H National Association of Engine and Boat Manufacturers Inc., New York; 12.C B. S. Folley; 12.3 *Boating Industry*. 12.8, 9 Westminster Dredging Group Ltd

Chapter 13. 13.5–6 Greater London Council; 13.7 Daphen Dunster.

Chapter 14. 14.1–3 Greater London Council.

1 The leisure background

The growing demand

The increase in the pursuit of leisure throughout the world since the Second World War has been dramatic and like most trends, this upsurge has drawn its impetus from the convergence of other strongly-marked factors of which the most significant are mobility, income, leisure time, education and population. These have all increased whilst the average number of working hours has dropped: 1.5, p. 16 takes these aspects and projects them forward to demonstrate their relative changes by the year 2000. Obviously there are dangers in such forecasting, however subtle the methods used, and the simple extension of present trends thirty years into the future can only be the crudest prophecy. However, economists in Britain agree that an annual growth rate as low as $2\frac{1}{2}$ per cent of the Gross National Product would result in the doubling of *real* incomes by the year 2000. Growth in science-based industry and increasing automation will also result in a significant increase in leisure time for the majority of people in fully-developed countries. This increase in time and affluence, coupled with the new desire for mobility with which to enjoy them, brings its own upsurge in the demand for that controversial piece of leisure hardware, the car, and the Road Research Laboratory estimates that the 12 million cars on the roads of Britain in 1970 will more than double by the year 2000.

Entry into the European Common Market is adding momentum to trends already apparent and it is not an exaggeration to say that this will mean a leisure revolution in Britain. Already thousands of skilled Britons are working in West Germany for more money and security, shorter hours and better holidays. With barriers removed British employers will gradually concede to Common Market standards and the consequential increase in both leisure and affluence is certain to have considerable repercussions on recreational services. It is essential to realise that a failure to plan now for recreational needs will not only create a chaotic situation in the leisure field itself but could do irreparable damage to the countryside and coast by the panic adoption of ill-considered facilities on unsuitable sites. The effects of these growth patterns on the landscape of the future has been put into context explicitly by Michael Dower,

"Three great waves have broken across the face of Britain since 1800. First, the sudden growth of dark industrial towns. Second, the thrusting movement along far-flung railways. Third, the spread of car-based suburbs. Now we see, under the guise of a modest word, the surge of a growth wave which could be more powerful than all the others. The modest word is leisure."[1]

1. Dower, Michael, 'The Fourth Wave: the Challenge of Leisure,' *The Architects' Journal*, 20 January 1965.

Comparative numbers of hours worked in manufacturing industries

Countries	1967	1974
Austria	43·6	39·5
Belgium	39·6	36·6
Cyprus	46·0	43·6
France	45·4	42·9
W. Germany	42·0	41·9
Greece	43·6	43·8
Eire	43·6	41·5
Italy *	7·9	39·7
Malta	46·2	N/A
Netherlands	45·3	43·0
Norway	37·8	33·6
Sweden†	161	140
	(1966)	
Switzerland	44·7	44·1
U. Kingdom	45·3	44·0
Spain	44·1	43·8
Portugal	45·3	N/A
	(1966)	
U.S.A	40·6	40·0

*hours per day   †hours per month

1.A   Over the last decade the hours in the European working week have fallen. At the same time, paid holidays have increased and most countries have at least a two-week minimum, while many are moving towards a three- or four-week minimum as in France and Sweden

Numbers of private cars in Europe and the U.S.A.

Countries	millions in 1967	millions in 1970	millions in 1974
Austria	0·9	1·2	1·6
Belgium	1·5	2·0	2·5
Denmark	0·8	1·0	1·3
France	11·5	12·4	15·0
W. Germany	10·7	14·3	16·9
Greece	0·1	0·2	0·4
Iceland	0·03	0·04	0·06
Eire	0·3	0·35	0·5
Italy	7·3	10·2	14·3
Malta	0·02	N/A	0·05
Netherlands	1·7	2·5	3·4
Norway	0·5	0·74	0·9
Sweden	1·9	2·2	2·6
Switzerland	1·0	1·38	1·7
U. Kingdom	10·6	11·5	13·8
Spain	1·3	N/A	4·3
Portugal	0·2 (1966)	0·45 (1969)	0·8
U.S.A.	80·0	88·8	104·2

1.B   An increasing number of European citizens enjoy the benefits of car ownership and have the ability to travel relatively quickly and conveniently away from home. Already surveys in Germany and Britain have shown that peak leisure traffic flows on summer Sundays can be greater than the weekday peaks for commuting and business traffic. The extension of motorway networks can only accelerate the effects of growing car ownership, although the congestion and frustration that results from the failure of

Total tourist receipts in foreign currency

$ millions

Countries	1968	1974
Austria	687	2,300
France	954	2,666
W. Germany	911	2,338
Greece	120	437
Eire	195	254
Italy	1,424	2,668
Netherlands	342	1,039
Scandinavia	436	1,511
Spain	1,213	3,209
Switzerland	592	1,793
U. Kingdom	678	1,956
U.S.A.	1,770	4,034

1.C Tourism has become very big business. Already it is the largest source of foreign exchange in Spain, Italy and Greece, and even in those countries where it is not the principal export it figures highly in maintaining the balance of payments. The above table shows a comparison in tourist incomes since 1968.

road building programmes to keep pace with car numbers, may bring a temporary slowing down of the growth of leisure trips. It is largely in the industrialised countries of North-West Europe that the car is most widely owned and used; and it is in these countries that rapid future growth is expected. It is predicted that France will have 25 million cars by 1985, while Britain will have trebled its present number by the end of the century

Increase in incomes and tourist expenditure 1971–74

Countries	% increase in personal incomes	% increase in tourist expenditure in foreign currency
Austria	+95	+95
France	+52	+60
W. Germany	+63	+55
Greece	+62	+48
Eire	+33	+25
Italy	+45	+45
Netherlands	+75	+85
Scandinavia	+50	+65
Switzerland	+60 (71–73)	+80
U. Kingdom	+40	+70
U.S.A.	+30	+60

1.D Tourism has brought a new source of direct and indirect income. Summer employment now exists in hotels and catering, and in recreation areas where there is a need for wardens and boatmen. There are, however, financial problems involved in the provision of leisure facilities, not least of which is the heavy investment, which has to be made speculatively, to establish the infrastructure for marinas

Second homes in Scandinavia 1974–1990

Countries	Second homes 1974	Estimated No. 1990	Annual increase
Denmark	180,000	500,000 (1 in 5 households)	20,000
Norway	210,000	500,000 (1 in 3 households)	18,000
Sweden	540,000	800,000 (1 in 5 households)	15,500

1.E The increase of second home ownership is a particular feature of the last decade. Although not yet extensive in Britain (somewhat less than 1 per cent of the population), in some European areas, notably Scandinavia, this is already a special tourist feature. The coast and lake shores are the most popular locations

Pattern of tourist movement 1974

Country of arrival	Total arrivals of tourists in millions	Receipts $ millions
Austria	10·8	2,300
Belgium	7·4	719
France	9·8	2,666
W. Germany	6·9	2,338
Greece	1·9	437
Italy	12·4	2,668
Netherlands	2·6	1,039
Scandinavia	15·1	1,511
Spain	30·3	3,209
Switzerland	6·2	1,793
Turkey	1·1	194
U. Kingdom	7·9	1,956
Portugal	2·6	513
U.S.A.	14·1	4,034

1.F Pattern of tourist movement 1974, in millions

Population characteristics

Countries	Population (in millions)		Density (persons/km^2)		% of population described as 'urban'
	1968	1974	1968	1974	1974
Austria	7·3	7·5	87	90	50
Belgium	9·6	9·7	314	320	66
Cyprus	0·6	0·6	67	69	*
Denmark	4·8	5·0	112	117	77
France	50·3	52·5	91	96	63
W. Germany	58·0	60·6	233	250	79
Greece	8·8	8·9	66	68	43
Iceland	0·2	0·2	2	2	*
Eire	2·9	3·0	41	44	49
Italy	52·7	53·7	174	184	48
Luxembourg	0·3	0·3	130	137	*
Malta	0·3	0·3	1,008	1,024	*
Netherlands	12·7	13·0	375	332	80
Norway	3·8	3·9	12	12	32
Sweden	7·9	8·1	17	18	72
Switzerland	6·1	6·4	147	156	42
Turkey	33·5	35·6	43	49	*
U. Kingdom	55·2	55·9	226	229	80
Spain	32·1	33·9	64	70	43
Portugal	9·4	8·7	103	95	23
U.S.A.	202·0	211·9	22	23	70

1.G For most countries the reasons for the increased tempo in leisure activity are not hard to find. National populations are increasing and becoming progressively more urban. There is more personal wealth, mobility and leisure time, and a greater awareness of the new opportunities for leisure brought about by more widespread education

* Figures not available

WOULD LIKE TO DO   HAVE DONE IN THE PAST

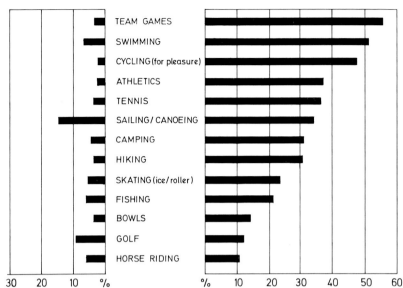

1.1 League table of leisure pursuits

1.1 There are no precise and reliable forecasts of the future pattern of leisure activity in Europe. At best, there are informed national estimates of the scale of growth envisaged but there is little information on the way in which participation in particular activities may be expected to grow. However, there is every indication that, for outdoor recreation, the north-west European countries are following, albeit some years behind, the trends already seen in America. At national level, at least a threefold increase in activity is expected by the end of the century. It seems likely that activities associated with motoring and water-based pursuits will continue to be amongst the most popular

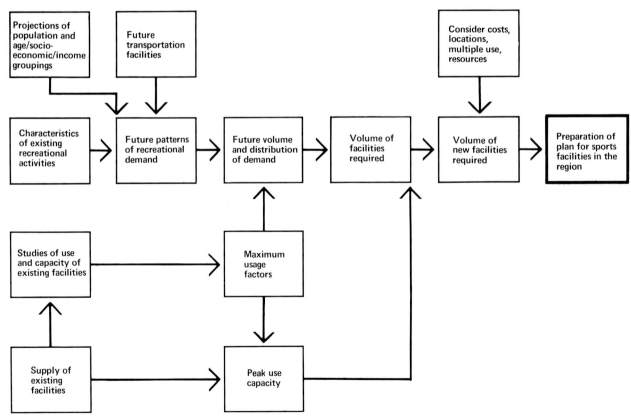

1.2 This chart presents a logical approach in assessing the requirements for sports and recreational facilities. Although the programme was designed for a region's total recreation requirements, it could be used, almost without alteration as a model for most marina projects, as a valuable part of the site selection process and a guide towards establishing a project's overall viability

The changing pattern of leisure

Within this overall trend, the demand for water-based activities has been positively explosive. As is so often the case the European experience followed a similar American pattern in which pleasure boating grew to the extent that the number of leisure craft in the U.S.A. showed a heavy increase between 1960 and 1970. By 1965 nearly 34 million Americans, one in six of the population, actively engaged in boating and in doing so spent dramatically each year. The growth in boat ownership in European countries during the same period was less impressive with Germany showing the greatest increase, France second and the U.K. third.

The Pilot National Recreation Survey taken from a total sample of over 3,000 people in Britain puts into perspective the relative popularity of sailing compared with other sports.[2] Having discovered those recreations pursued in the past it established those which people would most like to do in the future. From this table of active pastimes and their likely future demand it is certain that one of the strongest aspects of increasing leisure will manifest itself in sailing and canoeing 1.1. It also seems clear that because these demands stem directly from the more basic changes shown in 1.5 they do not imply some passing fashion but a deeply-rooted characteristic with a strong and increasing momentum. An interesting trend reflected in 1.1 is the desire for people to express their individuality in their choice of leisure. It is probable that four of the top five sports in the table had been available, if not compulsory, at school. It is even more probable that the bottom eight pastimes had not. The second column shows a relatively small number of people who wish to continue the pursuits of their schooldays and much larger numbers who wish to try, or to continue with activities which are largely individual, non-school and non-team.

From this changing pattern it seems clear that all forms of boating are going to become much more popular and occupy an increasingly high position in the league table of leisure pursuits. With this trend occurring within the overall increase in total leisure time and proportional expenditure, a situation is created whereby a planned programme is essential from the national to the local level. An awareness of environmental pressures and the need to conserve the available resources, and yet increase and improve the present facilities for boating of all kinds, can only be met by responsible and competent planning in both the public and private sectors.

The problem, which becomes more serious every year in all high-density countries is that as the popularity of water sports grows the space for its enjoyment effectively shrinks. The best of the accessible sailing water is already at full capacity during peak times and those inferior and less accessible areas become yearly more crowded. If the number of people taking to the water continues to increase at the present rate without a thorough and co-ordinated national plan there will be an inevitable deterioration in the boating environment and, in consequence, of the quality of recreation as a whole.

2. Rogers, H. B. *Pilot National Recreational Survey: Report No. 1,* British Travel Association/University of Keele, July 1967

Demands and resources

Except for occasional man-made additions the total stock of water available for leisure purposes is constant all over the world and it should be regarded as a natural gift needing to be preserved or allocated with great care and skill. Since 1850 more than half the English population has lived in towns and a new emotional need for recreation has grown as urbanisation and industrialisation increased. Now that more than 80 per cent of the population lives in urban areas, the demand for leisure activities in and around the cities has grown tremendously. The average man now has the chance of long weekends plus two major holidays a year and whilst the growth of European package-tour holidays has been phenomenal during the period 1965–71, there has also been a tremendous growth in second holidays usually spent in some part of one's homeland. In his search for recreation during such holidays the urban dweller has an inborn need to refresh his memory of those elements basic to life and here water has an instinctive attraction which could be seen as an opportunity for a more sophisticated recreational use of water resources both coastal and inland.

Given this situation where demand is rapidly rising and supply remains virtually static, the duty of all those concerned with the problem is to so plan the water resources that a balance is maintained between conservation on the one hand and satisfaction of a growing need on the other. This can only be achieved by formulating a positive policy and carrying it into effect. Such a plan must commence with an examination of the amount, location and type of water and its accessibility. The British Countryside Commission recognises two basic types of coastal recreation area—the 'natural' and 'intensive'.[3] The 'natural' reflects the need to preserve the natural environment and suggests that recreational facilities should only be provided where this can be done without serious damage to the present character of the area. The 'intensive' refers to areas where investment will be required for recreation aids of all kinds and also for access roads and car parking, the aim being to absorb the maximum amount of recreational use within a limited area with an emphasis on man-made facilities rather than the natural character of the coast. The development of such areas is most needed on those parts of the coast which have to accommodate heavy local and regional demands from the major centres of population, but they are also required in some of the main holiday areas. There are seven categories of water available for recreational use:

1 The sea
2 Rivers
3 Lakes
4 Harbours
5 Reservoirs
6 Canals
7 Water areas left after mineral extraction

Each contributes a different situation to the total water area and has its own constraints and opportunities. The principal task of authorities, developers and land owners is to ensure both the expansion of the supply of suitable water areas and the better use of existing water resources.

Having selected the 'intensive' areas the nature of the facilities to be provided must be considered.

The Countryside Commission recommends that because of the wide range of recreational activities that are often focused on quite limited coastal areas these should, wherever possible, be designed to allow for multiple use. The development of marinas and club-house buildings might offer the opportunity to provide facilities suitable for use in a number of activities.

3. The Countryside Commission, *The Planning of the Coastline*, H.M.S.O., London 1970.

It will be emphasised in later chapters that no marina, yacht harbour or even yacht club can survive economically by relying solely upon revenue from berth rental. Among other activities the encouragement of sports additional to sailing and cruising seems a natural step financially, as well as taking maximum advantage of the land and water resources of the area. 1.6 gives an indication of the compatibility of the more popular water-based activities. It would, however, be unwise to encourage clubs of conflicting interests within the same marina except where separated by the strict delineation of the water, and sometimes land, area or by some apportionment of time, during which the water is exclusively theirs.

As far as boating is concerned there is an important difference which should be noted between coastal and inland waters. Whereas in inland areas congestion can occur on the water itself the extent of coastal waters is in a sense unlimited and congestion occurs only at the access points.

Fixing the objectives

4. Greater London and South East Sports Council, *Sports Facilities, Initial Appraisal— Vol. 2 Water Recreation*, May 1969.

The Greater London and South East Sports Council speaks for many authorities in recommending the following action in promoting the best development of facilities.[4]

1   Encouraging the development of new water areas and their use for recreation
2   Supporting all action designed to control pollution
3   Increasing the recreational use of existing water areas by improving and adding to the slipways and access points to all forms of water
4   Extending the season
5   Providing additional moorings which can most easily be supplied by building marinas which would have the additional advantage of supplying other much-needed facilities, fuel points, changing rooms, boat chandlers and car parking

It is interesting to note that action under the last suggestion would help satisfy the needs outlined in all the preceding items. Examples of governmental encouragement in the development of marinas occur in publications of the Countryside Commission:

5. The Countryside Commission, *Coastal Recreation and Holidays, Special Study Report* Vol. I, H.M.S.O., 1970.

"The most critical problem is the shortage of storage for boats, especially moorings for keel boats. In the long run this can only be solved by the development of marinas in suitable locations."[5]

And from the same report comes this plea for public authorities to play a dominant role in the provision of facilities,

"Unlike private enterprise, public authorities are able to acquire and obtain public access to large areas of land where these are at present in multiple ownership. This may be particularly important where developments such as marinas involve the acquisition of parts of the foreshore and the adjacent land area."[6]

6. *Ibid.*

Advantages and criticisms

Apart from being the favoured method of overcoming environmental problems of conservation and pollution, marinas are by far the most convenient places for individual owners to keep their craft. Boat owners who have used a marina even for only a short while rarely wish to return to piles, swinging moorings or any other form of off-shore anchorage. Access to most marinas is good, being either centrally placed within its catchment area or strategically located at the end of fast road or rail links. It may therefore be reached from the conurbation within two hours or so.

The cost of berths is high when compared directly with casual mooring. In 1968 only 6 per cent of boat owners used marinas mainly because berths in them cost, on average, ten times as much as moorings elsewhere. But to help offset this considerable difference there is both the convenience and the hidden savings that yacht harbours bring. There is less need to lay up out of the water for the winter—thereby extending the sailing season. There is also a possibility of overnight accommodation on shore or aboard and this with quayside electricity and locker space makes out-of-season self maintenance a more likely proposition. The owner who makes maximum use of his marina may find that, with good service, less likelihood of damage, lower insurance premiums, free parking and other facilities, it may not *overall* be costing him more than casual mooring or trailing his boat. He will at the same time enjoy the convenience of the clubhouse, chandlery, fuelling points, showers, slipways, and hoists that a good yacht harbour has to offer.

Despite these advantages the concept of the marina is not without its critics. Opposition comes from conservationists who fear loss or disturbance to wild life and the natural environment; from local residents concerned about the inevitable increase in activity and from boating enthusiasts themselves, some of whom sense an artificiality about the whole idea. Others, whilst accepting or even welcoming the principle, point out that marinas are now responsible for the largest single increase in boating costs. Misgivings are to be expected where an innovation as recent as the marina imposes itself upon an activity as traditional as sailing and in areas as sensitive as coasts and rivers. As a control upon the more extreme and misguided proposals such doubts should be welcomed because schemes are often put forward advocating developments quite unsuitable to their location, scale or form. Were this not so there would be no need for the guidance and information which this book attempts to set down.

Marinas: the American experience and the British potential

As the United States coined the word marina and was the first nation to develop it as a viable recreational concept, some simple statistical comparisons are in order here to help indicate from the American experience the social potential for marina development in Britain.

USA		UK	
	3½ million sq miles		93,000 sq miles
	200 million population		55 million population
	56 persons per sq mile		589 persons/per sq mile
	12,383 miles of coastline		4,910 miles of coastline
	3 in of coastline per person		6 in of coastline/per person

From these simple comparisons it can be seen that America is about forty times larger than the United Kingdom, has about four times the population and therefore accommodates its people on average at about one-tenth of the density. It has approximately 14 ft of coastline per sq mile of area compared with 95 ft per sq mile in Britain and 3 in of coastline per person against 6 in per head in Britain. No point in Britain is more than about 80 miles from either the sea or a major inlet of the sea. In America large areas of land are 850 miles from any coast, although they may be nearer to rivers or the Great Lakes and climatic conditions for recreational boating vary much more dramatically than in Britain. The northern extremities of both coasts suffer severe winters which lock many marinas in ice and prohibit sailing of any kind, whereas in southern Florida the temperature only very rarely drops below 10°C (50°F), but hurricanes and tropical storms are a danger during their season. Britain, because of its geography, its empire and its tradition, is a sea-faring nation: America is not, although a larger proportion of its population live on its seaboards than in Britain.

In 1965 33·9 million Americans over 12 years of age actively engaged in boating. This represents one in six of the population. Boating ranks ninth in popularity of American sports and pastimes, the list from 1 to 9 being:

1 Picnicking
2 Driving for pleasure
3 Sightseeing
4 Swimming
5 Walking for pleasure
6 Playing out-door games
7 Fishing
8 Attending sporting events
9 Boating, sailing and canoeing

As evidence of the undoubted future growth in popularity of all forms of boating in Britain, the Pilot National Recreation Survey states that if the sailing categories are grouped into two classes, salt and fresh water, then the former would be the first overall in the list of activities people would like to take up (11 per cent of the total sample) and the latter third (7 per cent).[7] Together the two types of sailing are a very clear 'first' among people's preferences. Though weighted towards the younger age groups (18 per cent of those between 17 and 24 years of age would like to sail) both salt water and inland variants persist strongly into the preferences of the 44 to 65 age band, where they are more likely to be gratified. It is emphatically the expensive forms of sailing (power boats and sail boats) that attract the greatest interest and this is expressed, realistically, most strongly by the highest of our three occupational ranges, but it is also at its strongest among the best-educated. Those now sailing or intending to do so are therefore

7. Rogers, H. B., *op cit.*

1.3 Holiday apartments are frequently provided within American marinas. These at Sanford, Florida, belong to Holiday Inns of America

youngish, well educated and well-off, mostly male, car owning, with a good job and enjoying increasing leisure, longer holidays and a shorter working week.

A feature that is common to most countries with regard to the future of leisure pursuits is the increasing popularity of activities which are family orientated. At least six of the first nine sports and pastimes in the American popularity table fall into this category. Such pastimes embarked upon when young, have the added advantage of being able to be continued after marriage and whilst bringing up a family, except perhaps for a few years when children are very young. The way in which most marinas in America are now being developed recognises and caters for this changing pattern of recreation, the increase in boating interest, the increasing demands of regular boating enthusiasts and the family proclivity of the sport. There is, therefore, an emphasis upon good access, generous car parking, adequate boat servicing and sales areas, together with facilities to extend the boating season at both ends.

Trends in boat ownership

This then is the consumer profile for an extension of sailing in the immediate future and one which accounts for a rapidly rising percentage of our total population. Though it is probably unnecessary to plan for a 10 per cent national participation, even a more modest demand will quickly clog existing facilities, once realised. We have seen that as prosperity grows the time available for leisure increases. Those who were previously content (or through financial restrictions were forced) to follow their leisure pursuits on shoestring budgets soon find that because of this, so much of free time and money is spent on the preparation for an activity, that little time is left for the actual enjoyment of it. The length of journey that enthusiasts are prepared to make is also increasing which again shortens the time spent actually participating. In boating, the answer to this problem usually manifests itself *in the rental of a permanent mooring rather than trailing one's boat* and by paying for regular servicing and maintenance rather than doing this oneself. However, the need for maintenance has decreased dramatically with the advent of fibreglass hulls and nylon and terylene sails. In America the demand for 'instant boating' has resulted in the evolution of the increasingly popular multi-stack boat store.

There is, however, always likely to be a gap between the sale of boats and the availability of moorings and this situation is not dissimilar to that

1.4 Fibreglass hulls awaiting completion at the Cobia boat factory in Florida, which was producing 5,000 boats a year within three years of its foundation

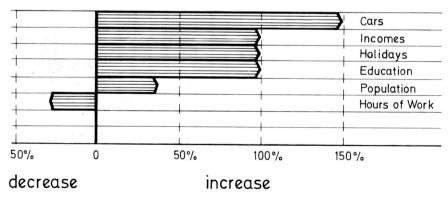

decrease　　　　　increase

1.5 Estimated changes in factors affecting recreation 1965–2000

of motor car manufacture and parking provision. Experience indicates that such gaps tend to widen rather than close as popularity grows, and the demand is not met because mass-production methods of creating the article cannot be matched in numbers or speed by the purpose-made nature of its refuge. Despite the remarkable American ability to fulfill quickly a demand in almost any section of the market, the need for moorings has lagged behind the increasing number of boats being sold and long waiting lists are common in the majority of marinas. The reason for this is not so much the dis-inclination by local authorities or private enterprise to enter the marina business, as the understandable time lag between manufacturing the boat and constructing its mooring occasioned by the different time scales of the two operations. These might more nearly equate if the comparison were purely between the boat and its mooring, but to take advantage of maximum financial returns a marina in most cases has to become a complex of amenities which, together with the type of engineering involved, has a lengthy design and construction period.

In the United States the type of boats and boating varies almost as dramatically as the climatic conditions. In the northern half of each coastline there is a noticeable preponderance of sailing over power boating and many carvel and clinker-built timber boats are owned and built: people really enjoy sailing, extending their activities from early spring until late autumn. In the southern States, however, fibreglass power boats are very popular and power predominates over sail. This is particularly so in Florida and southern California. In exclusive marinas near Los Angeles there are many large power boats which are used almost wholly by their owners as floating bars, and are only occasionally taken out of the confines of the marina.

The number of boats and boat owners has grown very rapidly in the United States, particularly within the last ten years. With the proportional rise in the number of boats now built in fibreglass the price of the small 14 and 16 ft outboard runabouts have in some places entered the 'impulse buying' market. The size of fibreglass boats now being made and sold is also increasing, 27-ft boats now being quite common and within the range

COMPATIBLE ACTIVITIES	ANGLING	CANOEING	ROWING	SAILING	SUB AQUA	WATER SKIING	HYDROPLANE / MOTOR BOAT RACING	MOTOR BOAT CRUISING	WILD-LIFE
ANGLING	●	✓	✓	✓	✓			✓	✓
CANOEING	✓	●	✓		✓			✓	
ROWING	✓	✓	●		✓				
SAILING	✓			●	✓				✓
SUB-AQUA	✓	✓	✓	✓	●				
WATER SKIING						●			
HYDROPLANE / MOTOR BOAT RACING							●		
MOTOR BOAT CRUISING	✓	✓						●	
WILD-LIFE	✓			✓					●

1.6 Compatibility of water sports

of many manufacturers. There is no doubt that even larger pleasure boats of this material will very soon be available as 72-ft boats are now being made in fibreglass as replacements to the shrimp-boat fleet in the Gulf of Mexico. In the light of this it is to be expected that boat ownership in Britain will continue to increase. Although at present this is being held back by the adverse economic situation, there is little doubt that when the present economic restrictions begin to lift the demand for boats will take a steep upward turn and with it a new era of British marina development.

Marinas: the Middle Eastern potential

Since the first edition of this book the growing prosperity of many Middle Eastern countries has resulted in an increased demand for leisure products. Whilst fishing and commercial harbours have been a feature of the area for many centuries, water sports have not been a traditional activity. This situation is changing however. Western technicians and consultants employed by oil and other development companies have imported power and sailing craft as work boats and for recreation. This has encouraged local custom and pioneered a market of considerable potential.

The last fifty years' experience in America and the last thirty years in Europe, have provided evidence that, given a reasonable climate and moderate coastal conditions marina development follows closely behind boat ownership, which in turn stems from increased prosperity and mobility (mostly car ownership). These criteria are now so relevant to the Middle Eastern situation it seems inevitable that the area will provide opportunities for marina development during the next decade. The countries demonstrating most potential are Lebanon, Egypt, Saudi Arabia, Dubai, Abu Dhabi, Qatar, Kuwait and Iran. Oman shows certain promise confined to its foreign population and Bahrain's growth in leisure-boat imports is only limited by its lack of marinas.

There seems little doubt that those concerned with providing consultancy services, boat handling and other equipment, floatation systems or marine construction generally will be busy in the region during the next decade.

With an area so huge it is not possible to generalise about coastal conditions, but as with all seaboards each site must be surveyed and evaluated individually.

The great majority of the area is desert or semi-desert which generally continues through to the seaboards. Annual rainfall is very low— 25–100 mm (1–4 in) but some coastal areas are rather more humid except in summer. Sailing in these waters does not generally offer the variety and interest of more temperate areas. Partly for this reason power boating enjoys a much higher proportion of total boat ownership than in the West.

Distillation plants are currently in commercial use throughout the region particularly in the Persian Gulf but they are expensive for the amount of water produced. The use of solar energy to support distillation is being developed but solar stills have proved successful only on a small scale.

1. The Leisure Background: Bibliography

Adie, Donald, 'Marinas', *The Municipal Journal*, 28 March 1971 (Docks and Harbour Supplement).

Burton, Thomas L. (ed.), *Recreation Research Planning*, Urban and Regional Studies No. 1, George Allen and Unwin Ltd.

Central Council of Physical Recreation, *Sport and the Community* (The Wolfenden Report), London 1960.

Clark, Colin, *Population Growth and Land Use* (Revised Edition), Macmillan, London 1968.

'Community Recreation: the Need for a Wider View', *The Architects' Journal*, 5 April 1967.

Countryside Commission, *The Coastal Recreation and Holidays, Special Study Report*, Vol. I, H.M.S.O., London 1969.

Countryside Commission, The, *The Coasts of England and Wales*, H.M.S.O., London 1968.

Countryside Commission, The, *The Planning of the Coastline*, H.M.S.O., London 1970.

Countryside Commission for the Countryside Recreation Research Advisory Group (C.R.R.A.G.), The, *Recreation News No. 1*, November 1968.

Department of Education and Science, *Planning for Leisure*, H.M.S.O., London 1969.

Dower, Michael, 'The Fourth Wave: The Challenge of Leisure,' *The Architects' Journal*, 20 January 1965. *Also issued as a pamphlet by The Civic Trust.*

European Leisure Recreation and Tourism: A Brief Report, Donaldson and Sons, Chartered Surveyors, London.

Gloucester County Council Planning Department, *Outdoor Water Recreation*, The Council 1968.

Greater London Council, *Research paper No. 2: Use of open spaces*, The Council, London 1968.

Greater London and South East Sports Council, *Sports Facilities, Initial Appraisal— Vol. 2: Water Recreation*, The Council, May 1969.

Hanson, Michael, 'Investing in Leisure: Parking Problems for Boats'. *Municipal and Public Services Journal* 21 March 1975.

Head, Derek, *Marinas: 1 Water Recreation*, Cement and Concrete Association, London 1975.

Hookway, R. J. S., *Leisure*, The Countryside Commission, London 1970.

House of Lords Select Committee on Sport and Leisure, *Second Report Evidence, Appendices and Index, Proceedings*, H.M.S.O., London, July 1973.

Leisure in the countryside—England and Wales, H.M.S.O., London 1967.

Leisure Industries Review 1973–74, Gower Press, Epping 1973.

'Leisure' in *Official Architecture and Planning* (special issue), August 1969.

Outdoor Recreation Resources Review Commission, *Outdoor recreation for America*, The United States Government, Washington D.C. 1962.

Patmore, J. Allen, *Land and Leisure*, David and Charles Ltd. 1970, Penguin Books Ltd. 1972.

Regional Sports Council, *Appraisals of demand*.

Rogers, Brian, 'Leisure and Recreation' in Cowan, Peter (ed.), *Developing Patterns of Urbanisation*, Oliver and Boyd, Edinburgh 1970.

Rogers, H. B., *Pilot National Recreation Survey: Report No. 1*, British Travel Association/ University of Keele, London 1967.

Scottish Tourist Board, *Firth of Clyde study phase 2: Recreation planning for the Clyde*, Edinburgh 1970.

Sillitoe, K. K., *Planning for Leisure*, Government Social Survey SS 388, H.M.S.O., London 1969.

Sports Council, The, *Planning for sport—a report*, The Council, London 1968.

Sports Council, The, *Sports Council Review 1966–69*, The Council, London 1969.

Tanner, M. F., *Coastal Recreation in England and Wales*, The Sports Council, London 1967.

2 Site selection

How much choice?

Anyone given the task of choosing the best location for a yacht harbour is presented with such constraints on the one hand and demands on the other that the application of even the most basic criteria will probably suffice to eliminate a large proportion of waterside as being unsuitable. In Britain one factor alone is capable of reducing the search area along the coasts by as much as 16 per cent. These are the stretches designated as Heritage Coastline by the Countryside Commission and they take up some 734 of the total 2,741 miles of the coastline of England and Wales. Most countries nowadays have stretches of coastline which are protected from commercial exploitation for environmental and amenity reasons and within these defined areas it is most unlikely that permission to develop even the smaller type of marina would be granted.

Other criteria as simple and obvious as ownership, or availability may further shorten the list of possible candidates, but having excluded clearly unsuitable or prohibited areas there will doubtless remain a collection of possible sites which will require consideration in a logical and comprehensive way. This early decision is of great importance, for to go ahead without examining all the aspects may mean overlooking some fundamental flaw which could lead to a costly and perhaps irreversible error.

It is said that Conrad Hilton, when asked to name the three most important factors leading to a hotel's profitability replied "location, location and location". This may be equally true of marinas for in matters of accessibility, service and multiple function the two are not dissimilar. In fact investigation into marina sites needs all the careful planning and experience that would go into a small town because, in a specialised sense that is what it is. In many ways it is just as self-supporting and yet probably requires even better means of transporting its 'population' to and from its chosen pastime.

There is one common example of marina development where no choice of location is necessary or even possible. This is the commercial harbour converted for recreational purposes. Small fishing fleets are becoming less economic and the container revolution is bringing about the closure of an increasing number of docks and harbours. It is sad to see these once thriving waterside areas becoming derelict but it is not always easy to see what economic use they can serve when their commercial life has ended. They are usually locked, reasonably deep (although often heavily silted) and contain, around a steep-sided quay, warehouses and ancillary buildings which are substantial, attractive but often difficult or uneconomic to convert to other purposes. Draining and infilling the water area is expensive and only results in 'made up' land, the weight loading of which is low and

2.1 'A suitable case for treatment': site at
San Diego, California in 1968

2.2 Inland view from the same site

2.3 Taken in 1964, this captures the
down-at-heel environment at Port Hamble,
Hampshire, where only the offices are
completed

2.4 Later the same year: the Hamble River
Marina site after the completion of Stage I.
The offices are on the extreme left

2.5 Site at Lymington, Hampshire, as it was in 1967 when the first sheet piles were being driven

2.6 The same view one year later as the Berthon Marina nears completion

generally unreliable for anything but the lightest of structures. Piling through the dock bed for a foundation first necessitates the removal of the silt and is not usually satisfactory or economic. Generally speaking, if a developer is anxious for profit or a local authority is looking for a housing gain it is more realistic to consider a clear 'clean' site elsewhere.

However, there still remains the problem of what to do with an old dock often located in a key position in an urban water-front area. This was the situation at St. Katharine's Dock by London Bridge; at the Haven in Exeter and the Brentford Dock, Hounslow. In each case the problem was overcome by re-using the water for recreational boats. These cannot by strict definition be termed marinas for they may not offer all the services associated with a modern purpose-built facility, but what they lack in convenience they often compensate for by their fine old warehouse buildings, customs houses, locks, quaysides and wharfs.

The best of these may be preserved and converted to other uses either connected with recreation or not. Others may be demolished to allow new buildings to rise or perhaps for open space. Whatever the case at least the water area is saved as a valuable amenity, continuing to earn money from boats and providing a valuable addition to the number of moorings needed to meet the rapidly-growing demand for berths. Some may disagree that a conversion from commercial to recreational purposes can be termed a marina at all which by definition is 'modern' and purpose-built, and that 'yacht harbour' may be a better description. The question is of more than semantic importance for a marina and a yacht harbour are different things each with its own problems needing to be handled in separate ways. The marina, whilst taking advantage of any physical attribute that nature has to offer, is nevertheless a new development on a 'new' site. The other is a piece of urban surgery requiring the sympathetic introduction of new forms into an old and familiar setting which in many cases may form the very heart of the town from which it may have drawn its life and prosperity for generations. The general design principles applicable to each situation are discussed in Chapter 5 but as to site selection the marina requires skills in geology, planning and economic geography etc. whereas with the adaptation of an existing dock or harbour it will be primarily for the accountant and the architect to determine its financial and aesthetic viability.

Breaking down the problem

In examining a new site for a marina development there are six main categories of investigation all of which will need assessment either by the developer or by consultants specialising in particular aspects once a development team has been appointed:

1   Legal work
2   Planning
3   On-shore considerations
4   Off-shore considerations
5   Engineering
6   Finance

As will be seen in the Check List on page 35 each is composed of many related sub-divisions. Some of these are naturally more important than others but all are worthy of some research. No location is likely to be found that fully satisfies all the demands posed by each of these topics. It is only by weighing the collected data that a well-founded and realistic judgment may be made upon the suitability of the site to accommodate the proposed development.

The importance of each aspect will to some extent vary according to the type of marina envisaged. In this respect the site considerations will need to be set beside those of the design, the intended facilities, the type of user, the available capital and so on, for only by doing this will one build up a picture 'in the round'. At this stage of strategic planning it is as well to keep a reasonably open mind about the components of the marina development —the facilities, mooring patterns, equipment and so forth. *It is better to choose the site on the evidence of really good geographic, engineering and planning data and subsequently marry to it the type of services and layout which suit it best than to predetermine the exact role of the marina and then search for the ideal location.*

Appointing the development team

Site selection is a complicated problem which it is tempting to break down into separate components, for by taking it apart and examining the pieces it may more easily be ordered and analysed. The danger of this lies in

2.7 Ramsgate Marina, owned and operated by the town corporation, was created by adding floating moorings to the inner harbour

2.8, 9 Brentford Dock, Hounslow, London.
A major problem was how to handle the
8·5 ha (21 acre) site. This was solved by
filling in half the dock and developing the
remaining 0·81 ha (2 acres) as a mooring
basin for small craft. Around this, 3-storey
housing rises from the Thames bank and
6-storey housing overlooks the narrower
River Brent

investigating each constituent in isolation and not allowing sufficient interchange or reference between them. These elements, being integral aspects
of the same problem, are so interrelated that their individual conclusions
must be brought together either by the developer or an agent on his behalf.
Only then may a value judgment be given about a site and an assessment
made of its potential related to other locations under similar investigation.

Each of the divisions under which a potential site will need consideration
will itself be composed of several interrelated aspects. Such are their complexity that professional advice will be required in the majority of cases.

It may be that one group or firm of consultants is competent to cover all
the engineering investigations both on and off shore and that a single firm
could embrace both the legal and financial questions. Professional planning
advice will almost certainly be necessary.

In considering these problems it is good advice to those bent upon
establishing the best possible location to assemble a team of advisers skilled
in all aspects *at the very earliest stage*. Their expertise will be needed

2.10 Taken 11·3 km (7 miles) downstream from Westminster on the south bank of the Thames, this photograph shows the location of the proposed 400–500 berth marina which will be adjacent to Thames-mead's central area. The locked entrance will reduce siltation to an acceptable level in an area of high deposition

eventually in any case and good advice given during these exploratory stages is particularly valuable, for it is only by comparing competent evidence that a sound decision may be arrived at on this critical question. No planner will be pleased at being approached for advice on transportation after trial boreholes have been taken. No engineer will enjoy telling a developer that his yacht harbour will silt up in six months when the land has already been purchased. The later the stage at which serious mistakes are made the more expensive they are likely to prove—particularly if they remain undiscovered until near completion or even after. *It is worth remembering that there are no errors in the whole field of development more costly than those of marine engineering.*

With larger marina developments, particularly in America, it has become usual to appoint public relations consultants. In informing local people and other interested parties, usually through the press, they certainly perform a useful role, but this should not be a substitute for the retention of qualified consultants. A yacht harbour development on a fairly large scale will, nowadays, nearly always result in a public hearing. Proofs of evidence will be required to cover all the fields of consultation. If the proposed plan has been covered from its inception by a professional team and these people can be called upon by the lawyers to prepare and deliver their testimony backed by well-executed illustrative matter, then the proposal will have very much more chance of success and will incidentally help to alleviate the misgivings of local residents if presented to them in a straightforward manner as early as possible.

1   Legal work

Legal Searches will need to precede serious work in other fields. A reasonably clean bill of health from the viewpoints of ownership and the planning laws will be expected if abortive work and unnecessary fees and other expenses are to be avoided. Certain permissions will no doubt be needed before surveying is started or trial boreholes taken.

The first approaches to land owners and occupiers are usually made by the lawyer as are those to the statutory planning authorities. These questions are dealt with more fully in the Planning Guide on page 35, but it should be emphasised here that investigatory legal work is most important in the early stages for as well as being suitable on technical grounds it will be necessary to establish as soon as possible the attitudes of all the parties responsible for the development's eventual approval. This will probably include the planning authorities at local, county and central government level and in Britain may need to include the opinion of advisory bodies such as the Countryside Commission and the Central Council for Physical Recreation. Skill and tact at the outset will gain friends and influence people and even if the site favoured by the developer fails to find approval, allies made now may prove useful by their recommendation of more acceptable alternatives. Legal work usually continues throughout the consultative and contractural stages and a legal representative is often included on any management committee retained to administer the completed development.

2 Planning

The planning aspects will cover a wide spectrum of considerations from the broad issues of national and regional policies and the evaluation of potential in terms of catchment, transportation and future expansion, to the more parochial and immediate questions of land-use planning, consents and the preparation of overall feasibility studies. The planner plays a vital role in site selection for he co-ordinates, controls and brings together the relevant information upon which an objective and logical decision may be founded. He is at the site selection stage what the architect is later on and both will be leading teams of several skills and professions, uniting and correlating their expertise to the client's benefit. Some of the essential economic or engineering decisions may have to be accepted by the planner. If, for example, the tidal range at a site is 35 ft on a rough coast, then a locked harbour may be obligatory and the planning carried out within this constraint.

There is a saying that defines architecture as one building and planning as two. This is an over-simplification which implies that an architect is incapable of handling a group of structures and that planning is solely concerned with the physical environment—both of which are untrue. The planner's role embraces the broad policies of employment, transportation and land use etc. and at a certain stage of decision he hands over to the architect who conceives the physical, functional and aesthetic relationship between the component parts of the total layout within the site boundary, bearing in mind the context in which they are placed.

3 On-shore considerations

The requirements needing to be satisfied in this category mainly occur within the land area of the proposed site boundary and are as follows:
1 The actual quantity of land
2 The quality of the land
3 Its present use
4 Its future potential

As one of the primary environmental arguments for a site's development will be to upgrade its usefulness, it will be easier to achieve an improvement the nearer the site is to dereliction in the first place. The survey of derelict land in England and Wales that was completed in 1968 by local authorities found 39,000 ha (93,920 acres) of land so damaged or polluted by industrial or other development that it was incapable of beneficial use without treatment. The total is higher today and would have been higher then if the terms of definition had included land currently *being* made derelict and which, whilst being so destroyed, had no condition of reinstatement in its present planning permission. How much of this land is coastal is not known, but there are many examples of quite disgraceful misuse of waterside areas —colliery waste tipping in northern counties, slate and limestone quarries in Wales and hundreds of miles of rivers and canals polluted and deserted. In Durham alone nearly 4 million tons of colliery waste are dumped on to the 'beach' or into the sea every year. Similar stretches occur in all industrialised countries and it would be unreal to suggest that they could be readily re-utilised as marina sites, but the possibility of an *eventual* transformation may be envisaged following a planned programme of progressive reclamation. Formulating a firm policy leading to a specific end product is more likely to succeed than vague alleviation with no final aim in view. Such a reversal would be costly, requiring the removal of the cause or an acceptable alternative method of waste disposal even before reinstatement could begin. Examples abound, however, where lengths of coast or waterside, whilst not polluted or derelict, are swampy, inaccessible, too shallow or in some other way unusable for most leisure pursuits and are in most cases visually unattractive. Such areas may be quite cheap to buy and relatively inexpensive to reclaim.

As the shore area may consist of 'made up' land from off-shore dredging, questions of its load-bearing capacity will arise as well as the length of time necessary for its settlement. The necessary height, strength and materials of the proposed bulkheads to retain the fill should be approximated. Consideration should be given to a sloped or reveted edge treatment to some stretches as these are less expensive.

Land reclamation need not always be built out as a straight addition from the shore line. A peninsular treatment in an L shape or two mirrored Ls are sometimes used at right angles to the shore line and turning to run parallel to it. Dredged from either side and built up as large land areas containing most of the marina buildings, they can act as their own breakwater with moorings within the enclosed water only.

The extent of land drainage is important. Programming the construction needs consideration *at this stage*. It may be that if expensive compaction techniques are to be avoided, the newly-formed land may need to consolidate for a considerable time before being built on. This ties up money and will need to be accounted for. There are occasions when large quantities of off-shore material will need to be imported as fill from elsewhere or conversely removed from the site and barged out to sea or trucked to distant tips. Such operations are of course expensive. The possibilities of expansion must be tested by as careful an examination of neighbouring areas, both waterside and inland, as is given to the proposed site itself. A recently developed site in Florida was given approval on a good geological and hydraulic report which had stopped short at the boundary of the initial area. When expansion was proposed the survey of adjacent land showed a highly expensive development situation which prevented growth for at least two years.

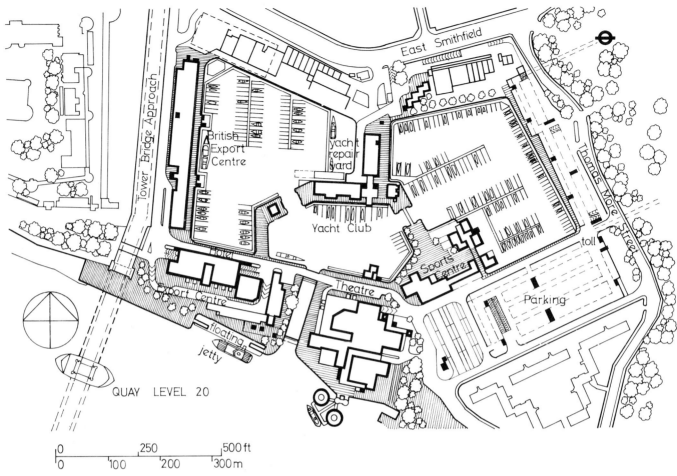

QUAY LEVEL 20

2.11 St. Katharine's Dock, London,
showing the berth lay-out in the West,
Central and East Docks. The Tower of
London and Tower Bridge are on the far
left. See also 3.39, page 61

2.12 The Ivory Warehouse from the West
Dock. This scheduled building is to be
converted to house the Royal Yachting
Association and other sailing clubs

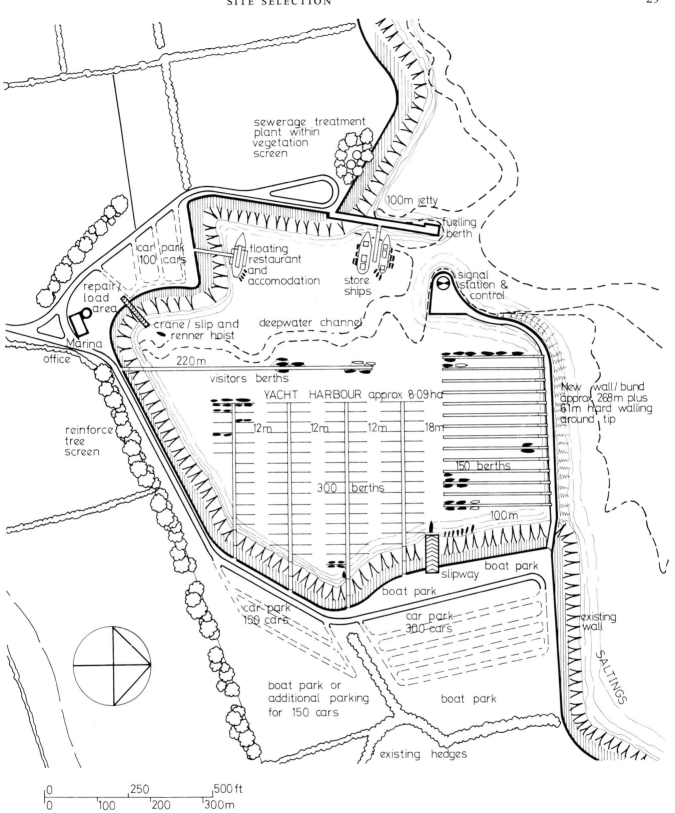

sewerage treatment plant within vegetation screen

100m jetty

fuelling berth

car park 100 cars

floating restaurant and accomodation

store ships

signal station & control

repair/load area

Marina office

crane/slip and renner hoist

deepwater channel

220m

visitors berths

New wall/bund approx 268m plus 61m hard walling around tip

YACHT HARBOUR approx 8.09 ha

reinforce tree screen

12m   12m   12m   18m

150 berths

300 berths

100m

slipway

boat park

boat park

car park 150 cars

car park 300 cars

existing wall

boat park or additional parking for 150 cars

boat park

SALTINGS

existing hedges

0   250   500 ft
0   100   200   300m

2.13 This proposal for a 450-berth marina on the River Blackwater, Essex, is unique in that virtually all the accommodation, as well as the craft, will be floating. Because of its location on an area of saltings in a rural estuary with few existing structures except farm buildings, it was thought appropriate to adopt a waterborne concept rather than erect permanent structures. The main reasons for the decision were environmental but constructional difficulties will also be avoided. It is interesting that the idea stemmed from the developers rather than being a planning requirement

2.14 Moody's Swanwick Boatyard on the
River Hamble in 1965

2.15 A similar view in 1969 showing the
first stage completed. The quay wall line
shows there was as much reclamation as
dredging

4 Off-shore considerations

The basic tasks under this heading are to:

1 Assess the existing natural conditions

2 Decide on the engineering methods to be adopted to modify or overcome them

The state of the tides, currents, prevailing winds and overall climatic conditions will need independent examination from the viewpoint of sailing conditions. Apart from this, however, the less obvious elements of wave refraction, swell, tidal currents and littoral drift may have a significant effect on the construction of the harbour and upon its efficiency when completed.

The services of specialists in coastal and submarine engineering will be necessary and their research may be quite lengthy, as equipment for hydrographic analysis will need to be installed and a study made of the area's history in respect of tides, flooding and siltation. These together with other engineering aspects are dealt with in Chapter 6 but it is worthwhile mentioning that in dealing with the selection of new sites an attempt to predict the levels of deposition within the harbour *after* construction will be of real value because the existing patterns may be changed and worsened by the envisaged development. An hydraulic model is the only really reliable way of calculation. Off-shore and on-shore problems are allied in many respects and investigation into them may be combined at the early stages and the general principles to be adopted combined into a plan upon which the future development could be based.

By indicating the type of marina and the overall layout of the land and water areas it will at least allow rough estimates to be made regarding the basic engineering requirements of depth, fill, length of bulkhead, type of lock, fixed or floating piers, number and type of piles and so on. This is not

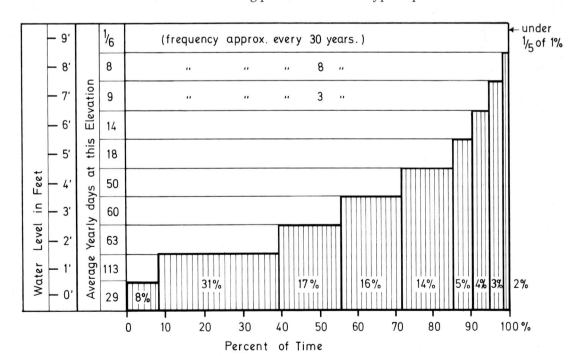

2.16 Tide records are a valuable source of hydrographic information. When the relevant information has been abstracted it may be set down either as a table or a bar chart. Interpretation of the result will affect the decision upon the site's overall economic viability and, if confirmed, the need for locking, the engineering criteria and the choice of berthing system. See also table on page 31

Table of water levels

Average elevation	0 m 0′	0·305 m 1′	0·610 m 2′	0·914 m 3′	1·219 m 4′	1·524 m 5′	1·829 m 6′	2·134 m 7′	2·438 m 8′	2·743 m 9′	
	Number of days observed										
Year											Total
1921	—	185	74	16	36	—	—	—	—	—	311
1922	50	66	101	10	30	23	1	—	—	—	281
1923	1	65	59	37	37	23	64	—	—	—	286
1924	—	39	39	47	73	8	2	8	22	—	238
1925	—	29	42	123	53	8	—	—	—	—	255
1926	—	—	37	63	62	34	31	29	—	—	256
1927	29	217	59	1	—	—	—	—	—	—	306
1928	60	81	1	24	51	16	9	15	—	—	257
1929	30	80	30	52	23	9	17	15	—	—	256
1930	—	—	1	131	83	23	6	4	16	1	265
1946	—	121	66	35	59	22	—	—	—	—	303
1947	—	58	32	32	68	23	9	15	15	—	252
1948	—	71	24	72	35	4	11	6	20	—	243
1953	—	23	6	3	17	28	20	23	27	10	157
1954	12	108	28	17	14	—	—	—	—	—	179
1956	—	1	6	5	15	11	20	—	—	—	58
1957	7	154	44	17	41	4	—	—	—	—	267
1958	14	135	8	44	5	—	—	—	—	—	206
1959	—	40	115	57	96	8	—	—	—	—	316
1960	—	36	84	32	28	50	45	19	36	—	330
1961	110	138	50	8	—	—	—	—	—	—	306
1962	87	109	64	43	—	—	—	—	—	—	303
1963	53	71	36	68	15	—	—	—	—	—	243
1964	32	52	40	57	5	3	7	18	—	—	214
Days each level	485	1879	1046	994	846	297	242	152	136	11	6088
Per cent each level	8%	31%	17%	16%	14%	5%	4%	3%	2%	Neg.	
Cumulative per cent	8	39	56	72	86	91	95	98	100		

a finished bill of quantities nor a complete layout plan but a feasibility study to give the developer some guidance. Perhaps alternative designs may be presented as well as alternative locations. Once the general location and layout of the harbour have been established it is in reclamation that the most important aspect of site selection is to be found. It is inevitable that engineering costs will form the largest single item of expenditure and within this category dredging is a major cost factor—quite often *the* major factor. In the typical example of construction cost expenditure on page 304 dredging and bulkhead amounted to 64 per cent of the total expenditure compared with 16 per cent for all floating piers and piling. These percentages pertained to a shallow lake-front development in the mildest of climates and could be even greater in tidal estuaries, strong currents, deep water or difficult sea-bed conditions. One of the single most important off-shore factors is that of tidal range. It will principally be upon this that the type of marina will be decided, whether it is a locked or tidal basin and whether, in either case, it has fixed or floating moorings or a combination of both. It will be a strong determinant regarding site selection *as such*, not only in appraising one location as against another but in some cases being the deciding factor in whether a marina is possible or worthwhile.

The engineering aspects of dredging are dealt with on pages 165–173 but as it is such an important consideration in site selection, it is necessary at this point to emphasise the role that a thorough survey plays in establishing the suitability of an area for development. The basic engineering structure of many marinas is constructed on the principle of dredge and fill where sheet piling is driven along the line of the future bulkhead wall and the material taken from the off-shore area to create the depth of water is transferred to the on-shore side to build up the land before it is drained and consolidated.

The workability of the material will therefore determine the ease with which this undertaking can be completed and the amount it is likely to cost. The type of plant necessary for the job and the length of time it is needed for must be estimated.

Each location will present different off-shore problems, in some climates ice formation, in others sufficient shade, marine boring organisms or extreme climatic conditions: for this reason it is important to retain experts who are experienced in the area or at least in similar conditions.

5   Engineering

Whilst the hydrographers and geologists diagnose the difficulties it is the engineer who proposes the means by which they may be overcome efficiently and economically. Even at this early stage of assessment he will probably need the advice of specialists particularly with the costing of statutory services, external works and buildings. Specialist firms are usually helpful in this respect and their expert knowledge is usually made available if they know the request to be genuine and may lead to their inclusion in the final list of tenderers. It is the engineer who is primarily responsible for much of the investigatory work. He will organise trial boreholes both on land and off-shore together with any contour survey work and recommend the engineering principles and overall framework of the construction. These will be set down in report form for each location examined and each alternative concept explored.

6   Finance

This aspect may be covered in the initial stages by the legal representative in co-operation with those responsible for costing individual parts of the future development, but as the services of both an accountant and a quantity surveyor (or his equivalent in other countries) will no doubt be needed eventually, it is wiser to retain their services at the outset. They may then

2.17 These wide canal-ways are a main feature of 'The Venice of America', Fort Lauderdale, Florida

2.18 Taken in May 1972, this photograph shows Bradwell Marina, Essex, as final preparations are being made to breach the sea wall before flooding the yacht basin. Near the mouth of the River Blackwater 98 km (58 miles) from the centre of London, this is Britain's first artificial harbour built behind an existing sea defence wall.

The basin, which already contains mooring posts, was excavated to a depth ensuring a 2·1 m (7 ft) minimum of water. It is estimated that half a million tons of earth were moved during construction. The completed marina complex will cover about 24.28 ha (60 acres), providing moorings for 267 craft, varying from 6·15 m (20–50 ft) overall. The marina has been developed by MAM (Marinas), a company formed within the Management Agency & Music group.

Plans for Phase I of the marina were announced only seven months previously and construction is continuing to include a yacht club, a boat yard and repair works. A chandlery, restaurant and dinghy park for up to 500 craft have already been provided.

contribute to the important and often irreversible decisions made at the exploratory stages.

Much of the costing will rely upon the financial estimates of the basic engineering work. Because site conditions vary so greatly so too will costs. This is why 'accurate' estimating is so important and so difficult. It is unfortunate that the work likely to occupy the greatest percentage of the cost is the hardest to evaluate. Other parts are reasonably straightforward —the land, buildings, and equipment being costed in a similar way to more conventional sites. Land costs may be as critical as planning permissions and first approaches are usually made by a legal representative who will establish the *sort* of price that is being asked.

As well as estimating the outgoings of construction an assessment of income is necessary, which in turn requires some idea of such things as length of season, possibilities of expansion and the type of facilities likely to be provided and their relative profitability. These questions are dealt with in greater detail in Chapter 12 but like other previously mentioned aspects a *rough* idea is needed at this early stage. It will probably be the case that no individual estimate of the complex items which together form the outgoings and income will in retrospect be entirely accurate—particularly in times of rapid price increases and unstable economic conditions— but it is often found that accuracy is helped by underestimates in one field being offset by overestimates in another.

These then are the principal factors for consideration before the location is finally decided. Each site presents its own problems and one of the more difficult tasks will be to evaluate the arguments between sites as well as within them.

In some cases, particularly with inland water, the search area may be so limited or the final location so obvious that the whole selection exercise is academic. Even when this is so investigatory work under the previously suggested headings will still be worthwhile for even after a decision has been taken in principle, most developers will wish to have some idea of the site's potential in terms of size, investment yield and cost benefit—even if the benefit is in social rather than financial terms.

Frequently, as will be seen in the next chapter, the persons taking decisions at this and subsequent stages will be members and officers of local planning authorities. Rarely will they be acting alone for they will more often than not be advised by officers of central government departments and by consultants of one kind or another. Whatever the circumstances and wherever the site this first stage will be an exercise in miniature, embodying most of the skills needed during the later periods of construction and development. If all the future team can be assembled early on and work together successfully during the selection of the site then a valuable foundation of goodwill will have been laid for the forthcoming development.

2. Site selection: Check list

Legal work

Examine planning acts relating to area.
Approach planning authorities
Check land ownership

        title
        negotiability
        covenants
        leases
        tenure
        property
        boundaries
        foreshore interests

Assist negotiation of land purchase and future options
Public inquiry work: acquire early any material useful for eventual proofs of evidence
Check wayleaves

      rights of way
      public access
      retention of footpaths etc.

(See also Chapters 12 and 14.)

Planning

Establish land ownership and title (check with lawyer)
Determine dates and phasing of land availability
Consider background information—future of region and locality
Examine

    regional studies
    economic planning council reports
    registrar general's statistics (population trends)

Measure potential catchment and assess area

        population
        social class
        sailing clubs
        boat ownership

Character of area: agricultural ⎫
      residential ⎬ growth or decline
      commercial ⎪
      industrial ⎭

    availability of amenities
    orientation and aspect
    approaches and views to and from site

Investigate potential for holidays

      weekends
      evenings
      tourism

Accessibility: evaluate in terms of time and distance for all modes of travel

Check nearby marinas for size, likelihood of expansion, are they competitive or comple-
mentary
Solicit local opinion and knowledge
Establish the need for and extent of present and future permissions re zoning
                                                                    planning
                                                                    byelaw
                                                                    estate
                                                                    legal
                                                                    coastal, river and
                                                                    harbour authorities

Contact national organisations (in Britain)
        Dept. of Trade and Industry
        Crown Commissioners
        DOE
        National Trust
        Countryside Commission
Future expansion, sufficiency of land area
Existing flora, fauna, tree and plant preservation, disturbance to wildlife
Discuss research material, programme, alternatives and format of report with members of
development team
Draft options on development packages most likely to succeed

Engineering: General

Draft out extent of all investigatory work thought necessary to determine site's viability,
covering: geological
        hydraulic
        climatic (prevailing wind, sea mists, fogs etc.)
        soil and subsoil survey (noting future foundations, quays, earthworks, retaining
        walls etc.)
Organize site survey, consider aerial photographs and see old records and maps (meteoro-
logical, geological etc.)
Check physical character of site: size
                                shape
                                orientation
                                physical features
                                type of ground
                                erosion
                                flooding situation
                                sight lines
                                drainage
                                smokeless zones etc.
Docks conversion: consider boating value ⎫
                        amenity value   ⎬ cost benefit study
                        historic value   ⎭
evaluate marina development against cost of infill
                                bearing pressure of harbour bed
structural condition of existing buildings and (with architect)
possibility of change of use/conversion
consider floatation system in light of existing dock walls
organise site and structural survey
Prepare material for report

Engineering: Off-shore

Assess length and type of bulkhead walls, breakwaters
Determine ease and amount of dredging
        area of reclamation
        type of plant and availability
        distance to tipping areas on land, at sea
        quality of dredged material for reclamation
        disposal
Consider suitability of basic marina type (off-shore, recessed, enclosed etc.)
Determine tides
        tidal range   commercial shipping lanes
        depth of water   any factors affecting sailing quality
        currents

Consider preferences for fixed or floating moorings
Examine siltation, erosion, ice
Consider benefits of locking, tidal basin, half-tide lock etc.

Engineering: On-shore

Contact statutory undertakers and others: gas
electricity
GPO
transport
police
fire brigade
parkways
Consider drainage and public health conditions, disposal of sewage
surface water
waste and garbage
Whilst awaiting site survey note condition and position of existing buildings
trees
walls, fences (condition and
ownership)
basements
tanks
roads
paths
tidal range
foreshore
high and low water line
Record suitable local materials
quarries
brickworks
fill, topsoil
nurseries etc.
Soil bearing pressure
ease of working
angle of repose

Finance

Clarify overall objectives and motives
Assemble basic economic facts
Consider likely land cost and/or charges now and in the future for the site and its neighbouring land
Review local labour situation: costs and availability
Effect of remote (or new) area (pioneering) versus popular area (competition)
Set social advantages against estimated expense (i.e. cost/benefit study)
Consider funding sources and likely interest payments against estimated income both short-term/long-term
Assemble factors for total programme budgeting
Profitability (or otherwise) of individual facilities
Consider all grant possibilities

See also Chapter 12

2. Site Selection: Bibliography

'The AJ Handbook of Building Environment' (Section 1: Climate and topography), *The Architects' Journal*, 2 October and 9 October 1968.

British Standards Institution, *Site investigations: B.S. Code of Practice C.P.2001*, The Institution 1957.

Building Research Station, 'Soils and Foundations' 1 and 2, *BRS Digests* 63 and 64, October and November 1965.

California City Planning Department, *Mission Bay Recreation Area*, The Department 1958.

Central Council of Physical Recreation, *Survey of Inland Water and Recreation*, The Council 1967.

Civic Trust, The, *A Lea Valley Regional Park—An Essay in the Use of Neglected Land for Recreation and Leisure*, The Trust, London 1964.

Colvin, B., *Land and Landscape* (Second Edition), John Murray Ltd., London 1970.

Countryside Commission, The, *Coastal Preservation and Development: Nature Conservation at the Coast. Special Study Report Volume 2*, H.M.S.O., London 1969.

Countryside Commission, The, *The Coastal Heritage—A Conservation Policy for Coasts of High Quality Scenery*, H.M.S.O., London 1970.

'Conference on the Future Development of Holiday Resorts,' (Organised by the British Travel and Holidays Association in Co-operation with the R.I.B.A. and The Architectural Association), *Journal of the Royal Institute of British Architects*, 13 November 1962.

'Developing Yacht Harbours in Great Britain,' *The Surveyor*, 21 December 1963.

Dower, Michael, 'Planning for Sport and Recreation', *Sport and Recreation*, July 1964.

'Final Report of the Council on the Recreational Use of Water Works', *Journal of the Institution of Water Engineers*, Vol. 17 No. 2, March 1963.

Folley, B. S., 'Inland Marinas,' Paper presented at the *Symposium on Marinas and Small Craft Harbours*, Department of Civil Engineering, The University of Southampton, April 1972.

Geiger, R., *Climate near the ground*, Harvard University Press, Cambridge, Mass. 1966.

Mite, James C., and Stepp, James M., *Coastal Zone Resource Management*, (Praeger Special Studies in U.S. Economic and Social Development), Praeger Publishers Inc., New York 1973.

Hewitt, Ralph (ed.), *Guide to Site Surveying*, The Architectural Press, London 1972.

Institution of Water Engineers, *Recreational use of waterworks*, The Institution, London 1963.

Kind, C. A. M., *Beaches and Coasts* (Second Edition), Edward Arnold Ltd., London 1972.

Land Utilisation Survey of Great Britain, Maps available from Miss A. Coleman, Geography Department, King's College, Strand, London WC1.

Lawson, Fred and Baud-Bovy, Manuel, *Tourism and Recreation Development: a Handbook of Physical Planning*, The Architectural Press, London, 1977.

Leller, Rene, 'The Biggest Holiday Development in the World', *The Daily Telegraph Magazine*, 21 August 1970.

Lovejoy, Derek (ed.), *Land Use and Landscape Planning*, Leonard Hill Books, Aylesbury 1973.

Manley, G., *Climate and the British Scene*, Fontana Books, London 1962.

Manley, G., 'Climate and landscape architecture, *Journal of the Institute of Landscape Architects,* May 1966.

Meteorological Office, *Climatological Atlas of the British Isles*, H.M.S.O., London.

Michigan Inter-Agency Council for Recreation, *The Present and Future Recreation Potential of the Great Lakes and Connecting Waters*, The Council, Lansing, Michigan 1961.

Ministry of Agriculture, Fisheries and Food, *Land capability maps*, H.M.S.O., London.

Ministry of Agriculture, Fisheries and Food, *Land classification maps*, H.M.S.O., London.

Ministry of Housing and Local Government, *Caravan Parks*, H.M.S.O., London 1963.

Ministry of Housing and Local Government, *Coastal Preservation and Development*, Circular 56/63, H.M.S.O., London 1963.

Ministry of Housing and Local Government, *Derelict Land and its Reclamation. Technical Memorandum No. 7*, H.M.S.O., London 1956.

Ministry of Housing and Local Government, *New Life for Dead Lands*, H.M.S.O., London 1963.

National Yacht Harbour Association, *Yacht Harbour Guide*, Ship and Boat Builders' National Federation, London 1963.

Olgyay, Victor, *Design with Climate*, Princeton University Press, Princeton, N.J. 1963.

Peterken, G. F., *Guide to the check sheet for international biological programme areas*, IBP *Handbook No. 4*, Blackwell Scientific Publications Ltd., London 1967.

Shwer, Max, 'Getting to the Grass Roots of Dereliction', *The Surveyor*, Vol. 140, 1 December 1972.

Tindall, F. P., 'The Care of the Coastline', *Journal of the Town Planning Institute*, Vol. 53, 1967.

U.S. Outdoor Recreation Resources Review Commission, *Shoreline Recreation Resources of the United States*, Publication No. 4, The United States Government, Washington D.C. 1962.

3 Categories of marina development

The word 'marina' is defined as meaning "a modern waterfront facility for recreational boats" and was coined in 1928 by the National Association of Engine and Boat Manufacturers Incorporated of America. The Association further described the term as being a facility offering services which have come to be part of modern boating: a place where boatmen may berth, launch, repair, fuel and provision their craft conveniently and be able to have a hot shower, dine ashore and be within easy reach of shops, communications and transport. Notwithstanding this description, the word is commonly and wrongly used as meaning any collection of moorings no matter how casually located, how small their number, how poor their quality or how under-provided they may be with even the most rudimentary services.

This misunderstanding of the word may be partly the reason why the word marina has a rather mixed image. Local residents fear them lest they bring the environmental intrusion of people, development, access roads and the inevitable cars and service vehicles. Many boat owners themselves criticise them for being cold and impersonal, and many are in fact too large, particularly in America, for such is the demand in areas of good climate and long seasons that expansion to over 1,000 berths is not uncommon. Such a number handled well need not be daunting, but too often the continual repetition of the same pattern is sufficient to deter the individualist. 3.2 shows the diversity of waterside areas at which marinas may be developed. Each, with varying facilities, will cater for distinct classes of boat and different groups of boat owners. Coastal sailing is of course the most popular, but inland sailing the fastest growing. Enclosed water generally offers safer boating ideal for sail training although control is often necessary to avoid overcrowding. Whilst each type of water area will pose different problems there are certain principles which can be applied to most categories.

Categories of marina

Marinas may be divided into several different categories and two of these based on location and ownership are set down in 3.2. To these may be added distinctions of size, sailing area, general quality of the facilities, fees and other charges related to the service provided and the kind of boat and owner. Each of these elements are directly or indirectly related and, in a well-balanced development there will be a sensible association between the principal factors of design and management which avoids extremes and pitches each aspect in relation to the whole, and at the same time follows a policy of overall progress on all fronts. Within these general principles the

3.1 This yacht club and office was converted from sailors' homes at Charlestown, South Carolina

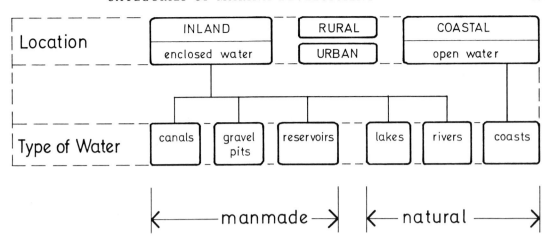

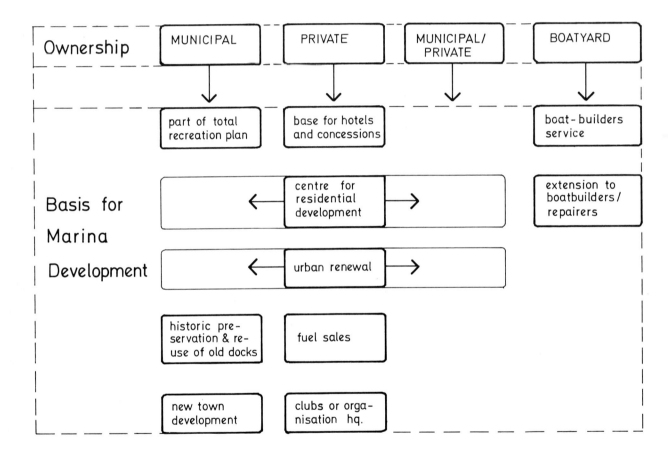

3.2 Categories of marina by location, above and function, below

function, shape and size of marinas varies considerably and is highly flexible, dependent primarily upon the factors mentioned above, each tailored to the constraints and opportunities serving the needs of the local or incoming boatmen. No development can succeed unless it appeals to the boating public for it is they who provide its source of patronage. The concept of a marina affords wide latitude. One boat owner may envisage a tiny natural inlet with a few moorings relying for essential provisions upon the nearby village: another may imagine a vast area of water containing 8,000 craft, several clubs, motels and every variety of boat-handling equipment, repair and service.

It is a mistake nowadays even to confine the marina concept to the terms

3.3 La Grande Motte living up to its name: the architecture attempts to sculpt interest into the featureless landscape. Cars get more imaginative treatment than boats

3.4 Pyramid apartments seen from one of the main walkways at La Grande Motte. The mooring charges range from 40p daily for small craft to about £3 for a large yacht. A canal connects this marina to the sea

of its own definition for, ever increasingly, marinas are being envisaged as merely a part, albeit an essential one, of the larger consideration of leisure development. At one end of the scale it may be a case of multiple-use with housing, hotels or commercial buildings being added to, or indeed being the *raison d'être* of, the marina. This occurred for example at St. Katharine's Dock, London, where a redundant dock area has been redeveloped, preserving historic buildings and retaining the water areas for use by leisure craft. At the most expansive end of the scale is what has been described as the biggest holiday development in the world—the Languedoc-Roussillon region in south-west France. Here there will be 20 new harbours for 40,000 pleasure boats along 120 miles of coast between Perpignan and Aigues-Mortes. By 1972 it was catering for a million tourists a year. Here marinas are taking their place within a tremendous scheme which includes 4 new airports, miles of motorway, a million new trees, huge areas for nature conservancy, a computerised irrigation system and no power or telephone lines above ground. By the time the scheme is complete 180 million ft³ of

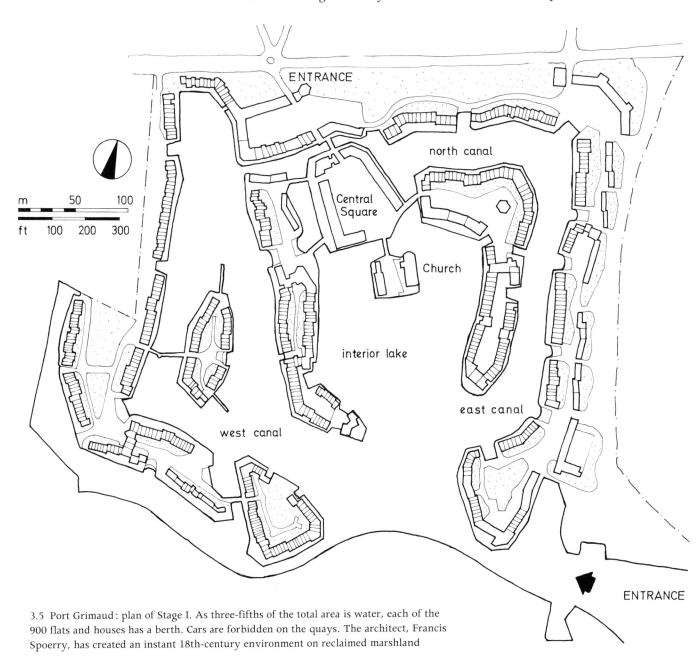

3.5 Port Grimaud: plan of Stage I. As three-fifths of the total area is water, each of the 900 flats and houses has a berth. Cars are forbidden on the quays. The architect, Francis Spoerry, has created an instant 18th-century environment on reclaimed marshland

3.7 This detail shows the simple and effective architecture

3.6 Cap D'Agde in 1971

material will have been dredged and moved elsewhere and the French Government together with private investors will have put nearly £400 million into this mammoth playground.

The way the French went about the task is a model for any developer. In its social awareness it seems faultless; as a business proposition cast iron and as an engineering feat it is Herculean. In many ways there is much for smaller developers to learn from such a colossal enterprise. What is in mind here is not just the building of a holiday coast but a new conception in leisure. For the plan to be successful the Government had to ensure there would be no get-rich-quick entrepreneurs flocking in in the wake of the basic engineering. They had at all costs to prevent ribbon development. The hardest thing was to hold off constructing the obvious and profit-making parts until the great majority of the infrastructure was finished. In this respect they did exactly the opposite of what has often been done elsewhere and that was to construct the roads, communications, sewage systems and amenities before laying one brick upon another. As well as the expertise and discipline needed to carry through such an approach it is essential to have the money or the certain means of raising it. Too little capital in the early stages is the greatest single reason for marinas failing financially and a common reason for their failure in the field of design.

3.8 A concept and its realisation: Le
Barcarès near Perpignan in model form

3.9 and an aerial view of the peninsula
during construction. With limited 1-to-1
moorings an intimate and informal
atmosphere is created

Associated uses

Situations where yacht harbours form only a proportion of a large development complex are becoming more frequent. The scale of projects continues to enlarge in almost every sphere and where increasing demands for disparate commodities are compatible, it frequently makes sense to join them, for as well as spreading the financial risk there may be a natural affinity between them in use, function and design. Housing, yacht harbours, hotels, restaurants, parks, offices, shops and some industrial uses—boatbuilding and repair for instance—may well be brought together to their mutual advantage, for an interest is thereby created between differing scales and uses which a skilful architect will exploit to achieve an interest not possible in a simple isolated marina.

1 Ancillary sports

It is water-based sports which are likely to blend most satisfactorily with marinas. Rowing, canoeing, water-skiing, swimming, fishing, SCUBA and all forms of diving can be introduced but care must be taken with regard to their compatibility and separation by location or timing may be

3.10, 11 Port Barcarès. This waterside terrace has been built since the aerial view was taken

3.12 The house and boat relationship could hardly be closer than at Leucate-Barcarès

necessary. Many other sports may be popular within the marina or nearby. Squash, tennis and bowls do not take up a lot of space whereas golf, although a suitable complement to sailing, could not be accommodated within the marina grounds. A course could exist however beside the marina site as at Pier 66 in Florida.

Boat owners and visitors often enjoy a change of activity and the total scope of the harbour is broadened considerably when ancillary sports are offered. Swimming, water-skiing, archery, fishing and Natural History are activities that require very little outlay in space or equipment on behalf of the management (or in cost to participants for that matter) and all are within the top 25 of the list of activities which people would like to take up seriously. If such a coverage was attained and added to sea sailing (top with 11 per cent of all adults) or inland water sailing (3rd with 7 per cent) any management would be able to offer a very large percentage of the most popular and expanding sports and pastimes. Secondary sports certainly are a draw that every management should consider, for those that are water-based become a relatively easily provided substitute to sailing on fine days and land-based activities are a valuable alternative on those embarrassing days when the weather prevents sailing and other water sports.

See also pages 79–84.

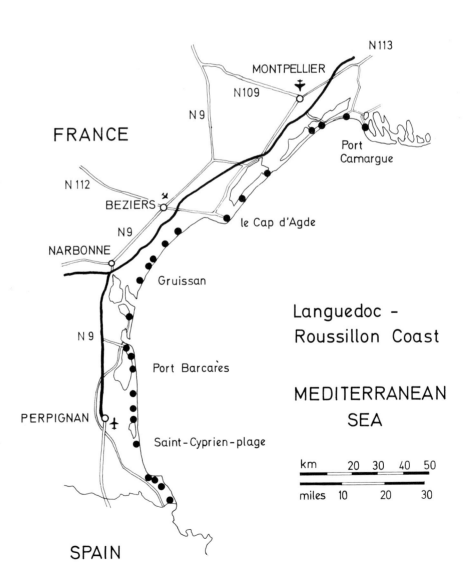

3.13 Along the 200 km (125 miles) of coastline from the Camargue to the Pyrenees 20 marinas are being created

3.14, 15, 16 Sharing with commercial users brings its benefit and its problems: fishing boats at Key West, Florida

2 Club or organisation headquarters

This is a specialist associated use which could be accommodated in such waterside premises as converted warehouses overlooking the marina. Whilst sailing organisations are an obvious choice, the accommodation of interests unconnected with boating could well prove satisfactory.

3 Community centre

In a municipal marina the clubhouse often becomes the centre for community interests which need not be confined to sailing or water-allied activities.

4 Conference/convention centres

Coastal sites are traditionally popular for congress meetings. The type of accommodation is of considerable variety and adaptability and an adjacent marina would, of course, attract water-based topics. There is an international shortage of medium-sized, flexibly-planned convention centres and accommodation and restaurant facilities could enjoy dual use with holiday needs. There are also appealing possibilities of combining conference, trade centre and exhibition facilities. The off-season custom thus generated would benefit the marina considerably.

5 Exhibition centres

Exhibition centres cover a wide field of use. They may be a prime element with their own ancillary uses such as restaurants, offices, banking and other commercial components, or they may be combined with, or part of, a trade or conference centre or attached to an industrial use or an individual manufacturer. Naturally those subjects related to the boat industry or leisure in general seem most readily suitable to form part of a marina complex, but incidental or independent topics may prove equally acceptable.

6 Historic vessels

The assembly of a group of historic vessels is a most attractive, appropriate,

3.17 At Thamesmead in the London Borough of Bexley a 12 ha (30-acre) lake was made as part of the first stage of building operations. Primarily a reservoir for surface water drainage, the lake is large enough for dinghy sailing and is linked to the open space network.

3.18, 19 Thamesmead residents enjoy a regatta on their lake. This photograph is the reality which stands up quite well when compared with the foregoing 'architect's impression' drawn 3 years before the lake's completion

and potentially remunerative possibility within a marina complex. It has already been successful at The Haven, Exeter, UK and is proposed within two marinas on the Thames—St. Katharine's Dock beside the Tower Bridge and at Thamesmead between Woolwich and Erith. The *Cutty Sark* (a tea clipper) and *Gypsy Moth IV* (Francis Chichester's round-the-world craft) are located at nearby Greenwich. In America there are collections at Mystic Seaport, Rhode Island and, most famous of all, at San Francisco where the collection of eight vessels is not only socially, educationally and economically successful in its own right, but acts as a well-known attraction to the city as a whole and the quay area in particular.

A collection may or may not follow a theme. At San Francisco the original craft had local connections, although other, unconnected, vessels have now

3.20 The *Balclutha* at San Francisco

3.21 The *Star of India* at San Diego

3.22 Aboard the *Star of India*

3.23 The *Eureka* at San Francisco

3.20–27 Historic vessels and marinas benefit each other. The harbour provides the access to old boats which are sure to be of interest to visitors. The *Balclutha* and the *Star of India* are splendid British ships restored by American enterprise

been added. Exeter draws its craft from all parts of the world and these are of general rather than historic interest. Some questions that may arise are as follows:

6.1   The choice of craft to exhibit

6.2   How they may be acquired

6.3   The cost of survey, acquisition and transportation

6.4   The cost of restoration and degree of historical accuracy

6.5   The type of dock (dry or floating) and setting

6.6   Staffing and maintenance

6.7   Likely income, overheads and profit

With craft too small for the public to enter, the means of collecting money may present a problem; a screened compound may also be necessary. With large cargo vessels the question of the holds arises and is sometimes solved by using all or part of this space to house an exhibition of artifacts and memorabilia connected with the vessel or its times.

3.24 A scow schooner and other vessels at San Francisco

3.25 Transport to the waterfront can be fun. This 20-minute service to Fisherman's Wharf is very popular in San Francisco

3.26, 27 Admirable and informative noticeboards

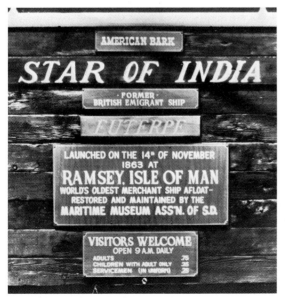

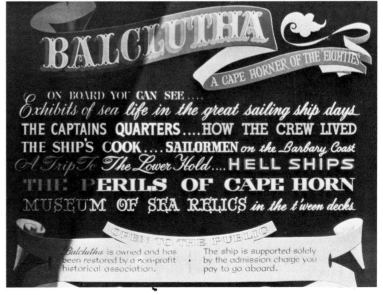

7 Hotels

These can either provide a normal service, support other uses or embody other activities within them. An adjacent marina would attract a sailing clientele to the hotel and there would be strong ties between the hotel accommodation and such facilities as the clubhouse, conference rooms and bars etc.

8 Housing

The provision of housing within a marina complex is an attractive proposition and the range of accommodation which could be offered in that situation is exceptionally wide. Flats, houses or maisonettes could be included for permanent residents on a freehold, leasehold or rental basis whilst temporary accommodation embraces such categories as:

8.1 Long-holiday premises
8.2 Weekend flats
8.3 Overnight rooms
8.4 Caravan sites

Apart from the more usual constructions mentioned above, housing can take many other forms such as villas, apartments, rooms, cottages, chalets and cabins.

9 Museums

This associated use is similar to the foregoing exhibition centre and equally wide ranging and flexible in concept, although their success lies in environmental rather than economic terms. Nautical subjects and those connected with transportation, local history or as an extension to a display of historic vessels seem particularly appropriate uses. A financial, land-use and town planning arrangement may be possible whereby the building may be staffed, maintained and the exhibits provided by (or provided in co-operation with) the local authority or historical society.

10 Offices

In many areas offices are the most remunerative associated use of all. Urban riverside sites, particularly obsolete or declining commercial harbours, are possible areas where offices could be acceptable to the local planning authority.

11 Restaurants

These can be very successful within the marina. The advantage of over-looking the harbour water can be considerable both by day and night: another clear case of mutual benefit to the restaurant and the marina.

12 Selected industry

Certain industries such as boat building and repair can be associated very satisfactorily with a marina. The first marinas proposed in Britain (in 1947) were a reversal of this arrangement, for they were envisaged as a means whereby a boat yard could offer convenient and permanent moorings to its clients. As the site requirements for both uses are similar and the commercial affiliation is so evident, the total locational and economic union of the two functions seems a natural consequence. Certainly the profitability of boat-handling equipment, chandlery, transport and many other mutually necessary but expensive components will be greatly enhanced and the appeal to many owners of enjoying the duel services of boat yard and marina on the same site can be a considerable draw. There are difficulties however: modern boatbuilding has largely lost its image of rigging and woodshavings and the noise and smell from a grp boat factory is considerable and, whilst not easily overcome, may be minimised by careful siting, insulation and shutting the works during the marina's busy summer weekends. Other industrial uses which could be suitably combined with a harbour for recreation craft are as follows:

12.1   Navigational instrument manufacture
12.2   Cabinet making
12.3   Sail making and canvas goods
12.4   Nautical clothing manufacture
12.5   Local crafts, artists' studios
12.6   Automobile service and repair station

Depending upon the circumstances existing on site or nearby, many other classes and types of industry may prove equally suitable even when unrelated to boating. Whether the industry is established or proposed it will be wise to check with the local planning authority to ensure that any new use—or the continuance of an existing firm—is likely to meet with approval.

13   Shops and commerce

The traditional harbour plan often places the shopping and market area just behind the quay wall. It is unusual to find shops *fronting* the water firstly, because single-sided shopping is rarely successful and, secondly, for reasons of exposure. Nevertheless, the relationship between a marina and a shopping area presents exciting possibilities both physically and financially.

14   Trade centres

There is a fast-growing interest in trade and commercial centres, their function being to offer display space, offices and excellent communication facilities. Several trades may be represented and visiting business men can be assured of seeing the latest commodities and having them explained by experts. Banking houses and, perhaps, insurance, transportation and other services are often available. As no goods are sold from the centre and visitors are fairly small in number and spread throughout the day, the problems of parking and road access are not difficult. Except for St. Katharine's Dock, London, where the relationship is indirect, no example of a trade centre and marina amalgamation is known, but there seems every reason to suppose that a trade centre for the boating industry covering craft, chandlery, clothing and navigational equipment could be a very attractive and successful proposition when associated with marina facilities.

15   Miscellaneous

The suitability of other uses will depend upon individual circumstances. Whilst the 14 associated uses outlined above appear to cover the uses likely to unite well with marinas in terms of economics and planning, there may be less obvious subjects which could be adopted with success.

Aquariums and dolphinariums are becoming popular water-based spectacles for example, and a zoo could be appropriate if it was not too

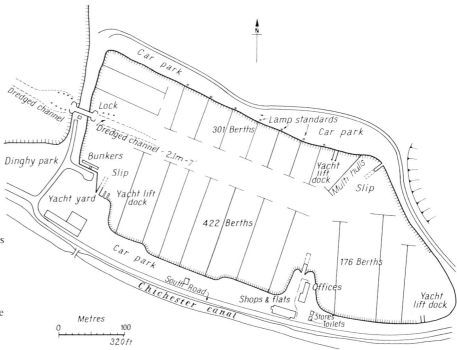

3.28, 29 Chichester Yacht Basin, Sussex, is the largest marina in the United Kingdom with just over 1,000 berths and a substantial waiting list. It is a fine, well-located example of the land-locked type, reliant upon the dredged channel from the River Itchenor and the very successful sector-gate lock design

large. Quite small menageries of 1 ha or less have a strong appeal to families with small children.

Whatever the relationship may be between the harbour and other associated uses *it is important for the marina management to be able to exercise some degree of control or influence upon any activity within or bordering the marina site.*

Specific examples of marina development

Permutations between uses are almost limitless and the origins of the more common existing marina developments are of interest for they have important implications for future schemes. Apart from development as a response to the specific sailing needs dealt with in Chapter 1, the nucleus from which yacht harbours grow is surprising in its variety and at least 10 different sources of potential development can be identified in the following specific examples:

1 As a base for catering and hotel development

The more sophisticated marinas in America, including both those developed by local authorities and private enterprise, appear to rely heavily on the income from concessionaire developers who are often operators of catering and hotel facilities. In Britain Forte's & Co. Ltd. was involved in the Chichester Yacht Basin—the largest marina in the country at the moment.

The example of Pier 66, Florida, also may be quoted, since the 150 moorings it boasts appear to have justified the construction of a multi-storeyed luxury hotel and restaurant topped by a revolving cocktail lounge.

2 As a centre for residential property development

Perhaps the most striking example of this in America is the Architectural Award winning 'Marina City' in Chicago. This multi-storeyed, twin-towered, luxury apartment building has a marina in its basement; craft are lifted by hoist or fork-lift truck and launched direct into the Chicago River. This process, from receipt of a resident's telephone call to launching, takes less than 10 minutes. Brighton Marina, which is being developed on England's South Coast will be the first major English example where a percentage of the land has been reserved for boat owners' homes and visitors' weekend flats. St. Katharine's Dock near Tower Bridge in London is a further example in this category and could also be included under categories 1, 4 and 8.

3.30, 31 Shore berthing is not nearly as convenient or attractive as wet moorings but it is very inexpensive. This converted tractor at Fairey Marine Ltd. launches the boat and then withdraws the empty trailer

3   As a visual amenity and facility in a total recreation plan

Most of the American National Parks illustrate this category, initiative being taken by the National Parks service itself which builds the facility and then lets it out on a concession basis. British examples of this include the Lee Valley Regional Park Scheme.

4   As a means of urban renewal and historic preservation

Mystic Sea Port at Rhode Island, USA, is a most inspiring example of both of these aspects of marina development, and involves the complete re-building of an old port. To a lesser extent Bucklers Hard on the Beaulieu River may also be included in this category. Here a large number of moorings and a landing jetty leading to a restaurant, hotel and grocery store have been placed in a setting of old preserved fishermen's cottages.

5   As an outlet for new boat types

The best British example of this category is the Fairey Marina Boat Park on the River Hamble. This is the first dry-land-storage marina, where 'Fairey' design craft are equipped with special trailers which enable them to be launched, retrieved and manoeuvred easily on the ground, where they are stored.

6   As an economic stimulant

The Maltese Government in an attempt to adjust the island's economy following the run-down of British troops there, hopes to rely heavily on the tourist trade in the future. To this end it has decided to make full use of one of the island's natural assets—its harbours, and to both rival and comple-ment the existing yachting facilities in the Mediterranean. Marsarnxshett Harbour is being rebuilt as a marina, with the Government financing the capital costs of all works involved, and sharing with private developers the cost of developing facilities on Manoel Island which lies within the harbour.

In the United States, a noteworthy example is the Muskingham Water-shed Conservancy District. Here a Public Works grant was obtained at the depth of the great depression, towards drainage and land restoration schemes. The result is now a completely self-supporting regional park of about 14,160 ha (35,000 acres) including several marinas. In Britain the Brighton Corporation hoped to improve the 'image' of the town by attracting a yachting clientele. More obvious examples are perhaps illustrated by a glance at the exceptionally high property values to be found in boating areas, particularly on the South Coast.

7   As a provision within a boat building or repair business

This is a very logical extension. An American example is the Thos. Knutson Ship Yard on Long Island, New York, which has a similar season to that in this country. The greater part of its income comes from repairs, but moorings are provided as a convenience in the hope that this will encourage people to patronise the yard. English examples are numerous and would include among others, Lymington Marina and Gosport (Camper & Nicholson).

8   As a club or organisation headquarters

This is a self-evident generator where, by common acceptance, areas of coast or river become recognised as boating centres. The next short step is

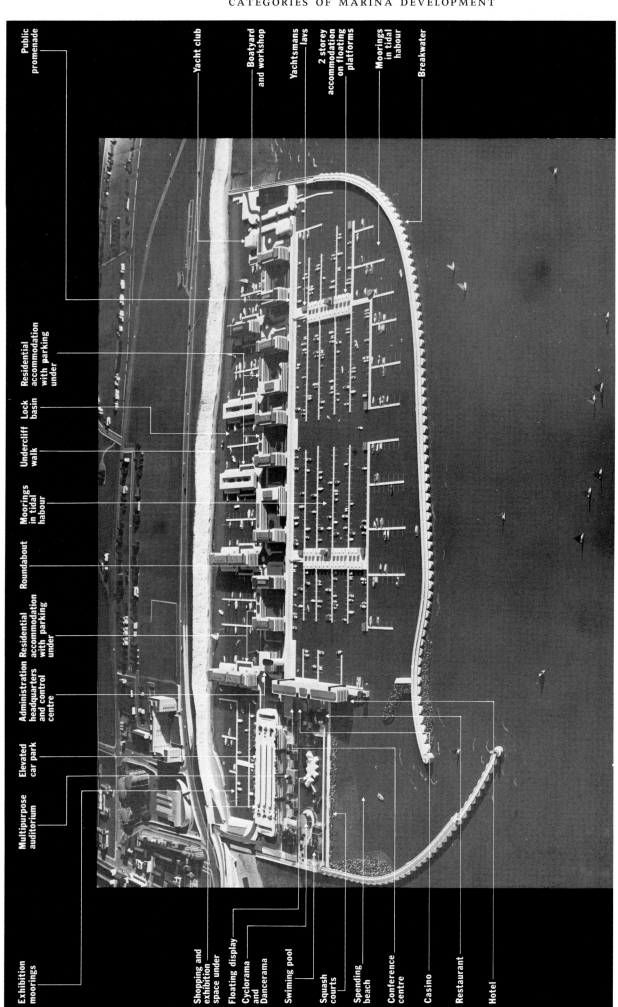

Exhibition moorings

Public promenade

Yacht club

Boatyard and workshop

Yachtsmans lavs

2 storey accommodation on floating platforms

Moorings in tidal harbour

Breakwater

Residential accommodation with parking under

Lock basin

Undercliff walk

Moorings in tidal habour

Roundabout

Administration headquarters and control centre

Residential accommodation with parking under

Elevated car park

Multipurpose auditorium

Shopping and exhibition space under

Floating display

Cyclorama and Dancerama

Swiming pool

Squash courts

Spending beach

Conference centre

Casino

Restaurant

Hotel

3.32 General view of a model of Brighton marina showing the principal facilities and components

3.33 Not a marina but an interesting artificial island shelter within a Dutch lake. The shape provides a long shoreline, a beach, areas for safe swimming and mooring and weather protection from all directions. The new planting will eventually increase the shelter and the attractiveness

to permanent associated on-shore facilities and an excellent English example of this is the proposed marina centres within the Lee Valley in North-East London. In the United States private club houses have frequently become the nucleus of larger public and private marina developments.

9 As an outlet for fuel sales

It is significant that in the United States, the large fuel companies are among the most prominent developers of marina facilities and an important example is Pier 66 in Florida, developed by Phillips Petroleum Company. It may be that British petrol companies will aspire to this initiative, from their present cautious assistance with fuel installation costs.

10 As an integral part of a new town development

At Thamesmead, the 'new town' within London, on the Thames between Woolwich and Erith, it is intended to develop a marina for 500–600 craft strategically positioned between the river and the town centre. There now exists a sail-training lake of 12 ha (30 acres) within the first stage of the project.

Reston New Town, near Washington, USA, embodies sailing lakes and moorings for over 1,000 craft.

Physical classifications of marinas

One of the principal factors governing marina type is tidal range. In the Great Lakes and other parts of America, in Dutch harbours and the Baltic coasts of Scandinavia, this is no great problem. But in Britain where a range of 6 m (20 ft) is common and extremes of 12 m (40 ft) are not unknown, it becomes the controlling factor in choosing from the following main categories, the locked harbour, the tidal basin and the haul-out marina:

1 The locked harbour
Locks are described in greater detail on pages 200–208, but here it is sufficient to say that they trap water in the harbour at a constant level or a designed range. One or several craft at a time may leave or enter the basin through the lock on either side of high water. The advantage is that by limiting the tide range the piles to which pontoons are secured can be much shorter and cheaper or, alternatively the whole mooring system may be of the fixed kind which is much more simple to construct and maintain. The disadvantage is that entry time into and out of the marina is limited to a few hours of free passage either side of high water or else restricted to the capacity of the lock.

2 The tidal basin
This is used where the tidal range is small or where the construction of a lock is impracticable. Piles are driven, the tide governing the length, and the pontoon system secured to them. In calm water and small tides it is possible to omit the piles and secure the flotation system with anchors and tie-bracing. With big-range tides piles will be very long and possibly therefore tied together for stability. The visual effect is strong. Given a 7·6 m (25 ft) tide and a need to brace the piles one has a situation where, at low water 7·6 m (25 ft) + 2·1 m (7 ft) (headroom at high water) + 0·3 m (1 ft) for bracing, equals a forest of 10 m (33 ft) 'goal posts' above water level at low tide. Add to this say 2·1 m (7 ft) depth of water, 0·6 m (2 ft) for silt and 2·4 m (8 ft) of pile driven into the bed and a total pile length of 15·2 m (50 ft) is necessary.

This effect contrasts with other circumstances where the long low pontoons are the only element above the water level except of course for the boats and their masts. It is most likely that economic and engineering criteria will determine which main system will be adopted, but it is important to realise that this decision will have a fundamental effect upon the marina's appearance. Both types have character, one restrained and quiet, the other exciting and dominant.

3 The haul-out marina
This is a giant version of the simple slipway. Converted tractors take the place of the car and trailer or the trailer and hand-winch. The disadvantage of timber boats 'drying out' on hardstanding has largely gone with the popularity of plastics. For club marinas the expense of employing staff to provide the service is overcome if 'volunteers' are available. Visually this type of marina is different again. Boats out of water present a distinctive appearance, they are taller and motionless and there is a danger of the haul-out marina looking like a nautical car park surrounded by chain-link fencing. It should be remembered that boats, like people and animals, don't look their best in cages. However, well handled they can be successful.

They are certainly one of the least expensive types of marina and boats up to 12 m (40 ft) long can be launched and retrieved in a few minutes at any state of the tide.

Secondary factors affecting design decisions

The motive behind the development of a marina is only one factor which will influence its eventual design and function. 3.2 shows the categories of water and type of ownership which will suggest different solutions in response to the different demands which are likely to occur within the individual headings. These primary functions are those over which the developer will have very little control once the site is determined. The ownership, sailing conditions and type of climate are, in the main, *given* conditions. They create areas, however, within which a choice of secondary considerations is possible and upon which early decisions will need to be taken. These subordinate factors include the following:

1 The number of moorings and overall size of the project
2 The number and quality of the facilities offered
3 The type of owner and boat catered for
4 The level of fees and charges for the services offered

Whilst these discretionary functions are themselves related it will clarify the situation if they are dealt with separately:

1 Size

It will be essential to have a fairly accurate idea, before construction begins, what the initial, intermediate and final size of the marina is to be. The information has a strong bearing on the whole concept of the project. The on-shore facilities will have to cater for the eventual number of boats, owners and visitors. This may seem too obvious to need mentioning, but cases are known where serious miscalculations of demand have led in some cases to custom being turned away and in others to the depressing sight of moorings lying empty for several seasons, tying up capital, demanding maintenance and depreciating even before earning their first week's rental. Mistakes like this are much less likely if a proper boat-ownership survey is undertaken to determine the current demand and the likely trends during the first few seasons; the average length of stay; the number of permanent rentals as against visitors' berths and the expected turnover of the latter. The methods of calculating this information and the effect of size upon shape, circulation and layout are dealt with in the chapters concerning the development of the marina plan and its special requirements.

One aspect often overlooked in assessing optimum size is the relationship between permanence of stay, the average spending per berth and the viable number of moorings. The point here is that, generally speaking, boat owners on the move spend far more than those resting at their usual moorings. If the marina is situated to attract overnight and weekend visitors or holiday families it is much more likely to draw more money per berth than its counterpart whose clientele are all regulars. A development there-fore with a high expenditure rate can prosper on fewer berths. Occasionally the sites themselves are so restricted that little choice is possible regarding size. In these fixed situations it is necessary to be sure of the project's feasibility, bearing in mind the impossibility of future expansion.

At the other extreme are locations whose size-potential appears almost

3.34 A lively scene at a Los Angeles marina. The locker-boxes sit above steel strengthening-angles

3.35 Within the covered moorings at Sanford, Florida

3.36 This roofed area at New Orleans covers only the pier, not the boats, and is more a sunshade than an umbrella. The I-section steel pile-columns, painted a light blue-grey, are certainly less oppressive than their timber equivalents in 3.35

limitless. Whilst this augers well for the early stages, care should be taken not to allow expansion beyond demand; to keep some slack in the situation for lean seasons (e.g. a short waiting list) and to remember the law of diminishing returns which is particularly applicable to marinas. Over-large harbours bring difficulties with lines of communication and run the danger of being psychologically and aesthetically unacceptable.

2 Number and quality of facilities

In parts of America marinas compete in the variety and quality of their associated activities and sometimes rather remote extras such as swimming pools and barbecue and ball-game areas are provided at the expense of essentials like boat-handling equipment and chandlery. Intelligent provision is largely a mixture of commonsense, demand and seeing one's way. Provision at the start must concentrate on essentials, particularly those not available nearby. Following this, the perceptive manager will soon sense the priority and quality of provision. The facilities demanded by owners of ocean-going vessels in the Mediterranean are fundamentally different from those of a sail-training centre for young people in the British Lake District.

3.37 St. Katharine Docks, London: these listed warehouse buildings and the adjacent water area are typical of scores of similar sites abandoned as uneconomic as the container revolution pulls commerce downstream to newly-built terminals. The dock area has been acquired by the Greater London Council for comprehensive development in conjunction with the Taylor Woodrow property company. All the water is retained as a marina and the scheduled buildings refurbished in a first-class example of public and private co-operation to create a high-class, high-density project

3.38 General view of the St. Katharine dock development

3.39 Key to the development:
(a) World trade centre. (b) Apartments and shopping centre. (c) Apartments, schools and health centre. (d) Conference centre and entertainment centre. (e) Hotel. (f) Yacht club and apartments (formerly Warehouse 'I'). (g) Chapel. (h) Apartments. (i) West dock. (j) Basin. (k) East dock. (l) River Thames. (m) Tower Bridge. (n) Tower of London

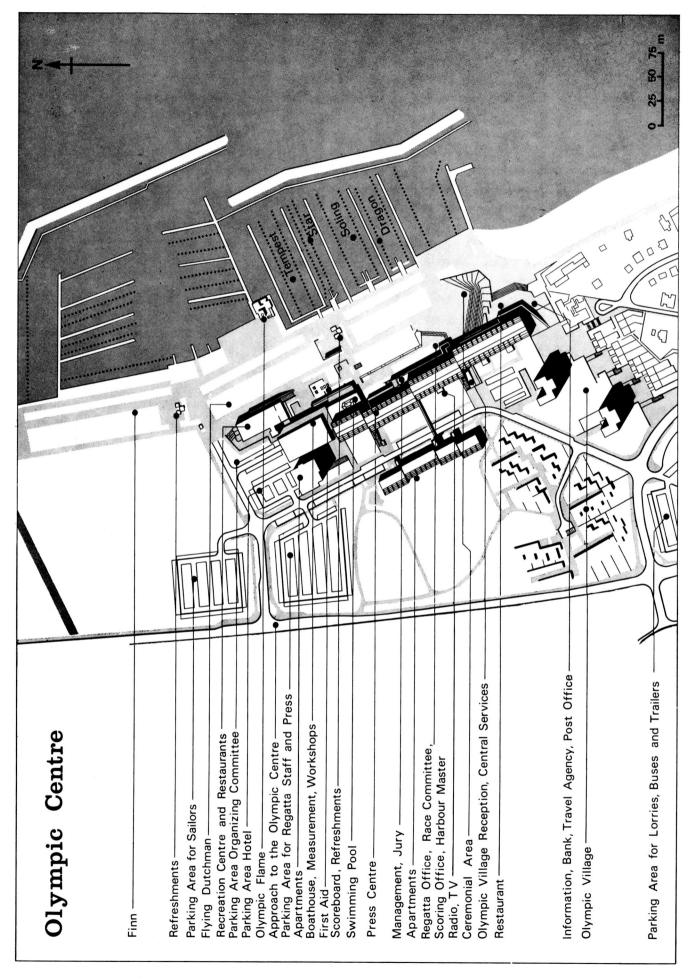

Olympic Centre

Finn

Refreshments
Parking Area for Sailors
Flying Dutchman
Recreation Centre and Restaurants
Parking Area Organizing Committee
Parking Area Hotel
Olympic Flame
Approach to the Olympic Centre
Parking Area for Regatta Staff and Press
Apartments
Boathouse, Measurement, Workshops
First Aid
Scoreboard, Refreshments
Swimming Pool
Press Centre

Management, Jury
Apartments
Regatta Office, Race Committee,
Scoring Office, Harbour Master
Radio, T.V.
Ceremonial Area
Olympic Village Reception, Central Services
Restaurant

Information, Bank, Travel Agency, Post Office
Olympic Village

Parking Area for Lorries, Buses and Trailers

3.42 The linear administration and press building with the Olympic village behind

3.40 The Olympic harbour at Kiel-Schilksee on the Baltic is a splendid example of what a specialist marina complex can be. Moorings for over 400 craft are provided, of which 173 actually sailed in the Olympic Regatta. The main buildings are set parallel to the quay and contain the central administrative offices, committee and jury rooms, boat measurement hall, restaurants, swimming pool, and communications and press centre. The latter occupies the top 3 floors of the main building in modern apartments each with a typewriter and television. All 3 courses were visible from the press boxes with 26 press boxes in addition.

32 bungalows and 2 apartment blocks provided flats for each yacht crew—team captains and managers having their own offices.

4,000 spectators were accommodated viewing the races from 14 steamers near the courses as well as on television. In the opinion of *Yachting World* (September 1972) the centre is 'absolutely unsurpassed by anything in the entire globe. Starting from scratch the Germans built a vast working monument which will stand as the finest collection of amenities for dinghy and keelboat sailors for years to come'

3.41 The Olympic Harbour at Kiel-Schilksee

3.43 Looking out into Kiel Bay across the Olympic harbour. The building in the foreground is the restaurant and recreation centre at the northern end of the complex

The former may require expensive boat-handling plant as an essential service and a night-club atmosphere for entertainment: the other will probably need a simple slipway and an inexpensive, self-catering hostel. The associated uses outlined in the categories on page 45 will also influence the type and extent of facilities. Generally speaking there are three main types of provision: that which serves the boats, that which serves the owners and that which, whilst it may be within the boundary of the harbour, is only remotely connected with boating or perhaps not connected at all. The first and second categories are found in the Check List of Marina Accommodation and Services on page 149. The third includes all the associated uses and is dealt with on page 45.

Sites with mixed uses or several functions within the same overall use are far more common now than they have been in the past. Marinas need be no exception to this trend which, when properly handled stimulates interest and contributes to the architectural character of the development. The reconstruction or conversion of existing commercial harbours may generate opportunities for many different associated uses, some perhaps retained from previous service and others accepted as either new uses for the existing buildings or brought in as new development within the on-shore boundary. An example of this is illustrated in 3.37 where at St. Katharine's Dock the historic warehouses have been converted for different uses and new river-front buildings designed to fulfil new roles in a functional and efficient way, at the same time respecting their closeness to Tower Bridge and the Tower of London. This thoughtful piece of urban surgery therefore contains within one site, most of the problems likely to be encountered in an urban dock redevelopment. The most important initial decision was of course to retain the water area. This not only made provision for the leisure-craft moorings so badly needed on the Thames, but preserved the character of the site and its historic surroundings.

Whilst the number and type of facilities needs to be flexible some programme of provision will need to be outlined at the early stages to ensure that the land area and capital are available when needed. The master plan should include a timetable to show when it is proposed to develop and implement certain facilities within a policy of overall growth. Careful planning will ensure that a compact circulation and economic layout accompanies the staged expansion. As this incremental enlargement may be taking place over many years the progressive manager will avoid a rigid approach. To continue to work one's way through a detailed and fixed programme that was founded years previously is to ignore later technology and more advanced concepts. This is particularly the case outside America where although the techniques may not be new the supplies of proprietary brands of constructional hardware and equipment are just not available. This situation is understandable when America has over 3,000 marinas and England approximately 60. Unless the supply of British boat-handling equipment and pontoon systems improves to keep pace with the increased demand for berths a situation will arise where developers will have to import American goods of which the variety and quality is quite remarkable. An alternative will be to import such goods from Europe, where far-sighted planning and development is already giving rise to the manufacture of very good marina equipment, purpose-made for the new Mediterranean harbours. This marina furniture is quite readily transportable: boat-hoist

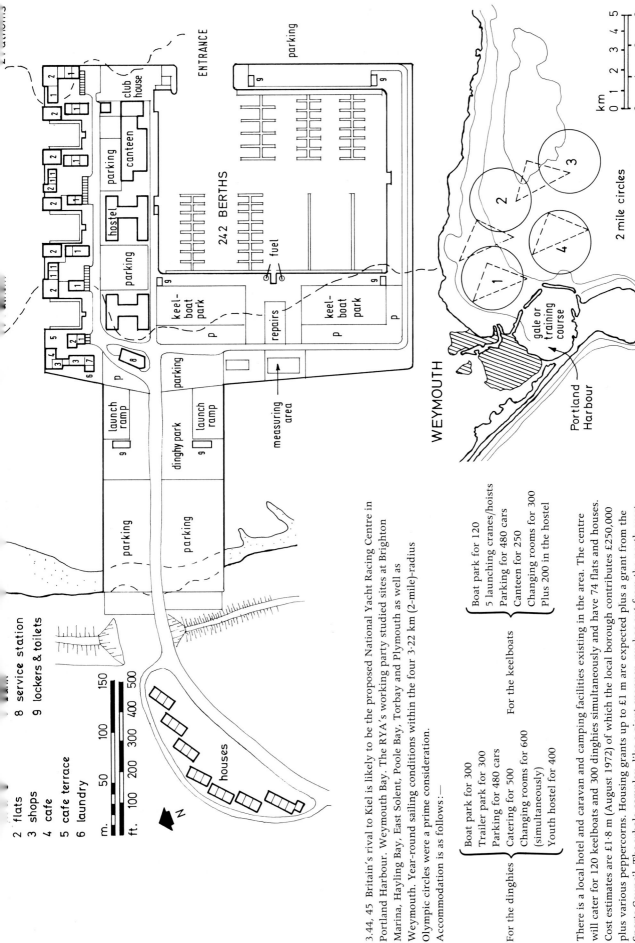

3.44, 45 Britain's rival to Kiel is likely to be the proposed National Yacht Racing Centre in Portland Harbour. Weymouth Bay. The RYA's working party studied sites at Brighton Marina, Hayling Bay, East Solent, Poole Bay, Torbay and Plymouth as well as Weymouth. Year-round sailing conditions within the four 3·22 km (2-mile)-radius Olympic circles were a prime consideration.

Accommodation is as follows:—

For the dinghies
{
Boat park for 300
Trailer park for 300
Parking for 480 cars
Catering for 500
Changing rooms for 600 (simultaneously)
Youth hostel for 400
}

For the keelboats
{
Boat park for 120
5 launching cranes/hoists
Parking for 480 cars
Canteen for 250
Changing rooms for 300
Plus 200 in the hostel
}

There is a local hotel and caravan and camping facilities existing in the area. The centre will cater for 120 keelboats and 300 dinghies simultaneously and have 74 flats and houses. Cost estimates are £1·8 m (August 1972) of which the local borough contributes £250,000 plus various peppercorns. Housing grants up to £1 m are expected plus a grant from the Sports Council. The whole complex, like a giant spanner, pushes out from the north-west corner of Portland Harbour. Space is still left for a 1·6 km (1 mile) radius heavy-weather and training course.

and storage equipment can be 'knocked down' and floatation systems stacked for travelling to be assembled in kit form on arrival. Buying good quality plant from abroad would appeal to many developers more than paying for purpose-made fittings—particularly when considering future expansion, spares and maintenance.

It is important that the actual buildings which house these facilities are considered in their relationship to the programmed evolution of the marina. The retention of the original architects (and for that matter the whole of the development team) is recommended to ensure satisfactory continuity and for their experience of the area and its planning and architectural concepts.

3 Catering for the boat and its owner

The original survey of the possible catchment area from which the new marina is likely to draw its custom should have provided information not only about numbers but also about the type of people and boats likely to become permanent or visiting customers. This information is likely to affect almost every aspect of the project for, whilst there are basic provisions common to every marina it is the variation of craft and customer that determine the character of the development. Normally, as in most business ventures, the management will try to give the kind of service that their average customer requires. The survey of potential craft and owners should aim to answer the sort of questions about providing the following facilities:

3.1 Average boat size, beam, length and draft (see page 98)
3.2 Range of sizes
3.3 Type of boat, e.g. power, sail etc.
3.4 Permanency of mooring
3.5 Number of visiting vessels, charter craft etc.
3.6 Length of stay
3.7 Accommodation required, e.g. overnight, weekend, holiday
3.8 Services required
3.9 Type of chandlery provisions and fuels
3.10 Amount of self-maintenance
3.11 Proportion of wet to dry berths and covered wet/dry moorings
3.12 Number of club members and type of club
3.13 Numbers of clients cruising, racing etc.
3.14 Requirements for children
3.15 Foreign boats visiting—customs
3.16 Likely average expenditure per boat and per head
3.17 Likely seasonal variation in numbers and activities

It may not be good economic sense to try to please everyone. If 95 per cent of boats in the catchment area fall below 12 m (40 ft) in length it may not be economic to cater for the few vessels over this size *if* they require special haul-out equipment or deep craft moorings. But if by so doing one is attracting sufficient numbers of a like kind to make the provision pay, the additional facility may then become a viable economic proposition.

It is sometimes a good policy, particularly in joint public/private marinas to provide a range of options in boat repair and maintenance; a thorough boat-yard service, perhaps in co-operation with an on-site boat builder and do-it-yourself facilities.

In considering types of owner, as distinct from types of boat there are two things worth remembering about sailing people. Firstly, even when they have plenty of money they usually are prepared to, and often positively enjoy, roughing it and going hungry. Getting wet, excited and even frightened is very much part of the fun and the millionaire yachtsman is just as prepared to do this as the dinghy sailor. Once ashore the basic needs are for somewhere to get oneself and one's clothes dry, something hot to eat, something (preferably alcoholic) to drink and somewhere to chat about the experience to like-minded masochists. Secondly, if the facilities which are provided are very luxurious and owners can afford to pay, this is splendid, but generally speaking, sailors are traditionally a remarkably undemanding group of people. What is more, they are excellent social mixers and this is particularly so with sail; power-boat owners tending to be rather more insular. The marina with a successful sailing club (or clubs) within its boundary has a great advantage. For many owners the social side of sailing is as important as the sport itself and a good, lively club is likely

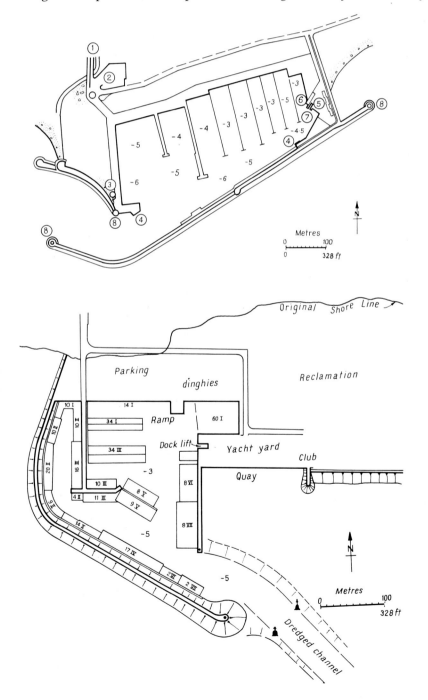

3.46 José Banus, Marbella, Spain: this marina, having a capacity of 915 yachts, is 30 nautical miles from Gibraltar and 35 from Malaga, and has been constructed on a sea front with no assistance from natural features. However, its cost, of a little over £2 m, is most reasonable and the berth charges are comparable with marinas in the Solent area. The sand dredged from the harbour has been pumped to form beaches to the west of the harbour
1 Entrance 2 Car service station 3 Marina office 4 Bunkers 5 Yacht yard 6 Slip 7 Yacht lift dock 8 Navigation lights

3.47 Larnaca, Cyprus: this is a yacht harbour for 300 yachts which has just been constructed in a country where yachting is in its infancy but where there is great potential. It is an example of the way in which dredged material may be used for reclaiming land around the harbour

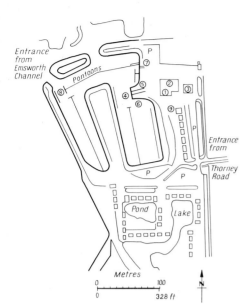

3.48 Emsworth Yacht Harbour is an example of a marina developed very economically by constructing a tidal basin with a dam at its entrance which submerges at high water. This allows ingress and egress for about 5 hours. The height of the dam is 0·5 m (20″) above Chart Datum. Further development is possible either by dredging or constructing a lock or a tidal gate.
1 Office and chandlery 2 Repair shop 3 Engine shop 4 Bunkers and quay (15t crane) 5 50t. slip 6 Grid 7 Slip (1:6) 8 Visitor's berth 9 Club house P Parking

3.50 Swanwick Marina is a typical river-side marina, developed as an extension to an existing yacht yard. All that had to be done to produce the yacht basin was to drive a line of steel piling and dredge an area of tidal foreshore. This was done in two stages and the berths were filled to capacity almost as soon as the pontoons were placed. See also page 29

to prove a marina's greatest single asset, not only as a place where members can exchange views, but *where a good manager can learn his customers' needs now and their requirements for the future.*

4 Fees charged and services offered

It must always be remembered that a marina has really only one function—to provide a *service*. The quality of this service and the charges made for it determine the value. The value given at any marina is weighed by owners against value given elsewhere and the outcome determines the yacht harbour's *attraction*. Provided an owner is sure of certain necessary services his next step is to compare and evaluate this with the competing harbour. Types of marina may be categorised by the value given (not just offered) as by any other criteria. Many owners stay out of marinas because they feel they are not worth the money and the wise owner not only moors at a marina but takes advantage of it. Good marinas are like good hotels—they are only worth the money, provided one can make use of the bath, telephone, colour television, lounge, bar, pool and sauna and meet people in a relaxed atmosphere. The sources of income of the average American marina are shown on page 301. This does not, however, show the range of facilities offered nor how varying the profits from any one facility can be between one marina and another. There may be a complicated interaction between elements whereby a low profit from one service remains viable because it draws people to use other high profit facilities—in other words marinas may effectively include loss-leader elements. In many harbours financial profit is not the prime motive. In the categories previously described there are several within which the marina *itself* is a fringe activity. Nevertheless, companies have a responsibility to shareholders as local authorities have to the public and no one wishes to see losses occurring where there need be none.

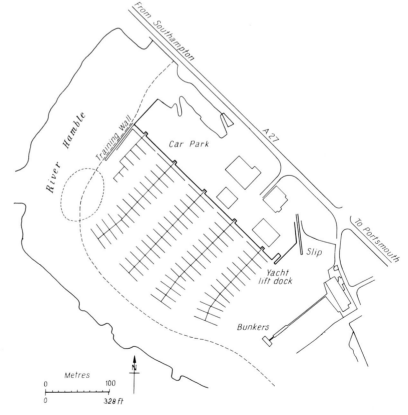

The outside influences governing a marina's attractiveness may be virtually unalterable factors such as climate and sailing conditions. What is possible is to offer incentives sufficient to lure visitors. In 1971 for example the French Government offered £15 flights from London and many of Europe's capitals to the Languedoc-Roussillon Marinas. It is now quite feasible for a moderately wealthy man to live and work in the Midlands of England and keep his boat moored at La Grande-Motte, Gruissan or Leucate-Barcarès, visiting it for long weekends and main holidays. This has already happened in the United States where New York businessmen keep their boats in Florida—a thousand miles to the south.

With these sort of influences being created by governments, competition is now international and with the advent of this polarisation across continents marinas are competing less at a parochial, just-along-the-coast level. If this is the trend for future promotion, countries with interesting, exciting and changeable sailing conditions have as much to offer boat owners as those with hotter and sunnier climates provided that sufficient safe, convenient and reasonably-priced moorings are available.

---

3. Categories of marina development: Check list

Type

Determine the land-use category of potential sites thus: Rural: urban
Open water: closed water
Inland: coastal
New: change of use
Natural: man-made

Investigate the main types: locked harbour,
tidal basin,
haul-out marina
(See also categories outlined in this chapter, pages 45–54.)

Geography

Consider site and potential marina type in terms of climate
length of season
quality and type of sailing water
profile of shoreline in section and plan
soil quality for piles, dredging, etc.

Economic Planning

Determine ownership/management category, e.g. private
municipal
boat yard
Is the locality or its catchment a growth area
a development area
a depressed area?
Are there plans for expansion under national strategy
regional development
local plans?
Is the site part of a complex
recreational plan
regional park
country park?
What are the economic opportunities locally
in the region
in the catchment area?
Find out the opinions and interests of hoteliers
holiday and tourist concerns
property developers
entertainment companies
Consider overall market research programme
Initiate a boat ownership survey as a basis for economic assessment (see this chapter, page 64)

Consider other surveys on demand for moorings

average length of stay
permanent berths
turnover at moorings
average spending per berth

Determine the socio-economic background of potential patrons
What category, size and quality of marina would be best under these circumstances?
What demands are likely to arise from this clientele? (see also Chapter 12)

Access Evaluate all modes of transportation locally and from catchment area(s)
Consider the effects that travel time/distance/mode will have on the marina type

Associated uses Is the marina likely to be (and remain) independent of other activities?
Determine likely scope and form of associated uses
Tailor marina to complement them in scale, type, patronage, quality, etc.
Research ancillary sports and assess compatibility
(See 14 associated uses outlined in this chapter, pages 45–54)

Size Consider the initial, intermediate and final size of development
Determine on-shore facilities commensurate with number and size of moorings at each stage

Facilities Prepare list of possible (desirable) facilities. Determine priority
Distinguish between initial (built-in) and eventual (added) provision
Separate likely provision into categories of that which serves craft
that which serves owners
ancillary services
associated uses

Design Evolve the design logically from the given conditions rather than imposing a predetermined plan
Allow physical structure to evolve from existing geography.
Tailor the scheme to be in sympathy with local environment

---

3. Categories of marina development: Bibliography

Allan, Tom, 'The Marina Scene', *The Guardian*, 1 June 1972.

Baumann, Duane D., *The Recreational Use of Domestic Water Supply Reservoirs: Perception and Choice*, University of Chicago, Department of Geography Research Paper No. 121.

Beazley, E., *Designed for Recreation. A Practical Handbook for all concerned with providing Leisure Facilities in the Countryside*, Faber & Faber Ltd., London 1969.

Black, A. N., 'French Marinas—St. Malo to La Rochelle" Paper presented at the *Symposium on Marinas and Small Craft Harbours,* Department of Civil Engineering, University of Southampton, April 1972.

'Brighton Marina' *The Financial Times*, 5 October 1976.

British Waterways: Recreation and Amenity, Cmmd. 3401, H.M.S.O., London 1967.

British Waterways Board, *Leisure and the Waterways Board*, The Board, 1967.

British Waterworks Board, *Leisure and the Waterways*, H.M.S.O., London 1967.

Browne K. and Ginsberg, L., 'Brighton Marina: An Alternative Plan', *The Architectural Review*, February 1975.

Burke Inc., Ralph H., *Small Boat Facility Survey and Marina Requirements (for Milwaukee County)*, Chicago 1957.

Dunham, James W., 'Design Consideration for California Marinas', *Journal of Waterworks and Harbours Division*: American Society of Civil Engineers, November 1960.

Forester, T., 'South Sea Bubble', *New Society* 3 October 1974.

Hockley, D. H. E., 'The Marina—in the Sailing Scene—and the Environment'. Paper presented at the *Symposium on Marinas and Small Craft Harbours*, Department of Civil Engineering, University of Southampton, April 1972.

Jones, Lewis F. R., 'Ship Shape and Botel Fashion'.

Nicholson, P. C., 'The Advantages of Marina Moorings,' *Boat World*, Sell's Publications Ltd., 1972.

Roberts, Ken, 'Water Sports for Us All: Leisure for the Disabled', *The Municipal and Public Services Journal*, 11 June 1976.

Wood, Halloin, Lehmann, Schoeneiker and Euston, *Waterfront Renewal: Technical Supplement*, Wisconsin Department of Resource Development, Madison 1964.

4 Inland situations

It is easier to generalise about coastal sites than about those inland. In some cases, such as estuarial water, it may be arguable whether a site *is* coastal or inland. With sites inland not only the location but the scale too may range from a quarry of a few acres to what is virtually an inland sea. The perimeter of the Great Lakes for example is nearly two-thirds the length of America's total coastline and their acreage is larger than many of the world's seas. As can be seen in 3.2 page 40, the coastal group is covered by the single description of open water, whereas the inland group embraces canals, reservoirs, lakes, gravel pits and rivers. Each of these in turn may be sub-divided by size, location and use.

One factor that will influence the function of the marina and the kind of craft within it is whether the water is entirely land-locked or whether navigable outlets give access to open water. With rivers this is nearly always so and the canal system often links, via navigable rivers, to the coast. Even in Britain some virtually land-locked waters are large enough for enjoyable cruising. The Norfolk Broads (which are man-made), the Lake District waters and the Scottish lochs are examples.

There is a total of 130,000 ha^2 (500 miles^2) of inland water in Britain, over 2,000 km (1,250 miles) of navigable rivers and about 3,500 km (2,188 miles) of canal, of which 2,880 km (1,800 miles) has access by way of existing towpaths. The use of rivers and canals as bases for linear park systems is growing and water-based leisure complexes are being formed in the Lee and Colne valleys near London and by the Trent near Nottingham.

Inland problems and opportunities

Many of the basic principles pertaining to coastal marinas will be equally applicable to inland sites. The differences are mainly those of catchment, accessibility, shelter and the limited water area. Generally speaking the coastal site, being on the seaboard will have a semi-circular catchment area simply because only half of its compass radii will be on land anyway. No such constraint applies to inland water where people may be attracted from all directions. The accessibility of many inland sites from heavily-populated areas is the factor which gives them their greatest value and often creates the greatest problems.

In practical terms of sailing area the water accessible from the coastal site is limitless. For most inland sites, however, the population pressures create demands upon water area which are greater than can be comfortably accommodated. This difficulty would sometimes arise even if sailing were the only activity to make its claim, but angling, canoeing, rowing, sub-aqua sports, water ski-ing, motor-boat cruising and racing and wild-life conservation and studies may all have good reason to claim a share of the water. Such demands will no doubt raise questions of compatibility outlined in 4.B, for even within sailing itself the different needs of racing, cruising and sail-training may require segregation by timing or zoning. As Nan Fairbrother observes in *New Lives, New Landscapes*[1] this is a problem largely solved by our natural gregariousness, for as a general rule the more popular the activity the more communal its enjoyment and in this lucky law of human nature lies the salvation of our landscape. Most of us, even from crowded cities, don't want to get totally away from other people; on the contrary crowds mean "we've come to the right place". Much of the general gaiety depends upon the cheerful company of our fellows in holiday mood and recreation areas are therefore an excellent way of collecting us in a single place and discouraging our spread, and also of providing a large return of pleasure per acre. One factor which helps to moderate the populous conditions on British inland water is the more equable climate which lengthens the usable time for recreation well beyond that of its coastal counterpart. These kinder conditions are ideal for sail-training, particularly in shallow lakes where a depth of 1·3 m (4 ft 6 in) ensures maximum safety, whilst allowing adequate depth for small-keeled craft. A roughly circular or elliptical shape is best for this as it allows tacking practice with minimum interference from other craft. Intrepid coastal sailors often refer in disparaging terms to inland water and class its advocates as the least adventurous of the boating public. This is not altogether true. People have to start their boating somewhere and where better to learn about boat safety and rudimentary boat handling than on sheltered inland waters? Thus the modern inland marina can fulfil a very special need in acting as the nursery slopes where the whole family can participate.

At present all the inland marinas in England belonging to the National Yacht Harbours Association provide approximately 2,500 berths. This number would need multiplying by approximately 4 to account for all the other berthing spaces not represented in this membership. The resulting 10,000 equals 1·4 per cent of the 700,000 pleasure craft in use in Britain in 1972. It has been estimated that 3 per cent of the population go cruising or boating but that this is likely to rise to 20 per cent in 20 years time, when it would involve 5 million craft. Assuming a pro rata increase on inland sites by 1992, a further 70,000 moorings will need to be provided on inland water. How is such a figure to be accommodated? There are several possibilities. As the need for water grows, so more reservoirs will be needed, more valleys flooded and more estuaries impounded. *It should be ensured that these new water areas are in future combined with recreation.*

As the White Paper on leisure makes clear "Arrangements usually are and certainly should be made at the planning stage so that the full recreational possibilities can be taken into account and the appropriate facilities included in the plans."[2] On this Command Paper the late Nan Fairbrother observed, "There are to be no more water boards shutting off tracks of fine country like the Lady Bower reservoir near Sheffield, where water and landscape equally are sterilised by edict. Purification by edict is surely obsolete now that water engineers can deliver to us, pure from our London taps, water which has already passed through the sinks and baths and washing machines of Oxford and Reading and goodness knows where else besides."[3]

1. Fairbrother, Nan, *New Lives, New Landscapes,* The Architectural Press Ltd., London 1970; Penguin Books Ltd., Harmondsworth 1972.

2. *Leisure in the Countryside,* Command paper 2928, H.M.S.O., London 1966.

3. Fairbrother, Nan, *op cit.*

Planning

It is essential to discuss any proposal for an inland-water recreational development with the local planning authority at the earliest possible stage —certainly before an outline planning application is submitted. River and water authorities too should be consulted before firm plans are made. These bodies will undoubtedly have attitudes, and probably policies, towards development which will set firm guide lines to those who propose to build or expand beside inland water.

With the problems of congestion and compatibility of uses that so frequently accompany inland sites the potential developer may find that planning consent is more difficult to obtain than in coastal areas. In the National Parks and areas of high landscape or conservation value, speculative development will not be allowed—at least as far as Britain is concerned —although combined ventures between the public and private sectors within the context of an overall recreation policy are not unknown and often very successful.

Local authorities themselves, perhaps too frequently, tend to drowse behind outdated policies which have not been re-examined in the light of the growing needs for water-based recreation. Whilst there are many planning applications which are brash and entirely inappropriate, others are refused for reasons which do not take into account the progressive attitudes needed to meet the challenge of contemporary demands. Such is the adherence to a status quo and the reluctance to accept change in any form that if there is something more difficult than getting permission for a certain development, it would be obtaining approval to remove it ten years later.

4.1 The proposed inland marina at Gossmore on the River Thames. It was rejected on the grounds of being detrimental to its surroundings

Inland water resources

Recreation on reservoirs

The steeply increasing demand upon inland water for recreational use has put pressure on water authorities to provide—or at least allow—the necessary recreational and social facilities at their reservoirs. The growing willingness for water authorities in Britain to recognise the recreational potential of British reservoirs stems from a 1969 Government Circular urging greater use of inland water as an amenity. With the Metropolitan Water Board this more relaxed attitude first showed itself at the Queen Mary reservoir and the new policy was virtually built into the Thames Water Authorities' new Queen Elizabeth reservoir at Datchet near Windsor. This 377,000 million litre (83,000 million gallon) reservoir not only forms an important part of the regional water supply strategy but also provides facilities for the following sports:

Sailing

The reservoir provides 192 ha (475 acres) of deep water for sailing. Except for a petrol-powered standby safety boat no power boating is allowed because of oil pollution, thus precluding power boating and water skiing. Clubs are established with a membership of 400. Casual users may launch their own boats on a daily basis or hire one of 6 Mirror or 7 Wayfarer dinghies. Instruction is available for clubs and the 1976 Olympic sailing squad held trials involving 300 sportsmen.

Fishing

The reservoir has a perimeter of 5·3 km (3·3 miles). Shore fishing is not permitted because algae bloom on the 1:3 embankment is considered dangerous. Wet and dry fly fishing is allowed from electrically powered boats on a casual-user basis. One or two persons per boat is allowed. The reservoir is stocked with 100,000 brown and rainbow trout. Periodic re-stocking is necessary as trout do not breed in still water.

Horse riding

Annual permits are issued for use of the 2 km (1¼ miles) ride in part of the 89 ha (220 acres) of perimeter land.

Sub-aqua diving

During 1976 trials involving small teams of divers were used to determine the chemical and bacteriological effects of this activity on the water. Deep-water training facilities are scarce and this reservoir bottoms at 22·8 m (75 ft).

Rowing and canoeing

There seems no health reason why these sports should not be enjoyed but the storage problem of long and fragile craft has not yet been solved. The 1973 Water Act put a clear duty on Regional Water Authorities to make the best possible use of their water and associated land not only for water supply purposes but for use as a major recreational facility. The exceptionally dry summer of 1976 will accelerate the provision of both supply and compensation reservoirs. particularly in the south of Britain. This increase in the total stock of inland water available for recreation comes as a most welcome addition at a time when water-based pursuits continue to increase in popularity until, at a national level at least, a threefold increase in activities is expected by the end of the century. Already other reservoirs are planned. As well as the sports previously mentioned amenity areas are being prepared as are viewing points, car parks, picnic sites and access to the rim of the reservoir.

Boat accommodation

Except for a short quay or a holding dock, water authorities do not generally allow permanent berthing within reservoirs. A boat park is the normal accommodation usually situated outside the surrounding berm. A reasonable access road is required for vehicles towing boats and, as security gates are almost always necessary, arrangements must be agreed between the water authority, the club and individual members.

4.2 This wooded cutting at the eastern end of Islington Tunnel on the Regent's Canal is surprisingly less than a mile from London's King's Cross Station. It is typical of many quiet stretches of water suitable for leisure boating

4.3 Canals can be lively too: a boat rally at the Uxbridge Cruising Club Marina

Supply reservoirs

There is great variation in the policy of different water boards on permitting the use of reservoirs for recreation. Boards in Gloucestershire, Somerset and Derbyshire have been the most progressive. Usage depends on the water's natural purity and the consequent amount of purification required. Recreational use can obviously be allowed if the water has to be purified, but in highland areas such as Dartmoor, where water can be taken almost straight for drinking, purification entails considerable extra cost. In any case, additional roads and car parks, buildings, slipways etc. are necessary and the water level may fluctuate.

Club membership is usually required for ease of control over fishing, sailing and canoeing. Water ski-ing, bathing and sub-aqua are rarely allowed.

4.4, 5 These 'before' (August 1968) and 'after' (Spring 1971) pictures show what can be done to recover areas of total dereliction. The central building is The Longboat, an Ansells pub on the Birmingham canal

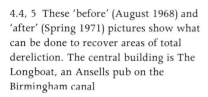

4.6, 7 The James Brindley Walk canal-side redevelopment scheme: more of the amazing transformation wrought by the City of Birmingham, England. The 'before' picture was taken in January 1957 and the 'after' in July 1969

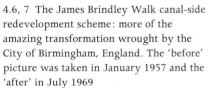

Compensation reservoirs

Most recreational uses are easily accommodated in these if the water is not to be drunk, for example on Snowdonia's Tryweryn compensation reservoir, which regulates the natural flow of the river.

Canal feeder reservoirs

These reservoirs, which maintain the water level in canals, are usually owned by the British Waterways Board, which may not, however, have riparian rights.

Canals

Canals offer great potential for relieving overcrowding in popular waterways such as the Thames and Norfolk Broads. They could be made to pay by encouraging commercial as well as recreational traffic, and by their use as a water supply for irrigation and industry and as a permanent water 'grid'.

The canal system of any country can certainly contribute a lot towards the total recreational provision of water if its potential is only recognised. In Britain, active promotion of boating, sailing and cruising is needed from the government, the new area waterways boards and the planning authorities through whose area they run.

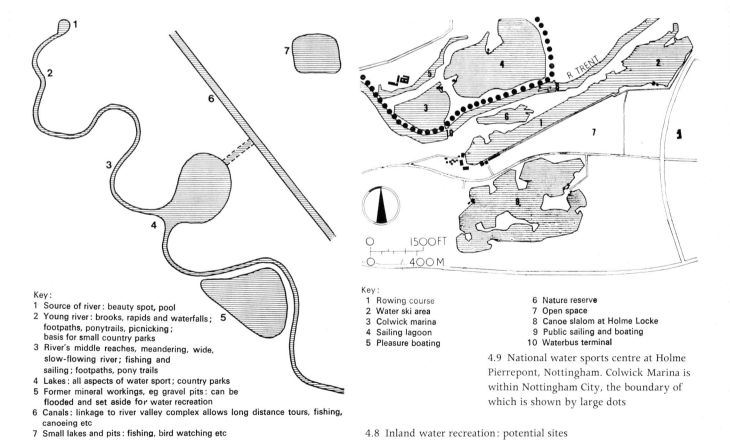

Key:
1 Source of river: beauty spot, pool
2 Young river: brooks, rapids and waterfalls; footpaths, ponytrails, picnicking; basis for small country parks
3 River's middle reaches, meandering, wide, slow-flowing river; fishing and sailing; footpaths, pony trails
4 Lakes: all aspects of water sport; country parks
5 Former mineral workings, eg gravel pits: can be flooded and set aside for water recreation
6 Canals: linkage to river valley complex allows long distance tours, fishing, canoeing etc
7 Small lakes and pits: fishing, bird watching etc

Key:
1 Rowing course
2 Water ski area
3 Colwick marina
4 Sailing lagoon
5 Pleasure boating
6 Nature reserve
7 Open space
8 Canoe slalom at Holme Locke
9 Public sailing and boating
10 Waterbus terminal

4.9 National water sports centre at Holme Pierrepont, Nottingham. Colwick Marina is within Nottingham City, the boundary of which is shown by large dots

4.8 Inland water recreation: potential sites

4.10 Inland sites can be large and luxurious: Chicago Yacht Club . . .

4.11 or small and simple: M-port, Florida

Recreation is a fairly new role for canals. They have always been well used by anglers but it was only in the 1950s in Britain that a wider public began to appreciate their leisure value with the growth of canal cruising holidays. This has been a significant factor in the retention and development of many canals which are no longer viable for commerce. In 1968 the Transport Act gave a new charter for inland waterways which acted as a springboard for a new concept of recreation and amenity on the canals. France has a wonderfully active canal system whose commercial and recreational roles both grow yearly. In 1966 this network consisted of 3,300 km (2,160 miles) of navigable rivers and 4,600 km (2,875 miles) of canals. Between 1966 and 1970 France spent 1,850 million francs (£130 million) on improving and extending its canal system. In the same period Britain spent £7½ million.

Marinas may be planned along canals wherever suitable widenings, docks, reservoirs or basins occur. These will probably be quite modest promotions with few facilities compared with their coastal counterparts, but the useful role that such outlets play when they are within, or easily accessible from, highly-populated areas extends far beyond their size.

Rivers

Freshwater rivers are controlled by the appropriate river authorities. Tidal rivers are often controlled by port authorities. They can be used for sailing, canoeing, rowing and cruising. Though water ski-ing does take place on rivers, it is discouraged as it causes bank erosion.

Natural lakes

England's relatively few lakes have attracted visitors for many years, and areas such as the Lake District are now seriously overcrowded during the holiday season, whereas Scotland and Wales have many remote lakes far from population centres.

Artificial lakes

Though it is unusual for large lakes to be created for purely recreational use, existing water areas—mostly lagoons in old gravel or clay pits—are enlarged and reshaped as at the National Water Sports centre at Holme Pierrepont, Nottingham.

Enlarged gravel pits

These are appearing rapidly along river valleys, and have great potential, especially in helping to form large new parks such as the Cotswold water park and the Lee Valley and Colne Valley regional parks.

Inland water recreations and their requirements

Sailing

Dinghies need a hard beach or, preferably, a slipway down which cars may reverse, allowing boats to be floated off trailers. There must be sufficient width to take a car and trailer safely and to allow room for persons helping launching. Slipways should extend to the low-water mark and are best set back into the bank to avoid obstructing navigation. Dinghies do not require moorings: for temporary tying up during sailing hours a slipway is adequate. Keel boats need moorings or—particularly to facilitate winter storage and maintenance—some form of lifting device. Mooring must be to a jetty or, if there is considerable change in the water level, a floating pontoon which will not foul the boat.

Lavatories and simple changing rooms and washroom accommodation are essential. Sewage disposal near reservoirs must be into a public sewer or by means of a chemical process. Parking space is needed for sailors' cars and trailers and for visitors' cars. Boat storage sheds are advantageous but not essential. A shed and/or trestles is useful for storing spares. A covered store for other small equipment can be incorporated with the other buildings.

Water areas less than 4 ha (10 acres) are used for competitive and recreational dinghy sailing, but the Royal Yachting Association recommends a minimum of 6 ha (15 acres). Boat density on inland waters may be in the order of 1 boat to 0·80 ha (2 acres) up to 1 boat to 0·20 ha (0·5 acres). The shape of the water area is not very critical but the minimum area must not include small bays. A bank consisting of long smooth curves or straights is desirable. Islands are acceptable if they are at least 46 m (150 ft) from the bank. The depth of water for sailing should be at least 1·50 m (5 ft), but 1·80 m (6 ft) is preferable. Shallows must conform to the same conditions as islands.

4.12 Harleyford Marina on the River
Thames: a neat built-in arrangement

4.13 This harbour in a German estuary is made from pre-fabricated parts and is
protected by its own integral floating breakwater system. This aerial view was taken at
Beaufort wind force 6

Rowing

Launching for this sport is by means of a ramp or steps. The landing stage
should be 18 m (60 ft) long for the sideways launching of 'eights'; if necessary, with a ramp and pulleys for hauling up narrow frontage.

The clubhouse should be sited preferably in a bay or sheltered inlet away
from traffic.

Training can take place in a rowing tank 12·6 × 7·6 m (41 × 25 ft) for one
eight. As shells are easily damaged by obstructions and swamping from
other craft, the water must be sheltered. If it is also used by power boats,
the water areas must be zoned or subject to timetables. The Amateur Racing
Association club competition standards require a stretch of water 1,500 m
(5,000 ft) long, not less than 1·83 m (4 ft 4 in) deep and 50 m (160 ft) wide to
accommodate 4 lanes. National and regional competitions, for example the
Federation Internationale des Sociétés d'aviron Standard C course, require
a stretch of water 2,000 m (6,600 ft) in length, at least 50 m (160 ft) wide to
accommodate 4 lanes, and 1·83 m (6 ft) deep. The Olympic or F.I.S.A. men's
championship course is 2,000 m (6,600 ft) long, plus 100 m (320 ft) beyond
the finish. It is 75 m (245 ft) wide to accommodate 6 lanes, plus 5 m (16 ft 5 in)
between outside lanes and the banks. The water should be not less than 3 m
(10 ft) deep.

Canoeing (general)

Canoes can be launched almost anywhere, but a landing stage is useful. In
addition, canoeists need a boat store and clubhouse. This sport need not
necessarily be confined to inland waters. Surf canoeing for example, takes
place in surfing areas and beaches may have to be zoned for safety.

Canoe touring

Water as little as 0·23 m (9 in) deep and 0·60–0·90 m (2–3 ft) wide provides
a possible route for this sport. Camping sites with portage facilities will be
needed along waterways.

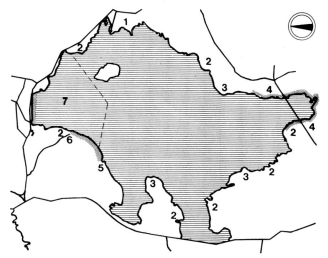

Key:
1 Picnic site
2 Entrance, parking and lavatories
3 Bird-watching hides
4 Nature reserve
5 Fishing hq entrance
6 Sailing clubhouse
7 Sailing area (summer)

Note: areas of dot tone at water edge indicate where fishing from banks is prohibited

4.14 Chew Valley lake (10 miles south of Bristol) Scale: 1:20,000

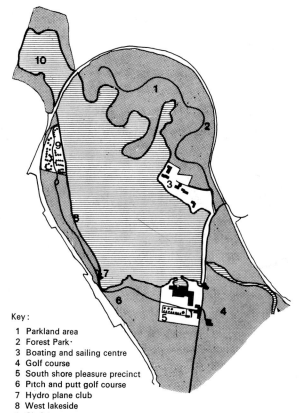

Key:
1 Parkland area
2 Forest Park·
3 Boating and sailing centre
4 Golf course
5 South shore pleasure precinct
6 Pitch and putt golf course
7 Hydro plane club
8 West lakeside
9 Chalet group
10 North pool

Note: dot tone indicates woods and parkland

4.15 Chasewater recreation centre at Cannock, Staffordshire. Its nucleus is a former canal reservoir among reclaimed waste land

Activity	Type of Craft	Approximate Water Area Needed in Acres	Equivalent in Hectares	Number of Craft	Total Acres	Total Hectares
Sail training	Dinghy	1·00	0·41	10	10	4·1
Plain sailing	Dinghy	0·75	0·30	20	15	6·1
Racing	Dinghy	0·50	0·20	16	8	3·2
Cruising	Motor boat 8 m (26 ft)	1·00	0·41	20	20	8·1
Cruising	Motor boat 8 m (26 ft)	1·50	0·61	12	18	7·3
Cruising	Sailing boat average 5·5 m (18 ft)	1·00	0·41	29	29	11·7
		Totals	107*	100	40·5	

* Multiplied by a diversification factor of 4 gives 428 berths.

4.A Sailing space needed for craft in 40·5 ha (100 acres) of water. With entirely enclosed water it is important to have *some* idea what area of water various craft need for comfortable boating conditions. This will vary according to the craft, the kind of sailing, the shape of the water area and the weather conditions. The above table gives an approximate breakdown for 6 kinds of boating, occurring together on a 40·5 ha (100 acre) lake 366 × 974 m (1200 × 3200 ft). Even at peak periods not all craft will be sailing at once. A diversification factor can therefore be used to estimate the maximum number of berths to ensure comfortable sailing conditions. On enclosed water in peak periods this could be about 4 (or 25 per cent usage) giving, in the above example 428 berths.*

The management may desire to accept delay in getting afloat for some owners at peak periods. It is a question of setting this against the considerable *under-use* of the water for the 95 per cent of off-peak time.

The example of course presumes that no other sport or activity is allowed on the lake, an unlikely (and selfish) assumption.

The quantity of inland water is sometimes given in millions of gallons. One million gallons will cover 1 ha to a depth of 447 mm (1 acre to a depth of 44 in).

'White water' canoeing
The correct conditions and unrestricted passage are required for this sport.

Canoe racing
The international distances for competition sprint canoeing are 500 m
(1,642 ft), 1,000 m (3,284 ft) and 10,000 m (32,840 ft). A 1,000 m (3,284 ft)
straight course of a minimum depth of 2 m (6 ft) and 45 m (150 ft) to accommodate 6 abreast is required.

Canoe slalom
This sport takes place on fast-flowing, turbulent water. There are penalties
for any divergencies from the straight course. The course should not be
more than 800 m (2,625 ft) long, measured through the gates. It may extend
several hundred metres down the river, or take a more serpentine route in
the restricted turbulent water below a weir. In the case of a hill river course,
it is a great advantage, even a necessity, if the river flow can be controlled
by manipulating sluices at a reservoir higher up in the valley.

Water ski-ing
The requirements for this sport are mooring and storage space for boats; a
clubhouse, and a ramp or hard beach, with jetties for dry starts.

As calm water is required, no designated area should be close to vertical
or concrete banks which tend to build up a rebound wash. For this reason
the sea does not provide the conditions necessary for competitions. Jumps
and a permanent slalom course may need to be maintained. The zoning of
water is important. Minimum dimensions of water for establishing a slalom
course are 640 × 182 m (1,950 × 555 ft) but 823 × 1,097 m (2,447 × 3,345 ft)
gives greater clearance for turning and accurate speed approaches. A jump
and figures course can be incorporated within a slalom course.

Power-boating
There are four main categories of power boats: motor cruisers, runabouts,
hydroplanes and off-shore racers. Of these, motor cruisers can usually be
accepted on inland water provided certain conditions of speed and sanitation are observed. Runabouts are small, general-purpose craft and when
used for day trips and picnics usually integrate well with sailing and other
pursuits. It is their use for racing and water ski-ing which is more difficult
to tolerate. Hydroplanes are the boating equivalent of the racing car and
must have a 'track' of their own. Damage to banks from the wash of fast
large craft is a continual problem dealt with by speed limits, setting a
minimum distance from the shoreline (which is rarely practicable) or treatment of the banks themselves. Their effect not only within the recreation
area but on nearby residents must also be considered.

Launching and mooring requirements on inland waters are the same as
for sailing dinghies. Larger power boats need the same facilities as keel
boats. Boats can be stored outside under their own covers but engines need
a lockable storage room. Fuel storage space is required. Lavatories, changing
and first aid facilities are necessary and a clubhouse is desirable. Car
parking should include provision for spectators. The minimum area of
inland water required for this sport is 6 ha (15 acres). A large bay in a lake
is desirable for 'pits' and shape requirements of the water area are similar

	Fishing	Swimming	Sub aqua	Wildfowl	Canoeing	Rowing	Sailing	Water ski-ing	Hydroplaning	Power-boats	Cruising
Fishing		X	X		PZ	PZ	PZ	X	X	X	PZ
Swimming	X			Z		Z	Z	Z	Z		Z
Sub aqua	X					PZ	PZ	PZ	PZ	PZ	Z
Wildfowl		Z						X	X	X	
Canoeing	PZ					PZ	PZ	PZ	PZ	PZ	
Rowing	PZ	Z	PZ		PZ		PZ	P	P	P	PZ
Sailing	PZ	Z	PZ		PZ	PZ		PZ	PZ	PZ	Z
Water ski-ing	X	Z	PZ	X	PZ	P	PZ		PZ	PZ	N/A
Hydroplaning	X	Z	PZ	X	PZ	P	PZ	PZ			N/A
Power boats	X		PZ	X	PZ	P	PZ	PZ	PZ		N/A
Cruising	PZ	Z	Z			PZ	Z	N/A	N/A	N/A	

Key—X incompatible: P programming; Z zoning; N/A not applicable

4.B Compatibility of water sports

	Lakes	Canal feeders and compensation reservoirs	Water supply reservoirs	Rivers	Canals	Sea
Fishing	X	X	X	X	X	X
Swimming	X			X		X
Surfing	X					X
Sub aqua	X			X		X
Diving	X	X	X	X	X	
Wildfowl	X	X	X	X		X
Canoeing	X	X	X	X	X	X
Sailing	X	X	X	X		X
Water ski-ing	X	X				X
Hydroplaning	X	X				
Power boats	X	X				X
Cruising	X	X		X	X	X

4.C Areas suitable for water sports

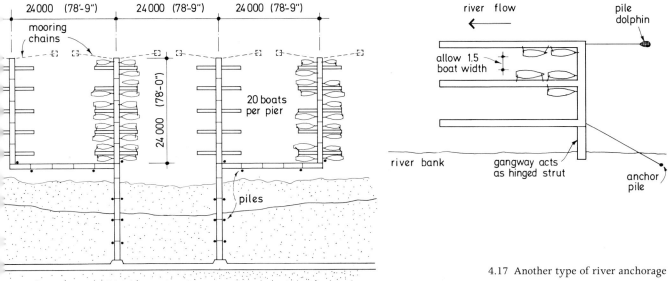

4.16 A mooring system for 80 boats built out from a river bank. The two main walk-ways are held between piles. Movement at the offshore ends is controlled by blocks and chains

4.17 Another type of river anchorage system

to those for dinghy sailing. The minimum water depth is 0·92 m (3 ft) and it must be weed free.

Cruising

Cruisers require moorings as a basic minimum and boat storage, repair facilities (with lifting equipment), a fuel supply, and chemical W.C. disposal points are also needed to a greater or lesser extent. Possible additional facilities include changing accommodation, lavatories, shops, restaurants, bars and car parking.

Because of great demand—particularly in south-east England—moorings are scarce and expensive. As traditional swinging moorings take up too much space, more compact methods must be used: either mooring alongside a jetty two abreast or mooring stern-on to a jetty with bows onto piles.

Bird clubs

The location of bird clubs (dictated by bird habits) is usually by 'natural' waters but some examples are by reservoirs. Specialised requirements are 'hides', and freedom from disturbance and therefore bird clubs are compatible only with angling and possibly sailing.

The Slimbridge Wildfowl Trust runs 'pop' versions, more like zoos, in Gloucestershire and Northamptonshire.

Angling

Angling is about the most popular water sport. Its varied facets allow it to be practised almost anywhere, but zoning is important. It is best restricted to banks on one side only, leaving large free areas for fish. Water disturbance caused by violent sports may drive fish into the free zones.

Swimming

The suitability of inland water for swimming depends on the level of pollution, obstructions and the other uses to which the water is put. Where necessary, lakes for example can be treated to make them bacteriologically safe, as by chlorination at the Serpentine Lido, London.

Underwater swimming or sub aqua

Water depth is the most important consideration in this sport, both shallow and deep water being required. The surface area is relatively unimportant. The sea is preferred for most activities but spear fishing is now allowed in inland water. Powerboats and water skiers are a danger to underwater swimmers and safety is obviously important. A code of practice for underwater swimmers published by the Ministry of Agriculture, Fisheries and Food is designed to ensure that divers do not interfere with fishing interests.[4] A clubhouse and storage facilities for equipment are desirable.

4. Ministry of Agriculture, Fisheries and Food, *Code of Practise for Underwater Swimmers*, H.M.S.O., London 1967.

Pollution and sanitation

Inland water demands even stricter pollution regulations than coastal sites. Under no circumstances can sewage or rubbish be allowed to discharge into enclosed water, canals or rivers. Adequate public lavatories ashore, sealing of heads, mobile suction units and efficient garbage collection are some solutions. Pollution from outside sources is particularly serious in enclosed waters with little natural flow. Nitrate seepage from adjacent hill farms or faulty sewage systems are frequent causes of algae growth on lakes and lochs. Driftwood is a problem in tidal rivers where commercial users exercise insufficient control and the river authorities take no steps to reduce the amount. It is not easily solved. Netting, 'sweeping' and special patrols help, but the best reduction comes by a change in attitude—don't let it get in in the first place.

Reservoirs are of three types: direct supply, regulating and rivers and lakes. The strictest control is exercised to prevent contamination and regulations often include the following controls:

1   No power-driven craft except rescue boats

2   Car parks sited away from water to prevent oil pollution

3   All buildings, land, jetties and roads must be owned by the water company and leased out to the individual clubs

4   No bathing or paddling except when launching boats

5   No tar or creosote etc. is permitted to be used. (This can be tasted in water even after chlorination)

6   Fully adequate sanitary arrangements must be provided and maintained

Design considerations

Liberated from many of the problems of tides, wave action and severe climatic conditions, the marina designer on inland water can form the development more freely and probably less expensively. Size and scale are likely to be smaller than the coastal counterpart, particularly with the basic engineering work. Some of the best architecture of inland docks and basins has a robust and functional elegance that has travelled upstream from the nautical tradition. Certainly in adapting inland commercial harbours to recreational use the vitality and assurance of much existing construction will set an example to be respected. The range is very wide, taking in the most rural and urban examples. In both cases the new construction should appreciate the past whilst accepting the future and aim to incorporate the inevitable changes with the least harm or even with benefit to the environment. The siting of the marina on the edge of enclosed water will be governed by the site conditions, access, orientation and aspect and in these respects is similar to the built-in type of coastal harbour which offers the maximum perimeter for the given water surface.

Linear forms such as rivers and canals present a different problem. They can of course be constructed away from the water's regime, being linked by a short channel or loch. In contrast to this is the river anchorage principle where the whole walkway system is strutted out from the bank by the gangway which acts as a hinge. (See 4.17.) Care is needed with engineering work likely to affect the established bed of inland water. In its natural state silt and organic matter will have formed a seal. Dredging and pile driving will puncture this, allowing water to permeate into nearby land, altering the water-table and causing damage to existing property and perhaps affecting the load-bearing capability of adjacent soil.

Edge treatment is very important and has a great visual and practical effect on any stretch of water. Inland water, except in estuaries, tidal rivers and some reservoirs, is unlikely to be subject to large variations in level. Where there *is* a range of levels the bank gradient will control the amount of exposure at low water. It is best to allow the soil's natural angle of repose to dictate what this gradient should be. Shingle is better than mud for maintenance and appearance although it can be quite expensive. Grass will grow well provided it is not under water more than about a quarter of the time and is protected from too much wear or pollution. Gabion mattresses or random flat stones set in pitch (rip-rap) serve as a hard-wearing edge where wave action is a problem.

Wave height may be calculated from the formula:

$$h = 0 \cdot 025 \sqrt{L} \text{ where } h = \text{ height of waves in feet}$$
$$\text{and } L = \text{ the greatest length of water surface}$$

Timber piles either driven as a continuous 'fence' or as spaced-out supports to horizontal planking make a successful light edge. If they act as a retaining wall then cable ties or deadmen anchors back into the land may be necessary.

Inland economics

The finance of inland marinas is likely to be very different from coastal experience. The size, on average, is much smaller and berths are even less likely to be profitable without support from ancillary services—either those integral to the marina itself or independent uses to which the marina is subsidiary. An important factor in favour of inland locations is the reduction in basic engineering costs compared with coastal sites. As against this craft are usually smaller, often trailed to the water and may never be permanently berthed at all.

It must be admitted that the opportunities for exploration and adventure are often limited. As one marina manager remarked "They go upstream the first weekend, downstream the next and the third weekend they say 'What do we do now?'" The amount likely to be spent on berthing is not high, neither are repairs, launching nor storage, the principal reason for marina berthing having been removed by the milder site conditions.

Sailing, given sufficient water area, is suitable on all inland water except perhaps canals and very narrow rivers. Cruisers are debarred from water-supply reservoirs and power boats are restricted to some lakes and canal-feeder and compensation reservoirs. *These constraints indicate that inland marinas are a better proposition where facilities may be shared as previously shown with other sports and recreations. This multi-use not only maximises the potential of scarce and popular water resources but improves the chances of financial viability too.*

There are thousands of acres of potentially excellent areas of water for recreation which lie derelict. Worked-out wet and dry pits could be turned over to boating without great costs being incurred. There is no doubt that the demand is there. In November 1967 the Greater London and South East Sports Council surveyed 56 wet gravel pits in west London alone. They found all of them intensively used, supporting 66 sailing and angling clubs with a total membership of almost 10,000 people. Abandoned canals too may be restored for pleasure cruising. This has been done with success on the Monmouthshire Canal. With social benefit as their motive rather than monetary profit, local authorities and official bodies can often create splendid water areas from the most unpromising material. Often the opportunity is rejected because it *appears* not to pay. The administrative difficulties are usually greater than the financial and technical ones and these must be overcome if the coming demand is to be even partially satisfied.

4. Inland situations: Check list

Planning

Contact local planning authority and river/water boards
Define catchment area in size
       shape
       population
       socio-economic categories
Consider transportation modes, ease of access
What is the total water (and other) recreation policy for the area?
Estimate demand for boating in context with other water pursuits
Determine degree of compatibility with other activities and organise programming and
 zoning accordingly
Examine degree of multi-recreation use
 how much sharing?
 what constraints?
Estimate special requirements for craft when sailing and berthed
Is water land-locked or is there access to open water?
Determine need for facilities such as changing accommodation
        lavatories
        shops
        restaurants
        bars
        car parking

Reservoirs

Determine water board's policy
What type of reservoir
      canal feeder
      compensation
      supply
What constraints imposed if boating is allowed?
Consider shape, size, depth, and edge treatment in terms of safe sailing—particularly for sail
 training areas
What launching and mooring facilities are possible?
Consider access
   parking
   restaurants
   other facilities
What are maximum/minimum water levels?

Canals

Contact local authority and regional waterways board
Is canal connected to open water?
Investigate access points
    launching ramps
    towpaths
    locks
Consider car-parking space
Will commercial traffic hinder recreational craft?
Is mooring to be along canalside
      canal basin
      feeder reservoir
Determine mooring capacity in terms of number
         size
         type of craft
Will boating opportunities create sufficient demand for berths?
Establish firm policy on houseboats and living aboard
Determine present and future policy on canal's maintenance
         navigability
         use for recreation

Rivers

(See also Planning and Canals above)
Contact river (or port) authority
Consider width
      tidal range
      salinity
      erosion and siltation
Will development affect the river's existing regime or constitute a navigation hazard by day
    or night?
Determine maximum flow in knots and decide type of anchorage system
Consider locational options for berths:
      within main stream
      partly recessed into bank
      fully recessed into bank
      on outside bend
      on inside bend
      within secondary loop, pool or ox-bow
      independent water or with connection to river
Determine distance to open water in km and sailing time
Estimate likely proportion of berthholders to visitors
What is authority's policy towards powered craft, speeds?
What constraints regarding water depth on keels
                bridge heights on masts
                river width etc. on sail training
                navigability (locks and other obstructions)
                debris on small craft or propellers
                speed of craft on bank erosion
                access to river bank for launching and retrieval
Check public health and sanitation requirements within site
Is cleanliness of river improving or deteriorating?
What basic type of mooring system is most suitable: fixed
                                floating
                                anchored
Determine design, pattern and materials for berthing or adopt suitable proprietary system

Lakes (natural and artificial)
and gravel pits

Integrate marina proposals with total recreation plans
Establish what other uses will share the water and determine their compatibility with the
    type of boating proposed
Estimate spacial requirement for craft when sailing and berthed
Calculate on-shore space requirements and optimum length and treatment of waterline
Will pollution be a problem in the form of algae
                          driftwood
                          methane
                          oil
                          sewage
                          weed
Check extraction company's responsibilities regarding reinstatement
Consider cost-benefit of enlarging, re-shaping or linking existing lakes/pits
Check grant availability for derelict land reclamation
                          leisure/sports promotion
Examine all safety measures for persons ashore and afloat

Sailing

Do water/space/wind/weather conditions inhibit quality of sailing?
Consider boat-stacker storage systems—investigate keel and mast problems
Decide launching methods: sloping shoreline
                      slipway
                      ramps
                      haulout
                      hoists
Provide safe and tidy storage for gear, masts, trailers, etc.
Calculate minimum water area for craft ashore/afloat
Apply a diversification factor to relate number of berths to number of craft out sailing

Rowing	Contact Amateur Rowing Association
	Contact FISA (Federation Internationale de Sociétés d'Aviron)
	Consider boathouses
	start pens
	finishing tower etc.
	Minimise unfair conditions across lanes e.g. gyrating currents, uneven crosswinds
	See Building Study of National Water Sports Centre, *The Architects' Journal*, 8 August 1973

Canoeing	Contact British Canoe Union
	See rowing above for facilities similar to and often linked with rowing
	Is a linked water network and/or access to open water available for canoe touring?
	Remember white-water canoeing demands either natural site
	improved site
	artificial simulation
	For canoe slalom conditions see preceding for white water

Water ski-ing	Contact British Water Ski Federation
	Carefully consider bank-edge treatment

Power boating	Contact Royal Yachting Association
	Consider fuel storage
	first aid
	safety equipment
	Will the site under consideration provide minimum water area of 6 ha, depth 0·92 m?

Cruising	Contact Cruising Association

Bird clubs	Contact Royal Society for the Protection of Birds
	The Wildfowl Trust

Angling	Contact National Anglers' Council

Swimming	Contact Amateur Swimming Association

Under-water Swimming & Sub-Aqua	Contact British Sub-Aqua Club

Pollution & Sanitation	Review all regulations affecting enclosed water
	canals
	rivers
	Consider lavatories
	sealing of heads
	garbage collection
	Also investigate debris
	litter
	driftwood
	oil
	creosote
	nutrients
	bathing
	siting of car parks
	See Water Resources Act 1963
	Ensure no discharge of effluent into underground strata
	Approach local authority re septic tank/cess pit emptying service

Design Considerations	See also Chapter 5
	Consider view of proposed development from opposite bank
	Select shape to give visual and recreational interest
	Investigate bank/edge/bulkhead treatment in terms of appearance
	cost
	function (sloping, retaining)
	maintenance

Consider car parks in respect of size
>access
>levels
>launching ramps
>landscaping
>screening
>surface treatment
>drainage

Examine lighting of grounds and water area
Design berthing layout
fixed or floating moorings
>anchorage system etc.
Research local materials and methods of construction

Inland Economics

See also Chapter 12
Determine catchment area in terms of transportation modes
>journey time
>popularity of water and allied pursuits
>class of craft
>socio-economic categories
>car ownership
>spending power, etc.

Evaluate drawing power/profit of ancillary facilities
Relate number of berths to density of craft per unit of sailing water

---

4. Inland situations: Bibliography

Arvill, R., *Man and Environment*, Penguin Books Ltd., Harmondsworth 1967.

Barr, John, *Derelict Britain*, Penguin Books Ltd., Harmondsworth 1970.

Beazley, Elizabeth, *Designed for Recreation. A Practical Handbook for all concerned with providing Leisure Facilities in the Countryside*, Faber & Faber Ltd., London 1969.

British Waterways Board, *Waterway Environment Handbook*, London 1976.

British Waterways Board, *Waterway Users Companion 1975*, London 1975.

Central Council for Physical Recreation, *Inland Waters and Recreation* (Survey of the west Midlands), The Council, London 1964.

Central Office of Information, *Sport in Britain*, Pamphlet 107, H.M.S.O., London 1972.

Chaney, C. A., *Marinas* (Second edition) National Association of Engine and Boat Manufacturers, New York 1961.

Countryside Commission, *Coastal Recreation and Holidays*, H.M.S.O., London 1969.

Department of Education and Science/Ministry of Land and Natural Resources, *Joint circular: Use of Reservoirs and Gather Grounds for Recreation*, H.M.S.O., London 1966.

Fairbrother, Nan, *New Lives, New Landscapes*, The Architectural Press Ltd., London 1970; Penguin Books Ltd., Harmondsworth 1972.

Gloucestershire County Council Planning Department, *Cotswold Water Park Draft Report*, The Council, Gloucester 1969.

Greater London Council, *London's Canal, its Past, Present and Future*, The Council, London 1969.

Greater London Council, *The Human Habitat*, The Council, London 1971.

Inland Waterways Association, *The Way Ahead for the Amenity Waterways*, The Association, London 1967.

Institute of Landscape Architects, *Lancaster Conference on Land and Water*, The Institute, London 1969. *Available only in the institute's library.*

Institution of Water Engineers, *Recreation on Reservoirs and Rivers*, The Institution, London 1972.

Institution of Water Engineers, *Recreational Use of Waterworks*, The Institution, London 1963.

Leisure in the Countryside, Command Paper 2928, H.M.S.O., London 1966.

Locher, Harry O. (ed.) *Waterways of the United States*, The National Association of River and Harbour Contractors, New York 1963.

McLoughlin, J., 'Control of the pollution of inland waters,' *Journal of Planning and Environmental Law*, June 1973.

Ministry of Agriculture, Fisheries and Food, *Code of Practice for Underwater Swimmers*, H.M.S.O., London 1967.

Ministry of Agriculture, Fisheries and Food, *Water for Irrigation*, Bulletin 202, H.M.S.O., London 1967.

National Yacht Harbour Association, *Yacht harbour guide*, Ship and Boat Builders' National Federation, London 1963.

Nottingham City Council Planning Department, *Colwick Park, Nottingham*, The Council, Nottingham 1967.

Nottinghamshire County Council Planning Department, *Holme Pierrepont, National Water Sports Centre*, The Council, Nottingham 1969.

Research And Planning Unit, *Greater London and South East Sports Council Information Sheet No. 9: Marina Development in the London Area,* June 1975.

'Water and planning,' *Town and Country Planning Journal* Special issue, June 1966.

Water Resources Board, *Water Resources in England and Wales*, Vol. 1: Report, Vol. 2: Appendices, H.M.S.O., London 1974.

5 General design principles

Having recognised the need for a leisure-craft harbour, preliminary investigations will have established where to develop, what type of craft and owner to cater for, when to commence construction, some notion of phasing the development and the initial, interim and probably final size. The engineering, legal and financial problems will have been explored and probably some investigatory work undertaken in the form of trial boreholes and soil testing. At this stage an outline master plan will have evolved, more for the purposes of obtaining the necessary minimum information than to determine an unalterable plan. The point is now reached where the professional expertise of the architect, engineer and others is concentrated upon the development of the marina plan in general and the solution of all the individual problems likely to be encountered in the design and physical development of the project. Whilst a main contractor will probably not yet have been appointed, the development team will nevertheless be outlining a work programme which may eventually be refined into a critical path network with individual bar charts for each main task over given periods of time. Because of the long duration of some survey and research work this will probably demand an early start, possibly before the final selection of the site is decided. This is particularly likely if advantage is to be taken of seasons or tides.

The principal elements of physical planning that the team will need to agree upon concerning the main engineering components would probably include the following:

1   Bulkhead and quay walls
2   Breakwaters
3   Locks
4   Piles
5   Dredging
    5.1   Type and amount of material
    5.2   Type of dredge
    5.3   Disposal of dredged material
    5.4   Volume of dredged material

Before dealing with individual cases it may be useful to outline a design approach which attempts to reconcile the marina with its setting and tries to see in what ways location is likely to affect the basic layout.

The design approach

The fundamentals of good design are common to all projects and usually involve relating the form to its purpose, which is a fairly accurate definition of function. The various categories of marina have already been shown,

5.1 The ultimate in multi-use: boat store, restaurant, skating rink, offices, car park and luxury flats (above) at Marina City, Chicago. The structure reminiscent of a stranded whale is a television studio

5.2 The boat store beneath the towers of Marina City. It was designed for 240 vessels, but only 130 can in fact be contained

pages 38–69, and each will require a different design approach stemming from the unique site conditions which prevail and solved only by a study of their singular problems. It is only possible here to generalise about questions commonly found within the prevailing types and to offer general guidelines towards their solution. The principal divisions affecting the design approach are those of cost, location and range of facilities—these aspects being, to some extent dependent upon each other. Whilst a talented design team will for the same cost produce a better final scheme than a less able group, certain design concepts such as cantilevers, open ground floors, changes in levels or a long irregular waterline will be precluded from the modest budget. The location itself may be very demanding in cost only allowing for the simplest layouts once the off-shore construction has been completed. The marina facilities will influence the plan form, both in their relationship to the water area and to each other.

The design approach must be strongly related to the local environment. A quiet lakeside set in gentle undulating countryside is asking for a soft approach, moulding the design to the setting and, by choosing natural

5.3 Notice boards conveying the marina's image deserve good design. These examples vary: from the garish at M-Port to the appropriately nautical at Nassau

materials allowing the organic form to grow naturally from the landscape as a tree grows from the earth. A converted dock might demand tough treatment where enclosed courtyards, iron bollards and simple materials are used to create a hard-edged urban atmosphere. The design, functionally and aesthetically will grow from these considerations of context and surroundings but will also need to take into account the public's needs and those of the developer. This can be difficult as preconceived ideas may need to be re-assessed and a critical examination necessary to bring all the inherent factors into line.

One of these will undoubtedly be finance which will have a fundamental effect upon design from the basic engineering approach through to the final quality of finishes. The roles of the accountant and the architect may be individual but they cannot be wholly independent. Brotherly love may be asking too much but a firm understanding based upon mutual respect will save both parties time and trouble and benefit the scheme and its developer.

Functional considerations

The land-to-water relationship

The percentage of land-to-water areas and the allocation of all the necessary components within that area is the central question in the planning process. The size of each of them, its position on the site and its relationship with other constituent parts are the three main problems to be solved. On page 104 the land and water areas are subdivided into their most commonly occurring functions and space allocated for each on a percentage basis.

The amount of practical advice that can usefully be given is limited because marinas vary so much in their siting, size and purpose. Even where basins are similar, great variations are possible. It is important to think carefully about the shape and form that the land and water make with each other, for this is half the battle in marina design. To a large extent geography and economics should determine the essential framework. It pays to design in sympathy with the existing elements, for taking advantage of the natural configuration will tailor the development to the site and bring aesthetic and financial benefits.

Practical planning starts from a broad statement of programme and works through the alternatives selecting and rejecting along the way. Much of this process is subjective but it will add to a general understanding to classify the factors and needs governing the location of each of the principal areas.

Spatial requirements and component parts

The profitability of most developments is very closely related to what can be got on the site consistent with good planning and a pleasant environment. Economy in layout on land and water will not only bring a direct monetary return but will also save by making best use of such costly components as jetties, quay walls, the dredged water area and the total land use. In general terms 5.37 and 38 can be used as guides in determining the allocation of land and water to some of the more commonly found uses. 5.38 averages out 4 prime divisions on a percentage basis, given a 1:1 land-to-water ratio. Any departure from this split will of course alter the relationship. 5.37 expands this idea by taking one marina and measuring each component as a percentage of the whole site. Here too the land and water areas are equal. The examples are intended as basic guidelines where only the marina is under consideration. The inclusion of an hotel, conference hall or any other added function will alter the balance.

5.4–7 **Land-to-water relationship:**

The land-to-water area in each case remains equivalent and constant but the shapes and relationships vary as the land wraps around the water. The off-shore marina has the shortest land/water interface but some land is 3 times further from the water than with the land-locked type

The geography determines the engineering
The engineering determines the profile
The profile determines the lay-out
The lay-out determines the architecture

5.4 **Advantage**
Minimum bulkhead wall
Minimum land take
Minimum dredging

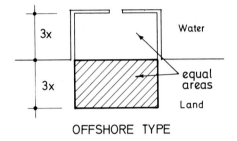

OFFSHORE TYPE

5.4 **Disadvantage**
Expensive in deep water
Vulnerable to weather, currents
Navigation hazards
Minimum enclosure
Silting by littoral drift

5.5 **Advantage**
Good for cut-and-fill economics

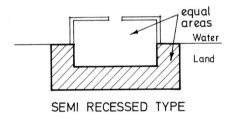

SEMI RECESSED TYPE

5.5 **Disadvantage**
Navigation hazard

5.6 **Advantage**
Uninterrupted shore-line
Large land/water interface
Considerable enclosure

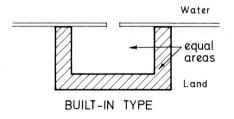

BUILT-IN TYPE

5.6 **Disadvantage**
Large land take
Length of bulkhead wall
Amount of dredging

5.7 **Advantage**
Maximum enclosure
Minimum interruption of shore-line

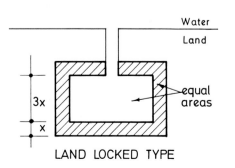

LAND LOCKED TYPE

5.7 **Disadvantage**
Maximum bulkhead wall
Distance from open water

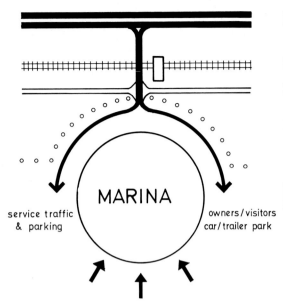

Motorway

Railway

Secondary Road

Pedestrians

Airport
Heliport

service traffic
& parking

owners/visitors
car/trailer park

Approach by water

5.8 Transport analysis is an essential part of the site selection process. If access does not exist or is inadequate it will have to be provided or improved by way of approach roads and reliable transport services

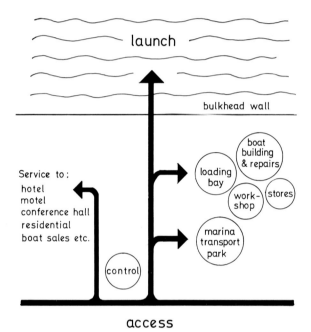

launch

bulkhead wall

boat building & repairs

loading bay

work-shop

stores

Service to:
hotel
motel
conference hall
residential
boat sales etc.

marina transport park

control

access

5.10 Some of the main activities which generally need to be included within a marina

5.9 This access road leads to Lyford Key Marina in New Providence Island, Bahamas

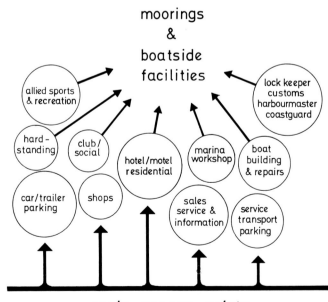

moorings
&
boatside
facilities

allied sports & recreation

lock keeper customs harbourmaster coastguard

hard-standing

club/social

hotel/motel residential

marina workshop

boat building & repairs

car/trailer parking

shops

sales service & information

service transport parking

main access points

5.11 How a draft amenity outline could be developed into a basic lay-out

1. Based on recommendations by Lt.-Cdr. G. J. Dunster, RNR, Member of the Institute of Navigation.

Spatial requirements for craft when berthed, parked or moored[1]

Craft may be kept, when out of use, by one of the following methods:

1 Parked on dry land or in multi-storey racks
2 Moored between piles
3 Moored to a permanent sinker ('swinging' moorings and 'trots')
4 Moored by a portable anchor
5 Berthed (marina moorings)

It is usual for craft longer than 5·48 m (18 ft) to be kept by one of the last four methods, unless there is a fairly sophisticated marina development with heavy-duty boat-lifts for hauling out and launching.

The purpose of this section is to give an idea of the comparative amount of space per boat required by these different methods. For the purpose of this comparison, a 7·62 m (25 ft) boat with a beam of about 2·43 m (8 ft) is used as an example.

Spatial requirements for manoeuvring

This can only be approximate, since the length to beam of craft varies, as do tides, water bed, individual marine engineering requirements, preferences for layout, types of hauling and the parking pattern. It does serve, however, to give a good indication of *comparative* spatial requirements. It is obvious that dry-land parking and marina-type berths are the most economical in space, and anchored or swinging moorings the most extravagant.

There is a factor which tends to lessen the apparent difference between these two extremes however, which is hidden by consideration of size alone, and that is manoeuvring space. In the case of the craft on swinging moorings or at anchor, there will be a percentage of their 'circle' which is free for manoeuvring, when they slip their moorings—in the case of berths and parked craft, no such 'hidden space' has been included in the calculations.

Nevertheless, the differences are still considerable, with freely-anchored craft occupying *nearly 100 times as much space as the same craft in a berth*. In practice, shortage of space has often forced moorings to be laid closer than this, but the closer they are, the greater the risk of collision.

2. Based on recommendations in Chaney, Charles A., *Marinas: Recommendations for Design, Construction and Maintenance* (Second edition) National Association of Engine and Boat Manufacturers Inc., New York 1961.

The following calculations[2] show how to find the spatial requirements of craft leaving their parking space or berth prior to being used or moved to another place. Again a great variation exists due to the differing handling qualities of craft and currents but broadly may be as follows:

Method	Approximate spatial requirements
Craft kept on land and being manoeuvred on a trailer with a car	186 m² (2,000 ft²)
Craft kept on land Being manoeuvred by hand or by a travel lift or hoist approximately 1½ times the boat length will be required	102 m² (1,100 ft²)
Craft kept in the water Require nearly 2½ times their length: e.g. for 7·62 m (25 ft) overall length	372 m² (4,000 ft²)

Whereas marina-berthed craft were economical of 'stationary' space, the above table shows that they are fairly extravagant in their requirements for manoeuvring space. However, unlike the other categories, they do not have any further requirements for launching ramps, etc. The comparative relationship is clear but the total space requirements appear exaggerated since the same manoeuvring space will of course be used by several craft, i.e., if 10 craft are parked or berthed alongside, the total channel which must be left for manoeuvring will be considerably less than 10 times the 371·6 m² (4,000 ft²) or 102·2 m² (1,100 ft²) shown as the requirement in the above examples.

		(Imperial dimensions are given in brackets)
	Min	Max
Land-to-water ratio	1:1	2:1
Density of boats per acre/hectare (wet moorings)	62 ph (25 pa)	162 ph (65 pa)
Density of boats per acre/hectare on hardstanding	25 ph (10 pa)	75 ph (30 pa)
Car to boat ratio	1:1	1·5:1
Density of cars per acre/hectare (2·44 m × 4·88 m 8′ × 16′ bays)	350 ph (140 pa)	520 ph (208 pa)
Ranges of boat length	4·8 m:13·7 m (16′:45′)	4·3 m:21·3 m (14′:70′)
Ranges of boat beam	1·8 m:4·3 m (6′:14′)	1·5 m:6·0 m (5′:20′)
Ranges of boat draft—Inboard	0·635 m:1·27 m (25″:50″)	0·483 m:1·65 m (20″:65″)
Outboard	0·305 m:0·559 m (12″:22″)	0·203 m:0·635 m (8″:25″)
Sailing boats	1·14 m:1·77 m (45″:70″)	1·01 m:2·16 m (40″:85″)
Average boat length	5·48 m (18′)	9·14 m (30′)
Percentage total parking area to total water area	20%	50%
Persons to boats ratio	1·5:1	3:1
Persons to cars ratio	1:1	4·5:1
Cars to boats ratio	0·5	2·0

5.A Guide to spatial requirements and likely size ranges

Aesthetic principles

The basic elements
In the same way that one may select a site by narrowing from a broad catchment area inwards towards the final position, so too may the main principles of design best evolve from recognising the context in which the development is located before determining the layout within the site boundary. A marina, as a separate entity, can never have the same appeal as an old harbour nestling in the heart of a fishing village. The harbour came first, chosen because of its natural protective setting which needed little improvement. That which *was* given over the years was probably massive, of local stone and increased the sense of enclosure and safety. The village and the harbour grew together, with boat sheds, pubs and houses crowding along its quays. The boats were part of the harbour and the harbour part of the village. Unless pleasure craft are directly replacing commercial boats in

5.12 Exploiting the charms of multiple use. A marina's industry, moorings, shops and clubhouse can gain more from mixing than from rigid segregation

5.13 Too-wide clearances bring a loss of scale and stretch the development area unnecessarily

5.14 These marina offices of traditional boarding have now been re-developed.

5.15 This old charter-boat fishing pier has been replaced by smarter moorings of much less character

5.16 A framed view at New Orleans

such a situation, marinas are not like this at all: quite the opposite in fact, for they are frequently sited on barren, featureless or derelict land. The relegation of the marina to rather dreary sites is likely to become more common as the demand for berths intensifies. By using poor sites for new boat harbours heritage coastal areas are, quite rightly, preserved from such development and the otherwise unusable and unattractive areas enlivened and employed for community benefit.

By their very function marinas are at sea level. There are likely therefore to be long-distance views both to and from the site particularly if the development is located in low-lying land which is often the case. The site may be considered as a point in the centre of a circle with views radiating to and from it. In a coastal setting sight lines will be from inland, along the shore-line and from the water. In inland situations, rivers, reservoirs or lakes there will be a prospect from the opposite bank and the effect of the new development upon the linear view from across the water will need to be considered. When there is rising ground or cliffs nearby one may have an almost plan view of the harbour from afar and an increased sense of enclosure from within the marina.

Off-shore views are important navigationally as well as visually. By tradition a sailor's attitude towards his home port is emotional and it is important to set it satisfactorily in its surroundings—both land and water. The night-time appearance too should be remembered. The topography of the area will play a large part in determining the harbour's layout for advantage must be taken of currents and tides, promontories, sand bars etc. and siltation and scouring avoided. Planning the basic shape of the marina within the demands of economic engineering is an exercise in good functional design and will do a great deal to ensure the right relationship between context, function and form.

There are so many different kinds of geographic situation in which marinas have been satisfactorily developed that it is impossible to describe all the problems likely to be encountered or to suggest solutions. There are however four basic characteristic forms, the off-shore, the semi-recessed, the built-in and the land-locked marina. Each has many variants and will become the more interesting for being adapted to suit the local terrain and conditions.

All types are found in coastal situations whereas the off-shore kind is less common in estuaries or rivers because of navigational hazards. In urban situations the built-in version is most common, allowing maximum land-to-water interface. It is worthwhile considering very carefully the merits and demerits of these broad categories. Some of them are set out in 5.4–7 and should be looked at in relation to the topography of the area at the site selection stage. The suitability of the format to its setting will be an im-

5.17 The large volume and cubic shape of most boat-stacker storage sheds requires careful handling, particularly on flat coastal terrain. On this account they may be denied planning permission. This example at Tampa, Florida, is not relieved by landscaping and planting nor helped by the giant notices. Inside, boats are stacked four high, and recovered by a forklift truck

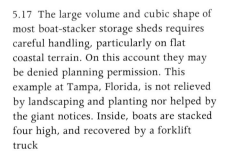

5.18, 19 Gordon Cullen's examples show what can be achieved. Compare this approach with the dreary and repetitive mooring patterns of pages 41 and 53

5.20 Security gates leading to floating piers at Seattle, State of Washington. The photograph shows the gloomy effect that tall dark piles can have

portant factor in the eventual design, cost and workability of the project. It also plays a major part in marrying the development to its surroundings.

The project team will need to bear in mind all the basic elements at the same time, in fact most of the items in the Check List on page 150 will need studying. Not all will be solved completely and there will be conflicts and compromises as in any other design problem, but the best solutions stem from the skilful application of full and accurate information.

The physical form

In the previous chapter mention was made of the importance of long-distance views of the site. The vista across low-lying land, from cliff-tops or from the water itself can result in marina sites being visually exposed and frequently being seen within a general panorama in juxtaposition to the entirely natural backcloth of water, sky, coastline or fields. This situation equally applies to inland or riparian situations with views across lakes and estuaries or from the opposite banks of rivers. The silhouette of the marina has therefore to be considered from all angles and this outline becomes increasingly vulnerable with the demands for high buildings such as flats and hotels.

The 'multi storey' boat stacker is particularly difficult to handle. Economically it may need to be about $21 \cdot 33 \times 30 \cdot 48$ m (70×100 ft) on plan and perhaps $12 \cdot 19$–$15 \cdot 24$ m (40–50 ft) high—often unrelieved by windows (see 5.17). Great skill is needed if such bulk is not to be a jarring intrusion in a natural landscape. Modelling the elevations, sculpting the ground, tree screens and soft, 'camouflaging' colours may all help to achieve a more satisfactory scale. It is worth remembering too that whilst reflections in still water can be quite charming, they tend to double a building's apparent size and it may be wise to keep large structures back from the water's edge if possible.

The liberating quality of water may make some strong architectural forms quite acceptable where otherwise they may appear extreme or inappropriate. In this category may be included floating buildings, cantilevers, curving or serpentine plan-shapes, building over the water on piles, buildings rising from the water or built upon an existing or reclaimed promontory, breakwater or jetty. The economy of built forms such as these need careful consideration—planning and ownership matters may need clarification but with the present land cost of really good waterside frontage these exciting and expensive structures may be feasible when set against more traditional land uses. More general features like balconies' flat sun-roofs, screens, pergolas and awnings will probably seem relevant or even necessary. Unusual problems invite unusual solutions which, skilfully

5.21 This splendid dandelion-clock fountain at San Francisco shows an imaginative use of water

5.22 Recognising the nature of the water surface by day and night

5.23 Smooth clear water reflects its image at Somer's Cove, Maryland

5.24 The seaward side of this natural breakwater at Nassau presents another of water's moods

5.25 Texture and function at Limone, Italy

5.26 Clear and simple brinkmanship at Iseo, Italy

handled are the very stuff of good architecture. At Marina City, Chicago, 5.1, 5.2 Bertram Goldberg created an unforgettable complex from what must have seemed an impossible brief.

The needs of marina users are complex and often contradictory. On one hand the open-air life, sunbathing and active pursuits demand terraces, lawns, panoramic views and exploitation of open spaces. Running counter to this is the need for enclosure and shelter which call for individual or communal seclusion in the form of courtyard plans, shuttered windows and intimate planning. Both are basic requirements for most people within one day and in sports as tiring as sailing, the need for active recreation during the day brings the need to relax after that recreation during the evening. The built environment should recognise and provide for these diurnal changes.

The qualities of water

The one material that is common to all harbour developments is, of course, the water itself, and some understanding of its many moods and qualities is as essential to the architect as a knowledge of its dynamics is to the engineer.

Water has a considerable range of character, even within the relatively narrow confines of marina development. On unsheltered coasts the off-shore side of the breakwater will encounter the pounding of ocean waves and, within the limits of safety there is no reason why the designer should not plan for people to enjoy the sound and sight of such a natural spectacle as they already do on many esplanades. Water within a tidal basin will be either choppy, rippling or quite smooth and untroubled. The vitality of light dancing on water is a splendid ingredient to be employed to advantage. Clear, flat water presents many faces. It can be transparent, revealing the bed of the harbour or, in the case of a canal basin, opaque and smooth like black glass reflecting a mirror image of buildings in the water. The sharp contact between contrasting materials has been termed 'immediacy' and the abutment of dissimilar elements is as exciting as the elements themselves. As Gordon Cullen points out in *Townscape*,[3] water provides the most obvious example of this because the transition between it and dry land offers the biggest of all psychological contrasts. Towns that live by the sea should live *on* the sea in the sense that the visible presence of the ocean should be apprehended from as much of the town as possible. The principle applies to inland marinas in exactly the same way. This does not always mean a full view of the water, but perhaps the glint of reminder or even a chasm of space closing the vista at the end of a street. For the coastal town the sea is its *raison d'être*—it is on the edge of the deep, it faces the constant but enigmatic horizon.

The same is true of the individual standing on the quay, except that for him the main tension is concentrated on the demarcation line between land and water. It is the emotional experience of this tension which gives the sense of immediacy which is best achieved by omitting railings at the line of vision. As examples of this Gordon Cullen instances Blakeney, Norfolk 5.28 where one may stand on the brink or even *lean out over* the water and Limone, Italy 5.25 where cobbles imitate wavelets both in their shape and shine, but are as hard as the waves are soft, a contrast which heightens the sensation of proximity.

3. Cullen, Gordon, *Townscape*, The Architectural Press Ltd., London 1961.

5.27 This wobbly quay wall near Emsworth Yacht Harbour, Hampshire is much more interesting than any dead-straight alternative

5.28 Interesting, functional and shapely edge treatment at Blakeney, Norfolk

5.29, 30 Two robust approaches to the design of breakwater steps

5.31 Uninhibited use of colour by the Job-Lot Trading Co., New York

5.32, 33 If there must be guard-rails, the simpler the better

5.34 Civilia, that low-rise, high-density city of revolutionary concepts, envisaged lake-side moorings in the heart of the town, with opportunity to navigate into the country's canal system and beyond

5.35 Exploiting the compressed view in Civilia: a slice of landscape is often more effective than a panorama

5.36 A horizontal 'slice' with a triple bonus: it reveals the view, uses the land and allows access

Access

5.8 shows a transportation situation in diagrammatic form. A satisfactory transport system leading to the harbour is of the utmost importance, and, if it is not adequate to serve the new development it may need extending or upgrading. Naturally, sites vary a great deal in their demands upon the existing transport network. At Brighton, Sussex, following a special Bill through Parliament for the development of the marina, which received the Royal Assent in 1967, it was found necessary to apply for a separate Bill for the road works which took a further three years. However, unless the development is to be very large or the existing system is very inadequate it is often unnecessary to spend large sums in improvement because observation of existing marinas confirms that the arrival and departure of owners and visitors is characteristically very dispersed even during the peak periods of a summer weekend. Relatively small roads may therefore serve quite large amounts of activity without undue congestion. This, however, applies only to the *marina* element and reappraisal will be required if other functions such as conference halls, shopping areas or entertainment centres are incorporated which will generate heavy and intense traffic. The proximity of railway stations and airports is also important if the marina is planned to have more than about 500 berths and transport to and from them should be made available.

Access into the marina itself is a problem between convenience and security. The number of entrances into the marina should be restricted to the minimum, for each will need supervision of some kind. It is also important to limit the number of entry points to the berths to assist control and security but owners naturally want short and easy access to their boats, the marina buildings and to things like garbage points, lockers and trailer stores.

Circulation within the marina

Having secured the means of travel to the harbour it is generally important to ensure the correct separation of vehicle types. This should visually take place before entering the site by clearly signing the separate entrances for owners and visitors on the one hand and service vehicles on the other. Bearing in mind that the former category may include cars (possibly trailing large boats), vans, motor cycles and cycles, it may be necessary in the larger developments to split these again within the grounds so that they may go directly to the launching bay, car park or club house without confusion or cross circulation. 5.10 shows some of the main activities which will probably need inclusion within a marina and it attempts to show what their physical relationship might be. The generic terms more or less correspond to the main headings within the check list on page 150 and may, within themselves contain as many facilities as are listed there. 5.11 takes this concept a stage further by suggesting how a layout could start to form from such a draft outline. Because permutations on even the basic needs are virtually infinite it usually helps to divide the on-shore elements into two: the service and social areas. 5.11 can be roughly separated in this way. The two parts will not usually be of equal area and they will overlap both physically and in function. They need not, and preferably should not be divorced. Owners will enjoy seeing boats being made or repaired and do-it-yourself facilities may be proposed. The plan must also be flexible

5.37 A typical allocation of on and off-shore space assuming a 50:50 land-to-water split. Differences between this and 5.38 are due to dissimilar American and UK standards particularly in parking allocation

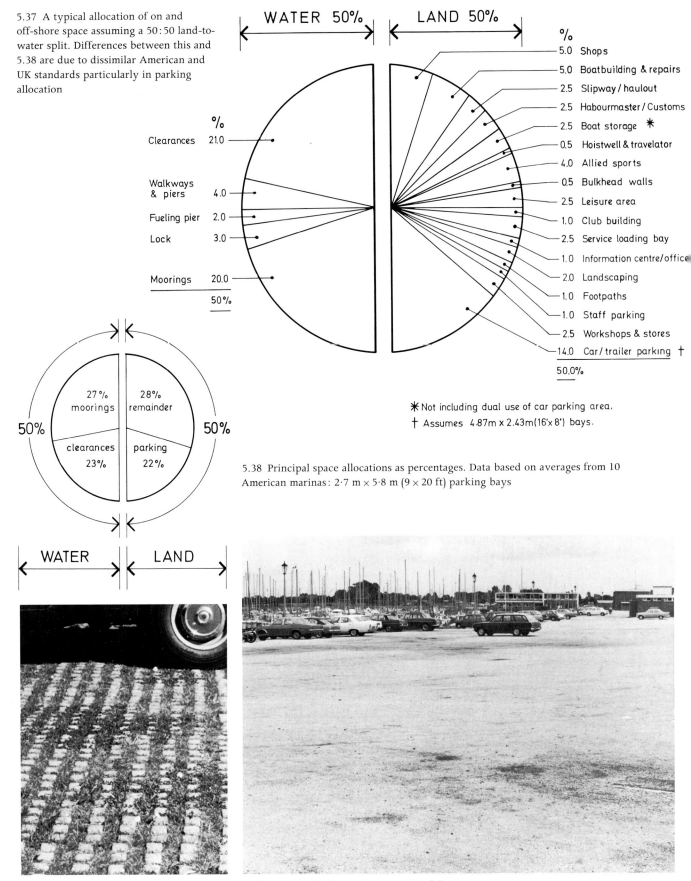

WATER 50% LAND 50%

	%
	5.0 Shops
	5.0 Boatbuilding & repairs
	2.5 Slipway / haulout
	2.5 Habourmaster / Customs
	2.5 Boat storage ✳
	0.5 Hoistwell & travelator
	4.0 Allied sports
	0.5 Bulkhead walls
	2.5 Leisure area
	1.0 Club building
	2.5 Service loading bay
	1.0 Information centre/office
	2.0 Landscaping
	1.0 Footpaths
	1.0 Staff parking
	2.5 Workshops & stores
	14.0 Car/trailer parking †

50.0%

	%
Clearances	21.0
Walkways & piers	4.0
Fueling pier	2.0
Lock	3.0
Moorings	20.0

50%

✳ Not including dual use of car parking area.
† Assumes 4.87m x 2.43m (16'x 8') bays.

5.38 Principal space allocations as percentages. Data based on averages from 10 American marinas: 2·7 m × 5·8 m (9 × 20 ft) parking bays

27% moorings 28% remainder
clearances 23% parking 22%

50% 50%

WATER LAND

5.39 This type of in-situ or pre-cast concrete slab filled with soil and seed, gives a grassy look but is very hard-wearing. Maintenance and mowing is the same as for ordinary grass

5.40 Car parks look worse empty than full. Why not plant trees? Vary the levels? Change the materials? None of these need be expensive: imagination is needed more than money

enough not to prohibit expansion along the shoreline. 5.11 also indicates those parts which need waterside positions and those which do not—an important decision, for usually more will be wanting this than can be accommodated, particularly with the off-shore type of marina with a minimum water frontage. When the planning and architectural work starts in earnest it soon becomes clear why early engineering decisions are so important. Unless, at that time, the team contained someone able to appreciate the significance of the resolutions taken, the architect may find himself presented with a *fait accompli* which may be quite wrong in imposing unnecessary constraints upon him. It is as if the quarry foreman rather than the sculptor were deciding the size and shape of the unwrought stone. The shape of the land and the shape of the water will determine the way they interrelate and this interface will have a great effect upon the whole character of the development.

On-shore facilities

The car park

It seems a pity to take up waterside frontage with car parking. Some access will be needed for launching boats arriving by land and the slipway and boat-well must be easily accessible without difficult manoeuvring. The main area should however be kept back for reasons of planning, appearance and, possibly, safety.

It is sometimes a good idea to allow some of the car parking area to be used as hardstanding for boats that are being maintained or laid up for short periods during the winter. As there are fewer cars and more boats out of the water at this time of year this will give dual use of the area and maximum economy. It also allows owners to drive their gear and tools alongside their boats.

The size of the car park will be calculated from the number of boats, the car-to-boat ratio and the sizes of bays and aisles. Page 98 sets out some alternative arrangements and dimensions. It may be advantageous to have more than one car park. Any idea which helps to reduce the impact of these boring places should be welcomed. Suggestions for screening, planting trees and treating the floorscape are given in Chapter 7.

It is useful and easy (although unmetric) to remember that a 'square acre' of 208 × 208 ft can contain 208 cars at 208 ft² each. This assumes 16 × 8 ft bays and 20 ft aisles. Authorities vary in their parking requirements. Some ask for 2 spaces per berth, others for much less. Dual use with other functions having different peak-use times may cut this ratio and dualing with boat hibernation during off-season periods maximises this land use.

5.41 Car and trailer parking bays at Detroit, Michigan

5.42 A far from unique mess. The trailer park at a large English marina

The trailer park

These should be near the hoist well and slipways or easily accessible from them. They are one of the most neglected aspects of marina design being frequently relegated to a jumble of metal on a muddy patch 5.42. They should be dealt with by providing both a special, separate area and by allocating part of the main car park to 9·14 × 3·04 m (30 × 10 ft) car-plus-trailer bays, thereby catering for the unhitched 'parked' trailer and the short-stay visitor. Rails or posts to padlock to are sometimes provided. Here too, screening, planting or a change in ground levels is often worthwhile.

Hardstanding

This is a well-surfaced area clear of all obstructions. Generally its position is between the boat store and the trailer park in the landward direction and the boat-handling equipment and the boat-collection bay on the moorings side. It must be easily reached from the access road. It will need to serve the slipway, haul-out areas and the hoist well and will continue uninterrupted until ended by the bulkhead wall. Its surface must be very well drained, non-slip and self-cleaning or easily cleaned. Concrete ridged with board-tamping, proprietary paving slabs or bitumen are good suitable materials. Sand, gravel, grass or stabilised earth are not. If a fairly dark material is used this colour and texture will contrast sharply with the smoother white surface of the bulkhead wall capping. This looks good, avoids glare and clearly shows the water's edge. The size and shape of the area are dictated by the need to manoeuvre boats, trailers and perhaps haul-out tractors around it, allowing for their turning circles without undue backing or shunting to and fro.

In large marinas or where layout has put a fair distance between the club and boat-repair premises there may be a need for two separate areas of hardstanding to avoid unnecessary circulation. The area needs to be well lighted at night, particularly along the edge, both for incoming craft and for pedestrian and vehicular safety.

The service transport area

This area will be directly accessible from the access road. It should be kept back from the water and preferably not visible from the moorings or club area except where it gives on to any launching bay. It may be used to link the boat building and repair yard, the marina workshop and stores (which they may share), the loading bay and the marina transport area. It must be large enough to receive a 6·09 m (20 ft) vehicle trailing a boat to the launching ramp. It may be adjacent to but separate from the service entrance to any hotel, conference, residential or boat sales component. It may need securing gates with a reception kiosk or office beside them. It will need a robust sub-base, a heavy-duty non-slip surface and be well drained with oil-trap gulleys.

Boat building and repair

Neither of these are essential to the operation of a marina but they can be most successfully incorporated either as a part of the marina's management or as an integrated but separately managed facility. A repair shop need not be very large whereas an economic boat-building yard may have to be quite

5.43 Repair-shop under construction at Beaulieu, Hampshire. The steel-framed, timber-clad structure blends well with the local landscape

extensive. As it can be positioned on the perimeter of any site it may extend or expand away from the marina. It will have direct access from the main road and, if reasonably small, may share service facilities with the marina transport area and workshop. Launching and retrieval may be into the marina itself or into its own bay. It may be noisy and smell of acetate if building involves grp and these disadvantages will warrant consideration in terms of the marina's management as a whole.

The marina workshop

This may be solely for the maintenance of the marina itself or be combined with boat repairs. It can have a waterside location to ease boat handling but may be kept back particularly if serving only as a maintenance unit. It will need a well-lighted under-cover area with good dry storage space, a small office, staff room and lavatories. Access may be through the service transport area. Delivery vehicles may be checked through the transport office and personnel through the reception office or kiosk.

Sales, service and information centre

These are often close enough to be brought under one roof except where boat sales are arranged on a concessionaire basis. In this case a separate building may be necessary for management reasons.

An ideal layout would be one which spanned the depth of the land area from the public entrance through to the water's edge, thus allowing the enquiries office to be near the entrance and shops and the boat sales to be accessible to the water. If, therefore, a narrow ramp can join the boat-display area near the marina entrance with a special or shared slipway to the water the best arrangement is achieved. This allows the boats for sale to be prominently displayed near the marina office where they may be seen upon entry and act as an advertising sign for the marina itself and yet be launched within seconds. Any salesman will confirm the value of a trial run.

It will depend upon management arrangements whether chandlery is included with boat sales, is treated as a separate element, or becomes a unit in the shopping parade.

The office should include the usual accommodation for the staff, a public space and enquiries counter, a manager's office, lavatories and perhaps a staff room. There is usually a large display map of the moorings with identification numbers and the owners' and vessels' names. This can be made an attractive feature. The information office should be readily accessible, with a high standard of attractiveness. A marina gives a *service* and a friendly and efficient impression from the buildings and staff is most important.

5.44 Strategically-placed weather-boards are an important aid to safe sailing

5.45–47 Three clean and functional marina buildings designed by the Bailey, Piper, Stevens Partnership. 5.45 at Swanwick. 5.46 Morgan Giles Ltd. at Warsash, Southampton 5.47 This good and exciting alteration to the Royal Albert Yacht Club Signal Station at Southsea, Hampshire, proposed in 1968, did not unfortunately reach fruition

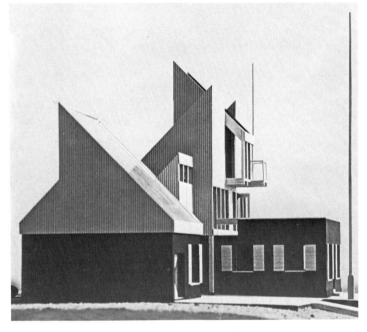

5.48 New residential development overlooking the California YC near Los Angeles. These low-rise apartments gain much from their clean roof-line and simple facade

5.49, 50 Two Californian marina developments. Such confidence in design and investment is rarely seen in British marinas but a score of such projects thrive in the Los Angeles area alone

Shops

The number and size of shops is generally governed by the extent of the moorings, the length of season and the nearness of outside retailers. There is a trend, following the American pattern, for chandlery firms, yacht brokers, agents and equipment specialists to re-open in marinas, bringing their business to the owner rather than remaining in towns and waiting for custom. Shops less reliant upon sailing are more difficult to establish and may close off-season or trade at weekends only. Things are easier in an urban environment, if there is a substantial hotel or residential element or if the shops can attract trade from the access road.

As 12.A, page 301 shows, over half the total income of an average American marina stems from sales and over one quarter of this is from chandlery and one-ninth from grocery, ice, vending machines, fishing tackle and bait. Shops are generally located near the entrance, the club, the car park and any hotel or residential component. If service transport adjoins the sales area this allows easier arrival and launching of boats for sale. The workshop staff if close at hand can maintain the craft on display if management policy allows. Orientation will require thought and a shop canopy as protection from sun and rain may be an advantage.

Hotel, motel and residential elements

To these elements may be added community centres, restaurants, conference centres, exhibition buildings and museums. Their location will depend entirely upon circumstances which are so variable that advice only in the form of broad guide-lines can be offered.

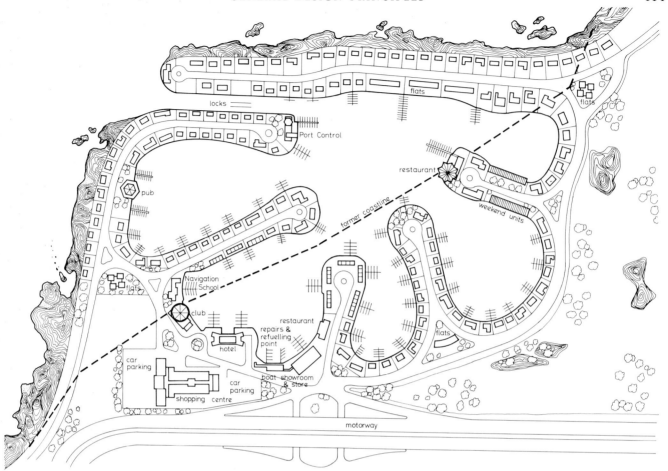

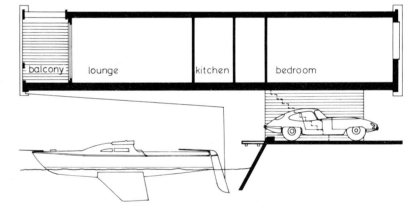

5.51 In this marina village concept, the architects Kaye, Firmin and Partners have achieved a high degree of both enclosure and waterside frontage. Edge-treatment costs are minimised by locking. Residential accommodation is envisaged at 20 per cent detached housing with private moorings, 20 per cent week-end units also with private moorings and the balance in small apartment blocks with moorings in the marina. Compare with the Port Grimaud plan, page 42

5.52 Week-end house unit. Within the village, week-end house units take the form shown here. It seems rather fortunate that the keeled boat has no mast

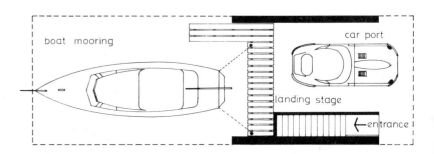

Whilst each of these facilities may gain from being within the harbour boundary, all of them may need to be self-contained to some degree. This will apply not only to the parts individual to them but to their parking and public and service entrances. They may have their own stretch of water-front and in the case of hotels and apartments their own moorings. It may be that the facility in question is the *raison d'être* of the marina in which case special circumstances will prevail. However, even if secondary in status and peripheral in location, any such component will have a strong influence upon the marina's character and clientele. It should certainly be close to the water, perhaps overhanging or projecting into it. Easy, covered access to any club-house, shopping or information office would be an advantage. The buildings will need careful planning to ensure the best conditions for aspect and orientation.

Views across the water may be featured and other advantages exploited by an enterprising architect. These kind of buildings are likely to be the principal elements on the site both in height and bulk and their relationship with other components, with the water the site and its surroundings including views from off-shore will need careful consideration.

Allied sports and recreations

One of the best recipes for success in marina operation is to cast one's net as widely as possible to attract custom.[4] All kinds of sports other than sailing may be encompassed, some of which—golf, tennis, bowls, archery—are not water-orientated and may be quite land-consuming. Others such as badminton, squash, deck games or trampolining are less so and may be quite readily accommodated.

Water sports are, of course, particularly suitable and fishing, swimming, water ski-ing and SCUBA diving may be candidates for inclusion. Sail training is sometimes included within the activities of the club-house or else set apart in a sailing school or separate club such as the Sea Scouts or

4. For allied sports and recreations on inland water see Chapter 4, pages 79–84.

5.53 Youth clubs of all kinds benefit any development by their vitality

5.54 Disabled people aboard a specially designed catamaran. Sailing can be a highly suitable activity for the disabled. Groups may be catered for in special or standard craft. Some disabled people are good swimmers and can sail alone in small unconverted boats. Marinas with their ramps and flat walkways are often ideal places for the handicapped. The amount of thought and money needed to cater for the majority of the disabled is very small when set against the resultant therapeutic effect

some other youth organisation. Areas of more passive recreation are also included under the general heading of sitting areas or sunbathing and viewing terraces. However general or specialised their needs may be they will require thorough research to provide the necessary layout, buildings, equipment and staff. It is important in planning their accommodation that they are not cut off from the main harbour but keep contact with it in both a physical and social context, thus providing for boat owners with a secondary interest and other sporting enthusiasts with the opportunity of becoming boat owners.

Water sports obviously need a waterside position, whether this is within the marina or not depends upon the sport and the circumstances. As well as providing links with the marina and clubhouse, allied sports will gain from *some* contact with each other, and a good plan will provide for this social integration and probably gain in economy of layout. As with any hotel or residential element the inclusion of special sports will need land additional to the average $1:1$ or $1\frac{1}{4}:1$ land-to-water ratio.

Car parking too may need to be separately annexed and additional to marina requirements.

In this part of the harbour there are opportunities in planning and architecture to provide an informal and relaxed atmosphere that is nevertheless efficiently designed for these specialised pastimes. Whilst the buildings and active areas will need to be hard-wearing and rugged the quieter parts can be landscaped with planting and trees.

Catering for the disabled

The need

The value of water sports as a means of integrating handicapped people into the life of the community is becoming more widely accepted. Until recently the idea of a paraplegic, complete with wheelchair, being the helmsman of a racing boat in testing conditions, would have seemed unlikely and dangerous. Now, however, it is recognised that water sports can contribute significantly towards the rehabilitation of the disabled. The facilities at most of the water recreation centres are not always suitable for those with restricted mobility because of steps, narrow doorways or fittings of unsuitable design.

More thought should be given at the early design stage to the problems of access for the handicapped. There is no reason why the basic requirements for complete accessibility should not be included in line with the British recommendations of the Chronically Sick and Disabled Persons Act. Good liaison between planning authorities, marina developers and regional sports councils should ensure that all water users can enjoy the facilities with the minimum of difficulty. There is no doubt that the process of learning how to sail has a considerable therapeutic value but for many handicapped people the problems start before they reach the water's edge. Access to *all* parts of the marina is essential if it is to be used by the whole community on an equal basis. Good access coupled with a few relatively minor aids will attract disabled users from a wide catchment. Most of these facilities need to be provided for general use anyway and the remainder are design problems which are most likely to be solved by thought rather than money.

Apart from providing a social benefit the return on the provision of facilities for the handicapped is likely to be good. The disabled are prepared to spend freely on their sport and having discovered an accessible and usable harbour they remain a very faithful group of users. Provision not only attracts the disabled themselves. A whole family with a handicapped child or infirm grandparent will all be influenced towards the marina which provides facilities for their less agile relation. No owner or manager should ignore them.

The car park

Disabled people usually have difficulty in travelling on public transport so their problems often start in the car park. Their adapted vehicles are usually taken as close to the destination as possible especially when arriving unaccompanied, perhaps with equipment and protective clothing to carry. The car park rather than the cloakroom is the base from which they often prefer to operate and some reserved spaces should be allocated adjacent to the main building. These should be wide to allow car doors to open fully and display a reserved space notice. Other requirements are surfaces suitable for walking aids and wheelchairs and convenient ramps or step-free outlets, but this is a requirement for the able-bodied too if trolleys are used.

Club facilities

The disabled user requires freedom of access to meeting rooms, pay desk, refreshment bars, toilets, changing rooms, shops, telephones, spectator terraces, information office and lounge. A public service lift should be large enough for a wheel chair. The access ramp to the front entrance should be not more than 1:12 gradient and not less than 1·6 m (5 ft 4 in) wide with a handrail height of 0·8 to 0·85 m (32 in–34 in) above the ramp surface. Door widths should be not less than 0·9 m (3 ft 0 in). Revolving doors are very difficult for handicapped people and raised mats are a hazard to wheelchair users. Fully glazed doors are often dangerous but particularly to the visually handicapped. Other features which make changing rooms more comfortable for disabled users are coat hooks, mirrors, wash-basins mirrors and benches at convenient heights. Shower room equipment should include non-slip tiles, a level floor, hand rails and a folding seat.

Sports facilities

The needs of the disabled regarding jetties, slip-ways and boat maintenance areas are common to most water sports. Getting in and out of small craft will always present some difficulty but with help from other water-users the problems can be reduced. Provided they have a ramped approach, fixed jetties are better than floating pontoons. They should be no more than 0·45 m (18 in) above the average water level. Wood jetties are better with their deck planks *across* the walkway which should be wide enough (at least 1·5 m or 5 ft 0 in) somewhere along its length to allow people to pass a parked wheelchair.

One of the slipways should be of not more than 1:10 gradient with its cross-ribbed surface continuing well below water level so that wheelchairs may go up to axle depth if necessary. Unavoidably pebbly beaches and soft ground are almost impossible to cross without assistance. Water skiing and sub-aqua swimming attract some handicapped sportsmen and women but no special equipment is needed other than good access to the waterside.

Equipment rooms, life jacket stores, canoe racks, bosun's lockers and workshops should all be accessible to disabled users. Special aids in the form of equipment on board are usually provided by the disabled themselves to suit their own disability or by their club or group if the equipment is of communal benefit.

Safety

Provided that local regulations are always observed water sports may be enjoyed by the disabled with considerable safety. Competence in water sports requires sound basic training during which people recognise their own limitations and avoid taking unnecessary risks. Disabled trainees are no different from anyone else in this respect. Their determination, enthusiasm and survival skills are often sharper than normal and many are good swimmers.

The clubhouse

The majority of marinas will include a clubhouse of some kind. Even where a separate utilities building is provided with lockers and showers for the visitor or casual user and paid for by a 'green fee' there is usually still a clubhouse as well.

As emphasised in Chapter 12, the financial return of clubhouses is very difficult to foretell for so much is subjective. Whilst many thrive and become the leaders of the marina, many others struggle, frequently change their management or perhaps contract to opening only at summer weekends. Whether the club becomes the financial and social heart of the Marina or a liability kept alive as a loss leader, the appropriateness of its design will be an important ingredient in its success.

Siting

The location of the buildings will naturally vary according to the type of harbour, but as a rule the club premises will be on the waterfront, not too far from any hotel, shops or apartments, close to any of the specialised sporting clubs and with an adjacent car-parking and hardstanding area.

Cars should be able to put passengers down by a covered entrance and the same may be said of arrival by water where space for a landing stage is necessary. Where possible an orientation which avoids the sun's glare when looking out to sea is an advantage.

Access from the club to the moorings, slipways and boat storage should be convenient. The building itself is often of two storeys with generous terracing, balconies and usable roof space. Unless this is separately located on the harbour wall or elsewhere, a starting post is sometimes incorporated with a clear view to and from the starting buoys. Entrance from the main approach road, particularly in large marinas, should ideally be separate from the main service entrance and this again should split to serve independant access for service vehicles on the one hand and members and visitors on the other.

Range of types

The scope and extent of the types of clubs within marinas is extensive. The following list embraces most of the current range, but new and specialist forms are added each season both in Britain and abroad.

5.B Types of club within a marina

1 Youth club	General sailing
	Specialist sailing class
	Youth organisation
	Private sailing club
	Local authority sailing club
2 Sailing club	Private (general)
	Private (specialist)
	Boat type
	Local authority
3 Allied sport	Rowing ⎫ etc. with sailing attached
	Canoeing ⎭
4 Cruising	Not racing, training or specialist class
5 Racing and sail training	From small club to Olympic standard
6 Social	From modest local club to those of international repute
7 Holiday inn/hostel/hotel/motel	Clubhouse attached to or within 'parent' building. Accommodation with specialist boating facilities. Usually with berths or small marina

The type naturally has a strong influence on design. Some categories may not even require an independent building, being integral with a larger structure. Nevertheless most of the recommendations given here will apply in principle if not in detail.

The range of accommodation

This will of course depend upon the status and financial resources of the developer and vary from very modest provision to building of international standing. There may be opportunities to extend the premises in step with the phasing of the total marina site. This is sensible economically and managerially but has to be determined at the initial design stage to ensure that the growth is sequential and that the building functions property at each stage with a minimum of contractural disturbance during extension.

The accommodation Check List on pages 149–150 give some idea of the provision for various situations.

5.55 An interesting amalgam of architecture: the club-house at Lymington, Hampshire

Basic	Desirable	De luxe	Specialist
Clubroom	Licensed bar	Snack bar	Gymnasium or weight
Lavatories	Snack bar	Dining/restaurant	training
Showers	Kitchenette	Kitchen	Coach's office
Dressing rooms	Gear/trailer	Library/chart	Provision for allied
Drying cupboards	store	room	sports
Lockers	Separate club or	Overnight	Press
Car/trailer park	washroom	accommodation	Officials'/stewards'
	entrances	Tennis	accommodation
		Badminton	Starter's tower
		Swimming pool	
		Children's/games	
		room	
		Committee room(s)	
		Manager's office/flat	
		Caretaker's flat	
		Viewing/sun roofs	

5.C Sailing club accommodation: a guide to levels of provision. These lists are cumulative; it is assumed that the full provision of column 3 would already include all from column 1 and selected items from column 2.

Sizes

In general terms a 500 berth harbour would have a clubhouse with a floor area of about 930 m²–1,400 m² (10–15,000 ft²). A reasonable average would be 1 m²–3 m² (10–30 ft²) per berth for a municipal marina and 2 m²–5 m² (20–50 ft²) for the private or club type. This perhaps seems generous but space must be allocated for lockers, showers, changing and drying rooms.

For economy of operation a flexible plan will allow for seasonal variations of occupancy. A large open restaurant and lounge capable of receptions and dances during the summer may contract to a snack-bar off season or on a week day.

Dressing rooms

It is inadvisable to lay down any hard and fast rules as individual requirements and levels of provision vary so much. There are however certain minimum basic standards which it is desirable to uphold especially in relation to dressing rooms and sanitary arrangements.

These should be proportioned to provide a floor space of not less than 1 m² (10 ft²) for adults and 0·75 m² (8 ft²) for children. For private clubs where numbered lockers are required within the dressing rooms the floor space should be appropriately increased. In clubs with basic accommodation it is usual merely to install sufficient seating around the walls with space below for boots and bags and hanging pegs above for each member. Linear allocation should be on the basis of about 610 mm (2 ft) run per person with a minimum width of 400 mm (1 ft 4 in).

Conveniences

Practicable sanitary arrangements will be largely dictated by local circumstances. If there are adequate public services of water supply and main drainage and the finances available will permit, provision may be made on a fairly generous scale. Where, however, such services are severely restricted or non-existent, as in many rural districts, the sanitary arrangements will consequently have to be curtailed accordingly.

5.56 Lockers, showers, toilets and telephone kiosks in one neat amenity block at Del Rey Yacht Club, California. The deep roof fascia conceals water tanks and piping.

Wherever possible the provision of conveniences should not be less than the following:

1 Male: 1 WC 1 urinal stall, 1 wash hand basin and 1 shower for every 25 persons

2 Female: 2 WCs 1 wash hand basin and 1 shower for every 25 persons

3 Provision for children and the handicapped should be considered

4 A drinking fountain should also be provided

It follows of course that where there are no public water supply or water-borne sewerage systems, these arrangements will not be practicable and earth or chemical closets will have to be substituted on a scale consistent with the facilities for emptying and sterilising. Washing facilities will also have to be scaled down in accordance with the method of supply and adequacy of water available.

The recommendations are offered as an absolute minimum with a view to avoiding undue extravagance, or the setting up of standards impossible of attainment in the majority of cases. It is realised that efficient sanitary installations involve an appreciable capital expenditure quite apart from the costs of running and maintenance.

All supply valves for basins and showers should, wherever practicable, be of the self-closing type to ensure the minimum use of water, and to prevent unnecessary waste.

Dimensions of conveniences should be not less than the following:

1 WC compartments: 1.372×0.737 m ($4'6'' \times 2'5''$) each

2 Urinal stalls: 0.610 m ($2'$) run per person by $0.305–0.427$ m ($1'–1'6''$) according to type

3 Showers: 0.914×0.914 m ($3' \times 3'$) each

4 Wash hand basins: 0.610 m ($2'$) run per place, by 0.457 m ($1'6''$)

5 Circulation space should be at least: 1.067 m ($3'6''$) between any group of conveniences

Drying rooms if provided should be separate for each sex of users. They should be fitted with convenient racks and be well heated and ventilated.

Entrances to dressing rooms should be direct from the hardstanding and not through the club room or lounge.

The boiler room for hot-water supply or heating will depend on the washing facilities included, and the method of heating most convenient. Where fuel has to be stored, adequate space should be allowed for the season's supplies.

Materials

The choice of materials for the clubhouse, as with all marina structures, is of particular importance for many sites will be exposed to high winds, salt spray and damp or made-up ground conditions. For these reasons it may be wise if the materials budget is somewhat higher than usual. Where harsh conditions are experienced for long periods it may be best to avoid the following:

1 Rendering
2 Untreated timber (except cedar)
3 Unpointed brickwork
4 Lightweight metals for roofs
5 Tile hanging
6 Very large panes of glass
7 Metal windows without sub frames

All metals are subject to corrosion or oxidation under atmospheric changes but exposure to salt water is particularly damaging. Salt water not only damages by direct contact but by accelerating deterioration by electrolytic action which occurs when metals of different galvanic rating are in close proximity.

The maintenance period for external materials on exposed sites is likely to be shorter than for normal areas, possibly 2 rather than the more usual 3 years. Internally, 4 years for public and 5 years for private accommodation would be normal.

Planning and construction

In planning the clubhouse and its environs it will be desirable to separate vehicles from owners and visitors cars and to provide entrances for goods and staff separate from the main club entrance. Entry to the dressing, drying and locker rooms from outside should again be separate and preferably away from the main entrance and close to the hardstanding and club moorings.

Circulation within the club is not normally very complex and the factors more likely to condition the plan are those of access, orientation and aspect in regard to its siting, and sanitation and plumbing in respect of its internal plan. There is an economic and functional need for the conveniences, showers and kitchen to be closely related and the establishment of these key elements may well determine the overall format of the plan.

All marina buildings are subject to the same structural town planning and building laws and regulations as any other. In strength of construction it may again be wise to budget somewhat higher than normal because of the harsh conditions of wind and salt water that such sites experience. For these same reasons and for town planning conditions it is unusual to see very high buildings forming part of the marina itself, although tall blocks may form part of such related uses as hotels or flats.

The building's structural frame must be decided in the light of site conditions, the building plan and the economics of the development. Clubhouse locations may well call for a non-standard substructure and long or short bored piles with connecting ground beams, plain or stepped rafts or robust retaining walls are all common alternatives to the more traditional trench foundations.

A continuous damp-proof membrane over the ground floor and a flexible dpc with cavity construction are usual for the external walls. These are often clad externally for further protection and warmth. Flat roofs are an advantage over all or part of the structure to provide popular sitting, sunbathing and observation areas.

5.57, 58 Simple concrete steps, uncluttered by handrails and balustrades at Havea, Valencia, Spain

Architecture and aesthetics

Most of the architecture associated with marinas and yacht harbours in the United States is quite good and some is very imaginative. In Britain there is, unfortunately, barely sufficient to provide a basis for a serious assessment although some examples illustrated here set an admirably high standard. Current development in the Mediterranean is often most disappointing. Little choice is offered between 18th-century romanticism complete with quaint Venetian bridges and the gimmicky ziggurats of La Grande Motte. It seems an indictment of the architectural profession that the biggest leisure development in the world should provide only these alternatives. *The coastline has always reacted violently to styles being imposed upon it and seems only at ease with forms and materials drawn from nautical traditions on whose foundations should now be built a satisfying vernacular.*[5]

Whatever the approach, the overall design policy should respect the area in which the development is sited both in local and regional terms and ensure that the built forms are in keeping. An excellent functional tradition is already present in the architecture of many coastal areas where the simple attractive forms of quays, and sea walls and the robust good sense of fishermen's cottages and marine artefacts stem no doubt, from nature's own unbeatable prototypes of water, sand and rocks along the ever-changing geometry of the shore-line. The designer's relationship with the developer can be vexing. Many operators thirst for eye-catching designs, 'themed' interiors or period lounges. Whilst these can sometimes be fun it is not easy to reconcile extreme styles with good architecture, and in the interests of harmony and sensible design it is usual nowadays to adopt contemporary design and not affect bizarre styles externally or internally.

Earlier in the chapter the whole development was analysed as a series of component parts. This is useful for identification and to illustrate the function of each, but once they have been collated and understood the design process must consider them in total, assembling and relating them so that no unnatural divisions occur but they flow one into the other with an ease that derives from a functional plan unified by good circulation and with colours and materials chosen to give a sense of unity, cohesion and strength.

Costs and economics

Whether a marina development is designed forwards from basic requirements or backwards from an analysis of investment yields is a critical question: the answer to it may well determine the size, layout, materials, construction and quality of the future scheme as well as its profitability. Whilst financial returns and overall development costs are important *they must only be a factor of collective control and not intrude as specific constraints upon individual elements.*

Clubhouses, due to their often doubtful contribution financially, may be vulnerable to being built down to a cost. This is poor policy and likely to sow the seeds of failure even at the conceptual stages. It is generally much

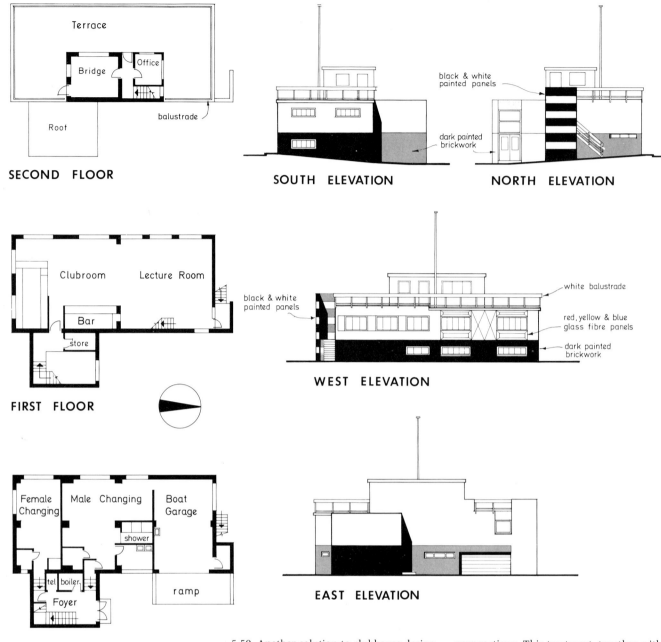

SECOND FLOOR

SOUTH ELEVATION

NORTH ELEVATION

FIRST FLOOR

WEST ELEVATION

GROUND FLOOR

EAST ELEVATION

5.59 Another solution to clubhouse design at the Stokes Bay Sailing Club, Gosport, Hampshire. The clubhouse was built in 1968 by converting and extending an existing war-time coastal defence station. The black and white banding, achieved by painting the brickwork, unifies the old and new sections. This treatment, together with the bold red, yellow and blue panels, gives the building a strong maritime flavour in keeping with its present and previous functions. The total floor area is 306 m² (3,200 ft²). Architects: Bailey, Piper, Stevens Partnership

better to phase the construction and allow the early returns to subsidise later additions.

Unit costs will vary according to the regions and the local conditions but in Britain the 1973 price would be expected to be in the order of £110–£130 per m² (£10–£12 per ft²).

6. For a detailed account of marina management see Chapter 13, pages 320–338.

Management[6]

The day-to-day running of the club cannot be examined here but in relation to the fabric of the building it is suggested that an annual maintenance fund is created to ensure proper financial provision for this purpose.

Shoreline facilities

Provision for customs and coastguards only apply to coastal sites and rarely will all the facilities dealt with below be included in one marina. Their needs are similar in some respects, each requiring a good view of the harbour and approach channels, good communication systems and buildings of warm and robust construction. They may, therefore, be located close to each other or actually together in the same complex. This will probably be at the end of the quay, sea wall or promontory in a tidal basin or beside the lock in a locked harbour. As well as providing their independent services they improve security with their surveillance.

Customs

This is only likely to be a small office—perhaps only staffed part-time. It will be on the harbour side of the lock or quay with suitable fixed steps or floating pier and ramp to boat-deck level.

Coastguard

Marinas can be quite suitable locations for coastguard stations if there are no naturally high features nearby. A tower is virtually obligatory for viewing, although the office or log room may be at quay level. The tower is usually reached from inside the office as outside stairs can be too exposed in bad weather. The journey to and from the premises must be reasonably sheltered and safe and well illuminated at night.

5.60 Lock-keeper's look-out at Chichester Marina, Sussex

Lock keeper

Accommodation for a lock keeper needs to be immediately beside the lock with the office at this level, but a look-out tower will improve visibility. A rough formula gives the visibility in miles as $1\frac{1}{3}$ times the square root of the observer's height in feet above sea level. The tower acts as a landmark for the approaching sailor and may be floodlit at night.

Harbour master

This office may be positioned on the quay or beside the lock or may share accommodation with the marina office and information bureau. It is more important that the harbour master is accessible to boat owners than vice versa. The office should be near the marina workshops and preferably close to the service transport and parking area.

Trolleys

Trolleys are popular for carting gear from car to boat. Types vary from 4-wheeled hand carts to the porter or golfer-type hand trolley. Key-holding owners take the trailer they need from a 'garage' beside the car park and return it after use.

Lockers

Individual lockers may be provided at each slip. These are convenient but tend to clutter up the walkways and should only be small, housing a few tools, paint and so on.

Larger stores to hold clothes, bicycles, etc., should be grouped near the car park, perhaps in a portmanteau building with showers, lavatories and trailer store.

5.61 Angle lockers at Hyde Street pier, San Francisco. These are always popular but they do encourage clutter on the walk-ways

Services

Provision of statutory services is dealt with in Chapter 8 on pages 239–266. Ducts and galleries are an important component within a fixed or floating system and they are best positioned on one side of the walkways terminating in a slip-side consul on the short side of each berth. It is wasteful to extend ducts along the finger piers.

Decking

This can be timber, asbestos, aluminium or concrete. In most cases it is applied to trimmers above the floats but can be the textured upper surface of the float itself. Minimum maintenance is most important as mistakes can cause continual expense. The surface should be slip-proof, well drained, not too abrasive or splintery and not too hot or cold to bare feet. Small units replace more easily than large ones. Straightforward 152×25 mm (6×1 in) timber planks in keruing or jarrah laid with 25 mm (1 in) gaps fulfil these conditions and take a lot of beating. Wood looks good against water particularly in rural areas. It is best left natural as creosote stains clothes and sails.

Fittings and accessories

Mooring fixtures may be single or double cleats, bollards, rings, traveller bars or many other forms, choice depending on cost and function in each circumstance. More important than the fitting is its position on the pier or walkway. They are easier to use when fixed to the top surface, but stringer-mounted fittings do leave the decks clear and uncluttered.

Most proprietary systems can be supplied complete with most of the accessories mentioned in the definitions list above, plus corner triangles, hinged junctions, angle offsets and so on 5.J, page 138.

Off-shore facilities

Treatment of the waterfront and walkways

The water area itself may be more clearly considered in terms of its relationship with the surrounding buildings than its position within the site boundary. As is seen on page 95 the shape of the land area, the length of the bulkhead and the average distance from the site boundary to the water will vary as the marina pushes into the land. The plan line of this interface can vary between a plain rectangle and a complicated series of facets and changes of direction. Departures from a straight edge help to relieve the tedium of uninterrupted bulkheads which seem so unimaginative, particularly in large marinas. There will need therefore to be a compromise between enclosing the largest water area within the smallest perimeter and an over-elaborate and expensive edge line.

The aim is to achieve an interesting harbour shape with some recesses and corners which will give the owner of each mooring a sense of individuality and enclosure without undue restlessness and at a reasonable cost. Where possible the form of the water area should stem from the engineering advantage that is taken of the natural contours of the site, but in an off-shore development on an unbroken coastline variety may be created by modelling the harbour walls, quays and breakwaters, by changes of level and by diversity of materials.

The margin between the water's edge and the buildings offers scope for

5.62, 63, 64 (below) Larger craft need larger gear for making fast. The bollard is very sculptural and the recessed links leave a clear walk-way

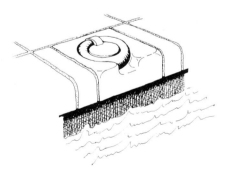

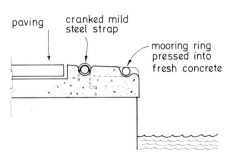

paving   cranked mild
          steel strap
                        mooring ring
                        pressed into
                        fresh concrete

imaginative treatment by way of shape, diversity and surface treatment. This is the water's frame and it may vary in width, constricting one moment and widening the next according to its function. Along the quay, buildings may in parts rise sheer from the water giving an immediacy and strength as they do on moles and cobs in many fishing harbours.

The height of the adjoining buildings and their distance from the water's edge may be used to vary the degree of enclosure. Many methods of margin treatment are possible but with regard to design in general it is recommended that this important area is handled imaginatively as an element in its own right and not left as a no-man's-land to be filled in as an afterthought. Orientation and aspect are important and so are views both across the harbour and inland from the level of the water, the quay and any upper storeys.

Thought must be given to the perimeter walkway—not only its continuity but its contribution as a linear 'precinct' with changing functions, free for once of vehicles and tuned to the pace of pedestrians. There are benefits to be derived from the ever-changing views that serial vision will present to the pedestrian.

Areas for sun and shade and protection from cold or prevailing winds are necessary because very uncomfortable microclimatic conditions can be created by an unfortunate juxtaposition of walls and openings. In urban marinas, particularly where craft will be manoeuvring to their moorings close to tall buildings, it may be worthwhile to test the model in a wind tunnel to ensure that difficult conditions have not been created or to correct them if they have.

With waterfront buildings there are opportunities for creating additional interest by constructing them or cantilevering them over water or, if they are light-weight structures, floating them on the water. These forms are expensive but may sometimes be justified—at least this creation of 'air space' saves on the use of land in the waterfront area where it is at a premium. An open ground floor exposes the hinterland to the water. Non-masted boats may even be launched or retrieved from a slipway 'through' the building to the hardstanding, car park or storage buildings behind.

The impression of a forest of tall piles at low water in a tidal basin has already been mentioned on page 58. These tall poles, some of which may be paired by a crossbar so that one walks through the 'goals' as one proceeds along the piers, create a mood quite different from the low flat pontoons and constant water level of a locked harbour. A uniform height above water for the bulkhead wall and quay is another consideration to be set against the varying amount revealed within the tidal range in an open basin.

It is evident then that the *form* of harbour and the *selection* of the fixed pier or floatation system will have a profound effect upon the eventual prospect of the water area. No one system is 'right', the decision must stem from engineering and economic considerations, the outcome then being designed in relation to the other components previously dealt with in this chapter.

Mooring patterns and harbour shape

An early decision on the relationship of the mooring pattern to the basic harbour shape will pay dividends later. Too often the engineering shell is totally fixed and the mooring system fitted in as best it may. This not only happens with small developments. At Brighton, 6 design and 18 cost options giving 108 theoretical and 78 actual alternative layouts were still under consideration not only after the infrastructure had been decided but after six months of construction work on the sea walls. If an architect only started to think of the room layout of a house when the builder had half-finished the outside walls, his method of working could be considered peculiar, and this is a reasonable analogy.

In addition to the regular slips, provision may be necessary for pontoons outside the marina, particularly with a locked harbour. Where a boat stacker is included boat collection bays are sometimes provided 9.19, page 275.

For some of the more common mooring patterns see 5.65. An important possibility which should not be overlooked is that some suppliers of mooring systems offer advantageous service terms by which they maintain and replace their equipment for a modest annual charge, including the removal of marine growth at regular intervals. Under such circumstances it is in the supplier's interest to provide a quality product to reduce the maintenance for which he accepts responsibility.

From an analysis of the catchment and the likely percentage distribution of sizes, the layout of the berths may be designed. The overall shape of the marina, its edge treatment and the positioning of the various facilities will all have a bearing on this. So too will the degree of protection offered by the harbour as a whole. A locked and well-protected marina allows a wide choice whereas more vulnerable areas will need to take account of tide and prevailing wind. The final choice will depend upon what importance is given to the overall density, ease of access and the return per berth which themselves depend upon the quality of the development and the percentage cost of the basic engineering work.

The layout of the berths will vary with the size of the craft accommodated. A marina for 7 m craft will generate a different pattern, density and appearance than one for craft of 12 m.

It is nearly always the case that small craft create a greater total length than large vessels. This fact—that smaller boats result in a greater overall length—seems surprising but is accounted for by their narrower beam, smaller turning circle and therefore less clearance widths for main and subsidiary waterways. Accurate prediction of craft length and the likely mix of permanent and visiting vessels is a skill which can have important influence on the harbour's economy and function. The options on varying the layout to suit the seasons or for maintenance work will increase if a flexible floatation system is used and if a deep or 'over dredging' policy is adopted (see Chapter 6 for calculating the dredge area).

Dredge depth and craft type are linked in mixed waters where sail and power are both popular. Predicting their percentage occupation is as difficult as identifying overall length. In areas where berth demand exceeds supply it is tempting to dispense with market studies, but if clever tailoring results in a well-fitted layout and minimum dredging then the savings can be re-invested or passed to berth holders in lower fees—or at least to stabilise them and ensure competitive pricing: it must be remembered that a sensible economy for the owner is an economy for the user.

If it does not put too much strain on costs the mooring arrangements could often influence the marina's shape to advantage. A marina is like a pub in that people don't mind its size provided there is a sense of intimacy and character in 'their' area. Many American marinas fail badly in this respect with rows of identical walkways in a graph-paper layout and no subdivision or creation of individual space or shape. Of course it is easier and may be cheaper to grid up a rectangle in this way but it is poor psychology and far from the snug embrace of harbour walls.

The water area

The density of boats per hectare varies considerably but is usually between about 62 and 160 (25–65 per acre). The sizes of berths will depend upon the craft and the mooring pattern. A pontoon system which is flexible has the advantage of allowing for seasonal variations. Berth positions are sometimes altered to allow shallow-draft boats to moor where silt has been deposited.

Main channels are usually about 15·2–18·2 m (50–60 ft) wide and subsidiary channels about 12·2–15·2 m (40–50 ft). Whilst berths and clearances may vary according to the boats accommodated in that part of the basin, the main channels must of course allow for the largest boats to manoeuvre safely in the most fluky conditions.

Types of boats and berths

In Chapter 2 mention was made of the catchment area of a marina and the type of boats to be anticipated. It is important that this information is as accurate as possible for two economic reasons: the first being the numbers of boats involved and the spending power of their owners; the second being the likely berth sizes, clearances and depth of dredge. *This last point is important for, unlike the other two, it is incapable of rearrangement.*

Yacht categories are governed by the Permanent International Association of Navigation Congresses (PIANC) which has devised a helpful shorthand on layout plans to use class numbers to categorise the berth sizes. These should be designed for the largest beam, length and draft in each class. Tolerances must be added to all boat dimensions to arrive at berth sizes and depths and these will vary with the berth's position, the mooring system adopted and the expected wave height in the harbour which, together with other factors such as wind and current will determine the maximum ranging the boats are likely to produce at their berths. The vertical tolerance for wave action should be half the wave height + 0·3 m in soft material and 0·5 m in rock.

Catamarans and other multihulls, due to their broad beam, present a berthing problem. This is best solved by allocating berths especially for them. End-of-pier positions, fore-and-aft moorings or (with a changeable mooring system) some extra-wide slips may be the answer. It may be necessary to charge a higher rental, for they are quite troublesome and occupy greater lengths of berthing frontage.

Fixed and floating piers

1 Definitions

Pier: A fixed or floating promontory giving access to moored boats and providing berths for them

Catwalk or finger pier: A narrow subsidiary pier

Guides: The means by which a floating pier is attached to a pile

Pontoon or float: The buoyancy system of a floating pier

Ref.	Type of mooring	Examples	Advantages	Disadvantages	Remarks
A	Stern to quay, jetty or pontoon, bows to piles	Chichester Le Grande Motte Rotterdam Kristiansund	jetty economy	not as convenient for embarking as alongside jetties or pontoons	
B	Ditto but bows moored to anchors or buoys	Deauville and the majority of Mediterranean marinas	jetty economy	not suitable with large tide range as excessive space required for head warps; danger of propellers being entangled in head warps	particularly suitable for large yachts in basins with little tide range where gangways can be attached to sterns
C	Alongside finger piers or catwalks one yacht on each side of each finger	Cherbourg, Larnaca (Cyprus) and many American marinas	convenient for embarking and disembarking		
D	Ditto but more than one yacht on each side of each finger	Port Hamble Swanwick Lymington	Ditto also allows flexibility in accommodating yachts of different lengths	finger piers must be spaced wider apart than in 'C' though this may be compensated for by the larger number of craft between jetties	fingers may be long enough for two or three vessels if more than three then provision should be made for turning at the foot of the berths
E	Alongside quays, jetties or pontoons single banked	Granville	Ditto		
F	Alongside quays jetties or pontoons up to 3 or 4 abreast	St. Malo Ouistreham St. Rochelle	economical in space and pontoons	crew from outer yachts have to climb over inner berthed yachts	
G	Between piles	Yarmouth Hamble River Cowes	cheapest system as no walkways, also high density	no dry access to land; difficulty in leaving mooring if outer yachts are not manned	not recommended except for special situations such as exist in the examples quoted
H	Star finger berths	San Francisco			

5.D Advantages and disadvantages of some types of berth lay-out: see 5.65 opposite

	A	B	C	D	E	
	sail & motor	sail & motor	sail & motor	sail & motor	sail	motor
Class						
I	32	30	26	40	9·7	11·3
II	37	35	32	53	10·5	12
III	44	42	40	62	11	13·5
IV	58	55	52	82	13·5	15·5
V	72	66	64	100	15·5	17·5
VI	100	90	90	140	25	25

5.E Recommended metric minimum clearances for the lay-outs proposed in 5.D for well-protected still-water basins

5.65 **Mooring layouts**

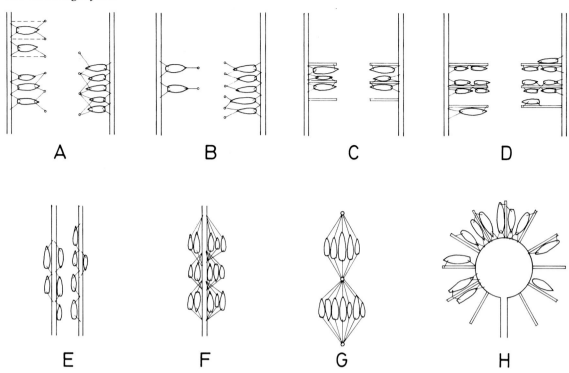

A   B   C   D

E   F   G   H

5.66 This double-banked arrangement is one of the most frequently adopted berth lay-outs. It is inclined to be rather repetitive if too many rows of identical piers are used, as at Shilsole, Seattle, but it is economic and simple. The Hamble River Marina, Solent, shown here, is unusual in having 'straight through' finger piers which master the main walk-way and are actually wider. The end-of-pier piles in roller housings are a neat and conventional solution. The timber decking adds, as usual, a warm and natural appearance

5.67 Below: Spatial requirements and likely size ranges (Reproduced by kind permission of Mrs. Daphen Dunster (née Collins) from *Boatplan: Facilities for Pleasure Craft*)

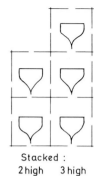

Stacked :
2 high   3 high

Method

a Parked = 7·620 m × 2·438 m + 1·210 m clearance on two sides (25′ × 8′ + 4′)
  = 8·830 × 3·650 (29′ × 12′)

Surface area (approx.)
32·33 m² (348 ft²) = 375/ha (150/acre)

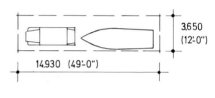

Parked with car, clearance on two
sides   = 14·935 × 3·658 (49′ × 12′)
stacked = (two levels)
      = (three levels)

54·63 m² (588 ft²) = 125/ha (50/acre)
16·17 m² (174 ft²) = 450/ha (300/acre)
10·78 m² (116 ft²) = 1125/ha (450/acre)

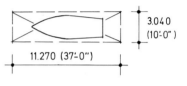

b Moored between piles
= loa + min. of 1·524 (5′) at each end of the craft.
0·610 (2′) beam clearance and 0·09 (1 ft²) piles in this example
= 11·278 × 3·048 (37′ × 10′)

34·37 m² (370 ft²) = 325/ha (130/acre)

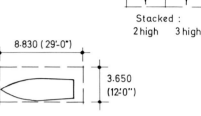

c. Swinging moorings or trots
Largely determined by the heights of the tides, which in turn determine the length of the cable or mooring line. Tidal range has been averaged at 4·570 (15′). The length of the cable is 3 or 4 times the depth, i.e. 13·710 (45′). The craft is then free to swing within this sector which has a
13·710 + 7·620 = 21·330 (45′ + 25′ = 70′) radius, giving a spatial allowance of

176·51 m² (15,400 ft²) = 7·5/ha (3/acre)

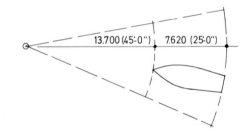

d Anchored (portable)
Length of cable in this case is not only determined by the depth of water but also by the type of anchor, boat and sea bed: it varies from 4 to 6 times the depth of the water. For this example assume 4·570 (15′) depth × 5 = 30·480 (100′) radius within which the craft is free to swing.

371·91 m² (31,422 ft²) = 3·75/ha (1½/acre)

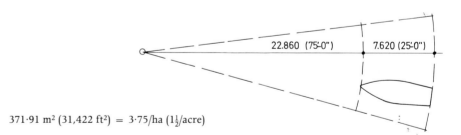

e Berthed
In a marina type of development or alongside a pier or jetty. The spatial allocation is similar to that for parked craft, allowing 0·152 (6″) either side of the beam for fenders and half the width of the catwalk in the case of a 7·620 (25′) boat this would average 1·219 (4′) on the length and 0·610 (2′) on either side = 9·144 × 3·658 (30′ × 12′) Allowing for clearances, piers, etc.

33·45 m² (360 ft²) = 302/ha (121/acre)
        = 125/ha (50/acre)

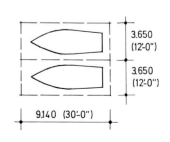

5.68 Craft moored between piles in the Hamble river

Class	Length overall L (metres)	Sub-classes
I	L < 5 (16′5″)	Motor boats Sailing yachts Motor/Sailing
II	5 < L < 8 (26′3″)	Motor boats with living quarters Motor boats without living quarters Sailing yachts with living quarters Sailing yachts without living quarters Motor/Sailing with living quarters Motor/Sailing without living quarters
III	8 < L < 15 (49′3″)	Motor boats Sailing yachts Motor/Sailing
IV	L < 15 (49′3″)	Motor boats Sailing yachts Motor/Sailing

5.F Classification of pleasure boats proposed by the International Commission for Sport and Pleasure Navigation of the Permanent International Association of Navigation Congresses (PIANC). For classification purposes it is assumed that craft above 7 m in length have living quarters and those less than 5 m do not

	Sailing yachts and motor/sailing			Motor boats and centreboards			Trimarans and Catamarans		
All dimensions are in metres		Berth			Berth			Berth	
Length overall L	Class	depth	width	Class	depth	width	Class	depth	width
L < 8	IS	1·5	2·8	IM	1·0	3·3	IT	0·6	4·5
8 ≤ L < 10	IIS	1·6	3·0	IIM	1·0	3·5	IIT	0·8	5·0
10 ≤ L < 12	IIIS	1·7	3·2	IIIM	1·0	4·0	IIIT	1·0	5·5
12 ≤ L < 15	IVS	2·0	3·8	IVM	1·3	4·6	IVT	1·2	7·0
15 ≤ L < 18	VS	2·5	4·5	VM	1·5	5·0	VT	—	—
18 ≤ L < 25	VIS	3·0	5·5	VIM	2·0	5·5	VIT	—	—
L ≥ 25	VIIS	4·5	7·0	VIIM	2·5	7·0	VIIT	—	—

5.G Yacht classification with recommended berth widths and depths. The depth and width of berths are assumed normal maxima for the class which will only be exceeded in a small percentage of cases. The dredged depth should be taken from extreme low water and increased appropriately to allow for wave action, dredging tolerance and siltation

							Languedoc-Roussillon (forecast)			
	Swanwick		Chichester		Emsworth		Large harbours		Medium harbours	
Class	S	M	S	M	S	M	S	M	S	M
I	13	5	30	14	46	10	30	45	45	65
II	25	9	24	15	30	4	24	24	30	24
III	19	2	4	6	4	2	23	20	15	9
IV	12	7	2	3	2	2	16	7	10	2
V	3	4	—	1	—	—	3	2	—	—
VI	—	1	—	1	—	—	3	2	—	—
VII	—	—	—	—	—	—	1	—	—	—
	72	28	60	40	82	18	100	100	100	100
	100		100		100					

All figures are percentages. Multihull craft account for less than 1%.
5.H Size distribution in 3 English marinas. A count by size of the yachts in the marinas at Swanwick, Chichester and Emsworth. For comparison the assessment of boat sizes expected in the Languedoc-Roussillon marinas is also given

Pier number	Length of pier	Number of berths each width (centre to centre of piles)						Total each pier
		2·743 m (9')	3·048 m (10')	3·658 m (12')	3·810 m (12'6")	4·267 m (14')	4·877 m (16')	
No. 1 west	73 m (240')	34	16	0				50
No. 5 east	92 m (320')	22	10	24				56
No. 2 west	80 m (260')	6	16	24				46
No. 6 east	85 m (280')	50	10	0				60
No. 3 west	85 m (280')				24	10	6	40
No. 7 east	73 m (240')				28	8	0	36
No. of berths each size		112	52	48	52	18	6	Total 288
Per cent of total berths (not inc. pier 4)		39·0%	18·0%	16·7%	18·0%	6·2%	2·1%	100%
Pier 4 (boatel wharf)	107 m (350')	0	8 At boatel: all small sizes	0	8	6	4	26 Grand total 314

Boat sizes

Note: larger accommodations at pier ends and unassigned bulkhead not included in the above count

Boats under 4·9 m (16')

Boats (4·9 m to 8 m (16' to 26')

Boats 8 m to 12 m 26' to 40')

Boats over 12 m (40')

Recommended number of berths for initial construction period. Future requirement to be based on records to be kept of local demands. Dry berthing will influence future totals.

5.1 Typical berth-arrangement chart. Most assessments of marina capacity include a chart of this kind and there may indeed be such a chart for each alternative layout.

Decking: The upper surface of a pier or walkway
Fender: The protective edge trim of a pier or walkway
Hinge: The means of articulation between floating units
Ramp: The usual means of access between fixed and floating surfaces
Floatation system: The entire floating equipment
Mooring pattern: The general arrangement of berths
Berth: A boat mooring station
Channel: The main route entering or within a harbour
Clearance: The distance between rows of occupied berths
Tolerance: The distance between a boat and its berth
Gallery: A duct carrying services
Stringer: The frame or edge of a pier or walkway
Anchor pile: A pile for mooring to
Traveller bar: A vertical or horizontal mooring rod fixed to a pile or stringer
Cleats, rings and bollards: Objects to secure boats to in their berths

The decision to adopt fixed or floating piers will be based on cost, tidal range, safety and convenience. A combination of fixed and floating piers is often satisfactory. Where conditions allow, a fixed system is considered to be better for it is often marginally cheaper, less costly to maintain, stronger, more durable and stable, and will certainly bear heavier loadings and withstand impact more readily than its floating equivalent. The advantages of a floating system are as follows:

1.1   Constancy of level between pier and water
1.2   Rearrangement of layout is possible
1.3   Alteration to mooring lines unnecessary
1.4   Less likelihood of damage to boats under tidal conditions

2   Fixed piers

These are open to a wide range of construction methods and materials. Steel, concrete and timber are the most common and combinations of these materials are usual. In a locked marina, a fixed system will be an obvious

5.69 Fixed and floating walk-ways at Fort Lauderdale. The decking paint has not been a success

5.70 Bow-to-quay mooring in New Orleans Marina. Shallow draft moorings in this position are economic in bulkhead construction

choice. With a constant water level, piles may be driven and capped off 610–914 mm (2–3 ft) above the surface, the beams and decking being constructed immediately above. This gives a neat, clean appearance to the berths, without a forest of projecting piles. The only changes in levels will be those necessary between the fixed pier and the quay, usually by fixed ramp.

3   Floating piers

In England particularly, there are some very odd collections of materials and constructions used to make up floating piers. Oil drums, steel and

5.71 This small dock is simple in design and can be constructed at the site in a fairly short time. The construction of such small docks is usually a part of a launching ramp development on lake shore or river. After a boat is launched, this type of dock provides a convenient landing place

5.73 Fixed concrete walk-ways at the Municipal Marina, New Orleans

5.72 Launching ramps constructed in groups of two or more should have some type of dock structure separating the ramps. One type of dock is illustrated here: the small finger pier. Structures of this type allow easy access to the boat during launching and hauling out operations and also facilitate boarding and unloading while the boat is in the water

5.74 Fixed piers at Sommers Cove, Maryland, re-painted for the forthcoming season

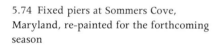

5.75 Crisply painted fixed walk-ways at Fort Lauderdale, Florida

painted asbestos piping are not uncommon. In any new modern marina, however, it is wiser to specify a proper proprietary system. Although initially more expensive, the advantage of being able to obtain spare parts, together with the flexibility that a good system will give, will benefit the developer in appearance and maintenance over the years. Another advantage not always appreciated is that considerable thought is given in the design of a proprietary system to the calculated balance between buoyancy and stability and, in this respect, it is generally found that the heavier concrete pontoon has advantages over most of the plastic materials.

The satisfactory design of floating piers is not as easy as it might appear, requiring quite difficult calculations for buoyancy, stability, tilt and so on and is best left to experienced professionals. Their physical properties are as follows:

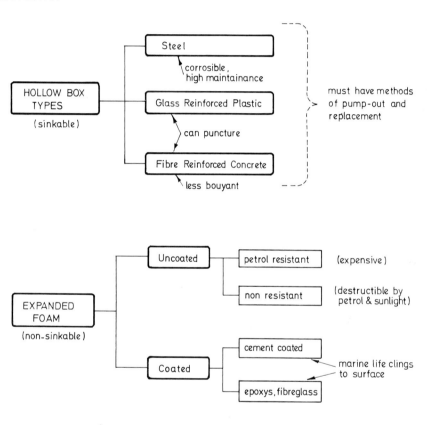

5.76 A four-roller pile-guide in the Uni-float system. This is a thoughtful solution although the cleat is misplaced. Compare however with 5.66

5.77 Poor materials and construction prove expensive in the long run

4 Materials: Concrete

This may be aerated or high-strength material. Sometimes the float is moulded around a collapsible former, or else made in two halves afterwards joined and sealed. One very good system in the United States is the Bercleve Unifloat. This very simple and attractive system combines the pontoon and decking in one unit by texturing the top surface of the float to give a cool and non-slip surface. All services including lighting, ramps and pile guides are supplied, giving a simple, good looking and unified system. Many different accessories are available allowing the designer to crank piers at any angle or to adopt T, L or pinwheel layouts at little extra cost. Concrete, as material for pontoons, is more reliable than it used to be. It had a bad name for cracking and leaking but advances in quality control have now improved its reliability.

Maker	Dimensions of standard pontoon unit(s)	Floatation unit	Dimensions of floatation unit	Deck	Width of deck	Remarks
Dibben Structural Engineers	12·2 m × 1·83 m	steel rectangular	1·83 m × 1·0 m (6′ × 5′6″ × 3′3″)	timber (keruing) on steel frames	1·83 m	
Walcon Limited	1·9 m (6′3″) width overall	expanded polystyrene encased in 19 mm (¾″) polypropylene fibre reinforced concrete	1·83 m × 1·2 m × 0·9 m (6′ × 4′ × 3′) wt. 9 cwts. draft (sea water) 9″, displacement (sea water) 128 lb/in	timber on steel frame	1·9 m (6′3″)	
Thos. Storey (Engineers) Limited	12 m × 2 m 12 m × 2·5 m 12 m × 3 m	polystyrene coated with plastic or concrete	2 m (2·5 m, 3 m) × 1·25 m × 0·65 m	hardwood timber in modular mats	2 m 2·5 m 3 m	Acrow-berth' system
United Floatation Systems		steel pontoon filled with polystyrene foam				
Magnum (Engineering) Ltd.	9·1 m × 2·4 m (30′ × 8′) 13·7 m × 2·4 m (45′ × 8′)	expanded polystyrene encased in elkalite	1·83 m × 0·53 m × 0·76 m (6′ × 1′9″ × 2′6″)	elkalite	2·4 m 1·2 m fingers	elkalite is made with high alumina cement and synthetic fibre
Concrete Utilities	7·3 m (24′) and over	expanded polystyrene encased in 13 mm (½″) ferro-cement		concrete timber or steel	0·9 m to 2·4 m (3′ to 8′)	Delivery within 100 miles

5.J Some types of commercially marketed pontoons

Plastic

Under this generic term are included expanded polystyrene, fibreglass and various kinds of plastic. Although it is possible to purchase floats separately from plastic manufacturers, here again a proper system is advisable.

None are affected by insects although crustacea adhere to them and are difficult to remove without damaging the cellular types. Expanded polystyrene can be affected by pitch, tar, petrol, oil and detergents. Where this is likely it is best to specify pontoons coated with glass-reinforced cement. Foamed plastics may be moulded to house services ducts or galleries. Whether these are lined with plastic conduit depends upon their positions, but care is needed at points of articulation. Uncoated foam may tend to crumble at the surface which, whilst quite harmless and not affecting buoyancy litters the surface of the water with white particles.

Fibreglass (grp) pontoons are increasingly popular, being impervious to attack from insects and most chemicals. They are long-lasting and whilst not easily damaged, should this happen are quite easily repaired.

Steel

Whilst steel has given good service in some locations, the incidence of failure in the past has been quite high. Ferrous metal in contact with water relies entirely upon its protective coating and any breakdown invites rust which, once inside the shell, is virtually irremovable. This applies to steel pipe boxes and drums. Wrought iron however resists corrosion better and is particularly well suited to galvanising.

Dimensions

Walkways are not usually less than 2 m (6 ft) wide and may need to be wider at their in-shore end when their length is more than about 50 times their width. Finger piers or catwalks serving 1 or 2 boats are usually 1·00–1·25 m (3–4 ft) wide. The decking surface may vary between 152 mm (6 in) and 610 mm (2 ft) above water level depending upon wave height and the average boat-deck level.

The maximum length of piers and walkways will vary with the system adopted and the tranquillity of the water. Unpiled floating walkways, the ends of which are only anchored, will experience considerable turning moments at the in-shore end and will have to be restricted in length. With a generously piled arrangement in still water of even depth the walkway could theoretically be of infinite length.

Ramps and stairs

The means of linking a fixed or floating system to the land area should be considered as part of the mooring arrangement. Ramps and steps are often not integrated properly and appear misplaced with odd materials and construction. Ramps are generally to be preferred as they cater for trailers. Stepped ramps are a compromise, shortening the total 'going' whilst allowing trailers to bump down if necessary. Free-standing stairs can of course be articulated too, with rollers or wheels bearing onto the floating walkway. Alternatively they may be fixed to the quayside to be drowned or exposed according to tide levels.

Sometimes complicated arrangements of floating walkways are adopted which climb or descend a fixed ramp according to the water's rise or fall. The cables and anchors involved and the need to ensure that the guides are clear of debris means they are by no means trouble-free and can be heavy on maintenance and initial cost.

The surface material looks better if it matches the rest of the decking but in any case it must be slip-proof. Handrails are necessary if the slope exceeds about 1 in 4.

Anchorage systems

The system chosen to hold the pontoons in position must stem from a thorough appraisal of the site conditions including:

1 Water depth
2 Bed material
3 Tidal range
4 Current
5 Wind conditions
6 Wave height
7 Cost and appearance

The first decision—whether to use piles or not—will stem from a study of these factors. If piles are economic they should be used as they usually provide the simplest, safest and most direct method although their appearance may be a factor against them. Their number, length and position in the arrangement will be an engineering decision and whilst no more should be driven than necessary there must be sufficient to ensure a stable system in the harshest conditions expected.

The three main ways of securing the float to its pile are:

1   The fixed-to-stringer type 5.79.
2   The built-in type 5.80
3   The use of traveller bars or guides 5.88

It is worth remembering that forces exerted horizontally upon piles become more severe at high water as the lever arm (the bed-to-fixing distance) increases.

Where piles cannot be used various cable and anchor systems are employed. This is usually where the water is too deep, the tidal range too great or the bed material unsuitable for pile-driving. Their disadvantage is the possibility of keels and propellers fouling the lines and the maintenance of the system. Their advantage is a clean, neat appearance.

5.78 Pontoon guides and bridging section, Los Angeles

5.79 Outrigger guides are best formed as strengthened extensions of the framing. This example would be inadequate for an exposed site

5.80 A 4-roller built-in guide. The timber batten reduces lateral movement but too tight a fit risks jamming

5.81 Below: layout and construction drawings to illustrate typical arrangements for both fixed and floating pier systems. Both types of pier cater for a tidal range of 1·829 m (7′)

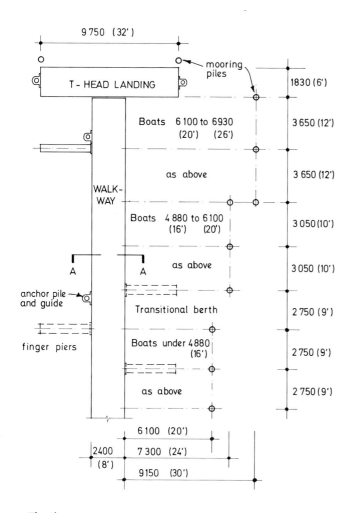

a Floating

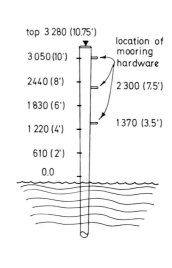

b Floating

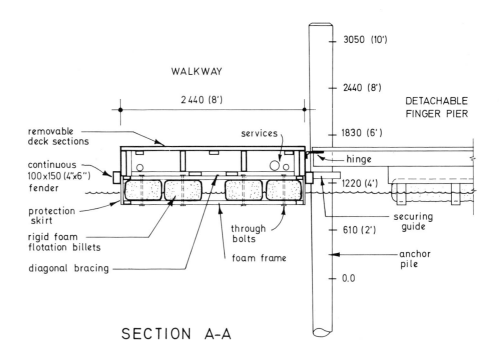

SECTION A-A

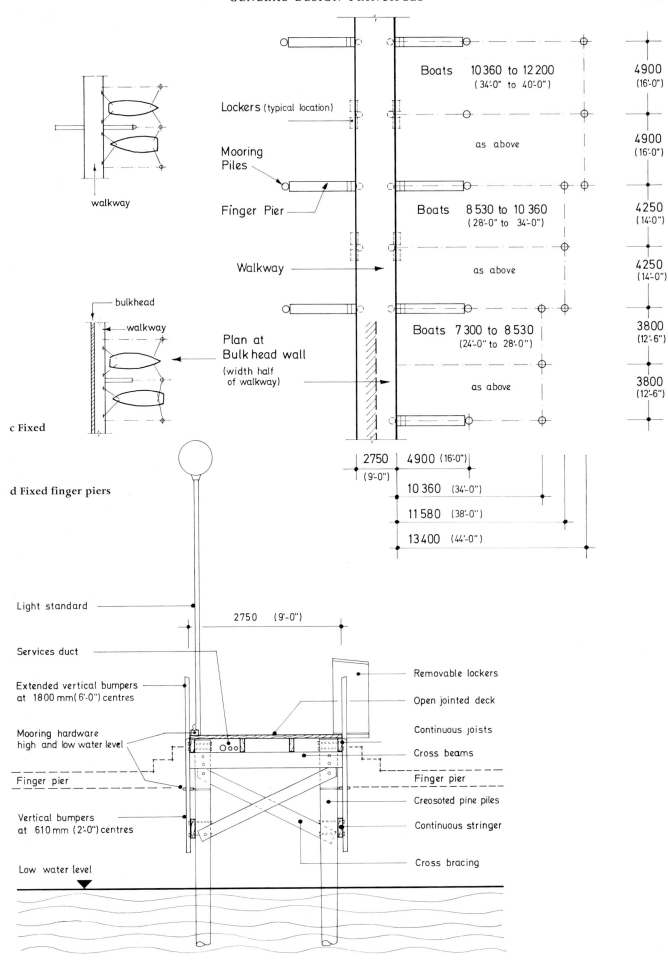

walkway

Lockers (typical location)

Mooring Piles

Finger Pier

Walkway

bulkhead

walkway

Plan at Bulk head wall
(width half of walkway)

c Fixed

d Fixed finger piers

Boats   10 360 to 12 200
(34'-0" to 40'-0")   4900 (16'-0")

as above   4900 (16'-0")

Boats   8 530 to 10 360
(28'-0" to 34'-0")   4250 (14'-0")

as above   4250 (14'-0")

Boats   7 300 to 8 530
(24'-0" to 28'-0")   3800 (12'-6")

as above   3800 (12'-6")

2750 (9'-0")   4900 (16'-0")

10 360   (34'-0")
11 580   (38'-0")
13 400   (44'-0")

Light standard

2750   (9'-0")

Services duct

Extended vertical bumpers at 1800 mm (6'-0") centres

Mooring hardware high and low water level

Finger pier

Vertical bumpers at 610 mm (2'-0") centres

Low water level

Removable lockers

Open jointed deck

Continuous joists

Cross beams

Finger pier

Creosoted pine piles

Continuous stringer

Cross bracing

e Fixed header piers

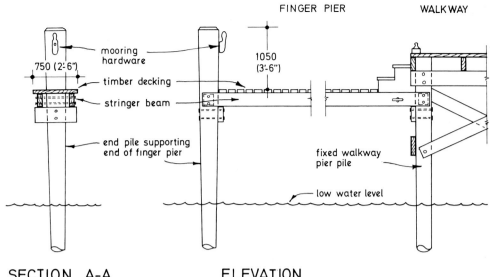

FINGER PIER   WALKWAY

mooring hardware

750 (2'-6")

timber decking

stringer beam

1050 (3'-6")

end pile supporting end of finger pier

fixed walkway pier pile

low water level

SECTION A-A   ELEVATION

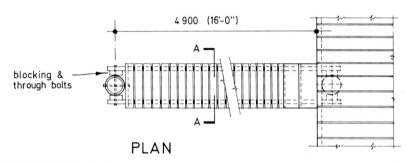

4 900 (16'-0")

A

blocking & through bolts

A

PLAN

5.82 Polystyrene floats with timber framework and decking

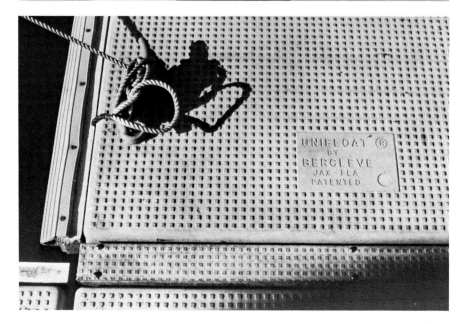

5.83 Close-up of the Bercleve Uni-Float Decking System at Sanford Marina, Florida

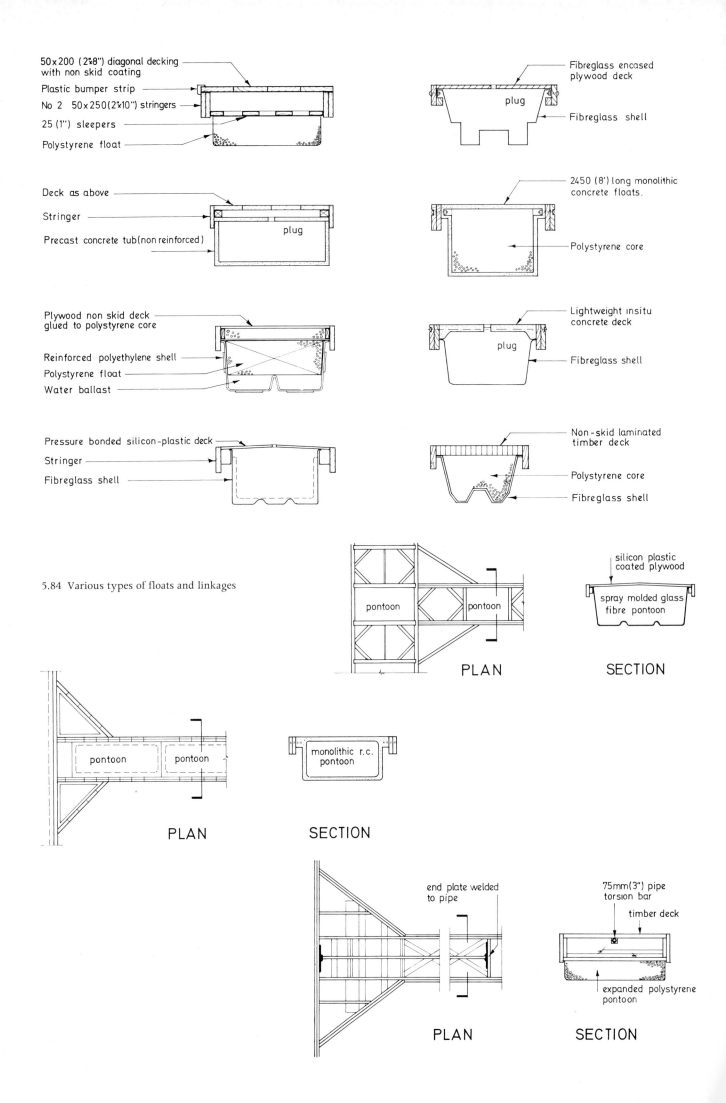

50 x 200 (2"x8") diagonal decking
with non skid coating

Plastic bumper strip

No 2 50 x 250 (2"x10") stringers

25 (1") sleepers

Polystyrene float

Deck as above

Stringer

Precast concrete tub (non reinforced)

plug

Plywood non skid deck
glued to polystyrene core

Reinforced polyethylene shell

Polystyrene float

Water ballast

Pressure bonded silicon-plastic deck

Stringer

Fibreglass shell

Fibreglass encased
plywood deck

plug

Fibreglass shell

2450 (8') long monolithic
concrete floats.

Polystyrene core

Lightweight insitu
concrete deck

plug

Fibreglass shell

Non-skid laminated
timber deck

Polystyrene core

Fibreglass shell

5.84 Various types of floats and linkages

pontoon   pontoon

PLAN

silicon plastic
coated plywood

spray molded glass
fibre pontoon

SECTION

pontoon   pontoon

PLAN

monolithic r.c.
pontoon

SECTION

end plate welded
to pipe

PLAN

75mm (3") pipe
torsion bar

timber deck

expanded polystyrene
pontoon

PLAN   SECTION

5.85–88 **Types of commercially-marketed pontoons**

5.85–87 An all-steel system by Walter Bower & Co., North Muskham, Newark, Nottinghamshire. The photographs show the steel floatation cylinders supporting the open steel-mesh
5.87 The drawing illustrates the coupling to the stabiliser bars and the perimeter walk-way

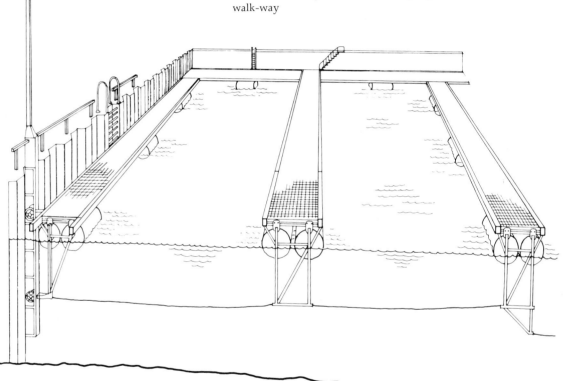

5.88 The RESA system of floating piers moored to piles. The picture shows the main jetty with breakwater profiles mounted to windward

5.90, 91 These sections show coated foam floats and shaped cement decking mounted on a superstructure of timber. Lengths are 12′ and upwards and 6′–8′ wide for walk-ways, and 2′, 3′ and 4′ for the splayed-end finger piers. Mooring rings and fenders are standard. Services and lighting bollards can be fitted if required

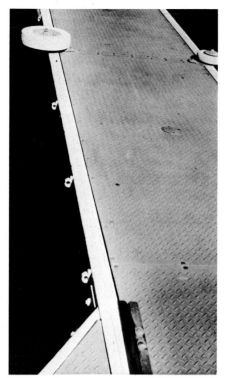

5.89 Anodised aluminium decking to a neat system of floating piers at Marina del Rey, Los Angeles. The decking surface remains remarkably cool in summer

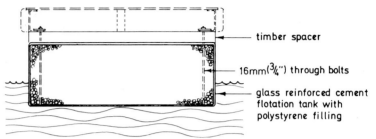

— timber spacer

— 16mm(¾") through bolts

— glass reinforced cement flotation tank with polystyrene filling

5.90 Cross-section of glass-reinforced cement floatation tank to show a typical arrangement for steel and timber decks

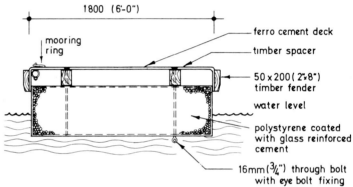

1800 (6′-0")

mooring ring

— ferro cement deck

— timber spacer

— 50 x 200 (2″x8″) timber fender

water level

— polystyrene coated with glass reinforced cement

— 16mm(¾") through bolt with eye bolt fixing

5.91 Cross-section of ferro deck and glass-reinforced cement floatation tank

5.92 Floatation system ready for assembly at Emsworth Yacht Harbour, Hampshire

5.93 Steel-framed ramp with timber decking at Emsworth

5.94 Aluminium ramp leading to a Bercleve uni-float system. This system, made at Jacksonville, Florida, is most impressive in its neat appearance and flexibility

5.95 Long ramp at the Port Haliguen, Quiberon, Brittany, France. The quay wall designed after the floatation system

5.96 If it can't float, it has to be raised

5.97 More complications at Los Angeles— and it all has to be maintained

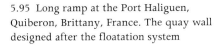

5.98 This man re-paints the pier decking every season with a paint-pumice mixture. Aluminium, plastic or even the right choice of timber would have saved such labour and expense

5.99 Even at Fort Lauderdale mistakes are made: ferrous metals make poor decking

5.100 'Styrofoam' (polystyrene) floats supporting asbestos decking at Santa Barbara, California. Fire-hose fenders always get ragged and are difficult to maintain

5.101 Timber decking at Nassau. The preservative stains the timber an attractive smoke grey

5. General design principles: Check list of marina accommodation and services

Note: this list is immediately followed by a broader check list covering the chapter as a whole.

Social activities	Clubhouse, boat-owners' lounge, public house, bar, snack bar, restaurant, offices, committee rooms, starters' post, lookout, viewing terraces, sunbathing Reading room, navigational library, weather forecast board, chart room, television, children's play space, crèche, paddling pool
Shops	Food and general stores, tobacco, stationery, etc. bookshop chandlery, clothes hairdresser, beauty salon barber's shop sauna masseur chemist laundry, launderette
Services and information	Marina office, information centre, caretaker's maintenance workshop, storage and staffroom banking post office, Giro visitors' information service (e.g. doctors, restaurants, entertainment) flagpole, windsock, weather and tides information kennels
Allied activities	Customs house harbourmaster's office coastguard, weather station and information radar, communications mast Sea Scouts lock-keeper's accommodation police, security station
Boatside facilities	Storage lockers lavatories (public and private) showers, baths drying rooms, cabinets bottled gas service electricity, lighting and power plug-in telephone service dockside laundry service tannoy system litter bins mail service
General services	Gas, main, bottled or in bulk storage electricity, lighting and power to piers and grounds (see safety equipment) sewage and refuse disposal water supply telephones centrally controlled security system
Boat services	Boat building, repair, maintenance yard, material store new and second-hand boat and engine sales and hire launching and hauling equipment (fixed and mobile) hardstanding launching ramps and slips dry storage of boats covered moorings (wet and dry) information board of local services brokerage, insurance, marine surveyors divers' service fuelling station or tender

Allied sporting activities: provision and instruction	Rowing SCUBA, skin-diving equipment and instruction water ski-ing, ski-kiting swimming fishing tackle (hire and sale of bait) sail training tennis, badminton and squash courts
Allied accommodation	Hotel, motel, holiday flats, public house, holiday inn
Transportation areas and services	Car parking and service (fuel, repairs and hire) trailer bays and hire bus bay transport to and from local centres carts for stores and baggage motor cycle/bicycle sheds (open and covered) boat trips and coach tours marina staff electric runabout marina workshops and transport areas
Safety equipment	First-aid post and observation platform fire-fighting equipment, fireboat life-saving equipment and instruction warning or flood lights to breakwaters, lock and harbour entrance general security system, fences and lighting de-icing or aeration equipment weather and tides information
Miscellaneous	Casual recreation area (e.g. picnic and kick-about areas) swimming pool vending machines, ice dispenser paved and grassed areas landscaping gardener's stores and sheds etc.

5. General design principles: Check list

General design procedure

Outline proposals	Collate survey material, data studies and all information relating to site, setting, context and environment Determine principal elements in broad outline only. Go no further until these are agreed with all development team At this stage, formulate design programme. Check procedure against Management Diary methodology (see Chapter 13, pages 305–306) Begin draft of critical path network
Basic marina form	On the basis of the site survey and hydrographic data decide on the fundamental type: off-shore semi-recessed built-in landlocked
Evaluation	Produce outline zoning plan by relating information from site appraisal to the land-use functions and user requirements Relate the framework of the plan to the anatomy of the surrounding environment
User requirements	Do not just presume user requirements. Discuss with berth holders at comparable locations (see Chapter 5, pages 106–117) for some basic user needs
Assets and liabilities	Establish which site elements are assets to be kept and which are liabilities to be dealt with

Design potential
Evaluate design potential of the site and its application to the proposed development
Prepare in outline the alternative design approaches.
Classify the factors and needs governing the location of principal elements
Review completed alternative design drafts. Discuss with client and development team and prepare guidelines for final selection
Obtain agreement with client and development team on the final selection from alternative design drafts

Report to client
Prepare report to client:
appraising the site and the problems
giving conclusion on the relationship of site to the proposed development
stating intended methods of proceeding with the project
summarising the financial and management situations
giving expectation of planning approval and probable programme
stating an opinion on the projects feasibility

Context and setting
Draw together all environmental information on the area
Continually review the outline proposals against background of local setting and consider:
orientation, aspect and prospect
distant, medium and close-up views *of* the site and *from* the site (both land and water)
consider night views, silhouette, and lighting proposals

Site Planning

Communications
Apply user requirement data to appraisal conclusions to produce communication pattern linking areas on the zoning plan

Access
Ensure that layout meets access requirements for deliveries, furniture removal, ambulances, refuse collection, launching and haul-out of largest craft

Checks
Use association charts and linear programming to check zone relationships and validity of circulation. Relate zoning and circulation to the outline building proposals to ensure unity of internal and external flow patterns

Services
Agree with statutory and local authority engineers on relationship of buildings, circulation and outline proposals for services, drainage, roads, waterline, edge treatment, locks, river works, etc.

Environment
Study likely effects of noise, overlooking, effects on wildlife, river regime, coastal equilibrium, micro-climate and adjust proposals to mitigate adverse effects

Research background
Keep abreast of research work, engineering and hydraulic studies, microclimate data. Tailor design to embody such evidence and be prepared to amend where necessary

Masterplan
Prepare draft master plan on the basis of the preferred alternative. Arrange meetings with local authorities to discuss and agree the approach in broad terms under such headings as planning criteria, access roads, environmental acceptability, basic services, bye-laws, fire regulations and daylighting

Local experience
Check the practicalities of design concept with experienced specialists—coastguard, lock keepers, harbourmasters, local fishermen, etc.

Statutory approvals
In light of the preceding data and following favourable reaction to appraisal report (see Report to Client above) submit for outline planning approval at this stage if not already applied for

Scheme Design

Components
Ensure that the elements included within outline masterplan include all of client's initial requirements and anticipate future needs and expansion opportunities. Run through Check List of Marina Accommodation and Services (pages 149–150)

Land to water ratio	Ensure that the apportionment stems from an allocation of space according to client/user requirement and control the physical and aesthetic manner in which this is handled. The latter must evolve from the former

See page 95 for relationship
See page 108 for spatial requirement and 5.A, page 98
Compile planning, design and environmental context material in preparation for possible public inquiry. Assemble notes useful in preparation of proofs of evidence. If in doubt, discuss with client and lawyer

Creation of Spaces

Masses and voids

Consider siting of individual buildings and engineering structures in relation to
existing topography
contours
existing trees and other vegetation
coastal or river profile
enclosure
drainage pattern
shelter
screening, etc.
Site groups of buildings in relation to each other to
define external spaces
create vistas and close views
avoid wind tunnels
avoid turbulent sailing conditions, dangerous harbour entrances
frost pockets, etc.

Harbour shape and existing contours

Relate mooring pattern to harbour shape. Do not allow engineering infra-structure to be totally fixed until function and design criteria are established. Try to give interest, variety and individuality to each berth or group of berths:
avoid grid-iron layout
favour serpentine perimeter and creation of enclaves

Visual linking

Consider linking of certain spaces, vistas and views. Evolve these from pedestrian circulation. Other spaces to be visually or physically separated: consider
privacy
surprise
concealment

Exposure/enclosure

Need:
Physical boundary, legal boundary, privacy, security, control, demarcation, wind, shelter, shade

Degree:
Complete, partial or slight enclosure
Physical but not visual
Visual but not physical
Filtering, aural

Permanency:
Fixed, temporary, moveable

Character:
Insuperable, domestic, opaque, slatted, psychological

Construction:
Walls, fences, mounds, planting, buildings

Levels:
Consider enclosure in relation to ground slopes

Openings:
Consider openings for pedestrians only, vehicles, and means of closing them
Design enclosures remembering sun, shade and wind turbulence

Shelter	Consider exact siting of sheltering elements and relate to levels. Select materials in relation to function, durability, colour etc.
	Prepare planting plan to illustrate form, density and species
	Consider need for model and wind tunnel tests

Movement and Control

Circulation

Consider movement of craft, vehicles, owners, visitors, staff and objects between
  zones
  buildings
  entrances
  ramps
  berths etc.
Examine intensity
    volume
    periods of movement
Think about movement at all levels
  ramps
  parking elevators
  underpasses
  escalators
  tidal movement
Deal imaginatively with static spaces such as
  paved areas
  parking spaces
  hardstanding
  laying-up areas
  loading bays

Control elements

Plan fences, screens, rails, gates, ramps, hoist areas, margins at water's edge, screening mounds, hazards, signals, notices

Perimeter walkways

Check existing
  easements
  rights of way
  towpaths
  riparian access
  coastal paths
  fishing and mooring rights
Consider continuity of access with alternative or replacement routes
Give foreshore and marginal areas an imaginative treatment

Ground modelling

Particularly with inland sites existing features often need shaping e.g.
  pools
  reservoirs
  lakes
  derelict pits
  quarries
Consider rising ground for
  views
  spectator banking
  hollows for privacy
  protection
  concealment (parking)
Consider fixed ground and changing water levels
Remember flooding and high water
  control of surface run off
  economy of sloping banks as against bulkhead walls
  contrast of land undulations against flat water surface

Edge treatment	Basically vertical, sloping or both, with choice dependant upon
	tidal range
	function
	cost
	water/wave forces
	Slopes give
	easy launching for light craft
	little protection and in tidal areas
	a varying land/water interface
	Vertical faces give
	constant land/water areas
	may mean hoists
	floating ramps
	perimeter hazards
Surface treatment	To encourage movement consider
	hard, level surfaces to vehicular standards
	hard non-slip surfaces to pedestrian standards
	semi-hard surfaces for casual walking (including grass)
	concealed hard surfaces, e.g. fire paths, parking, etc.
	To discourage movement consider
	hard, rough surfaces: ridged, stepped, cobbled
	semi-soft surfaces: ballast fill, thick gravel and sand
	soft surfaces: rough grass, planting
	Promote interest by
	modelling
	changes of level
	patterning
	colour
	bold, robust, nautical character rather than weak appearance
	avoidance of large flat areas of plain surfacing material
Continuity: serial vision	Marinas are primarily pedestrian places. Gear environment to pedestrian speed and scale or slow-moving (5 m.p.h.) craft
	Plan for interest on footpath routes and contrast visual constrictions (alleyways, stopped views) against courtyards, long vistas (along coast or across water)
Mooring (systems)	Design concurrently and relate closely:
	pattern of moorings
	harbour shape
	density of craft
	berth flexibility
	Explore all proprietary systems for suitability plus durability, stability, spares, maintenance, appearance, safety and cost
	Sketch or model the effect moorings and pile anchorages on (particularly) eye level views from harbour and local views *of* harbour
Mooring (components)	Decide on individual parts, e.g.
	service ducts
	articulation between pontoons
	decking, pile guides, ramps
	electric consoles, taps, fittings
	accessories, fenders, lighting, cleats

Form

Architectural considerations Likely to be low-rise for reasons of planning, stability and ground conditions. Consider use of roof space for viewing, race control and leisure. Avoid layouts that encourage wind funnelling or turbulence

Explore advantages of cantilevers for
balconies
press and public viewing
shade etc.

Explore advantages of pillars for
open-ground floor
launch-through advantages
visual links
open plan
flood precautions
boat storage or car parking

Explore advantages of building for
extension of land over water
area
drama
environmental impact and immediacy

Explore advantages of floating structures: may prove expensive but free of land costs

Continually check the design against unique site conditions. The best solutions stem from skilful application of full and accurate information

Structure Explore light-weight framing and cladding materials, raft foundations, wind resistant structures, uniformally distributed ground loading, pile bearing capacity and water-proofing methods

Materials Establish traditional materials, consider suitability, durability, texture, colour and cost
Check availability from local sources

Site furniture Determine the location and form of seats, tables, litter bins, signs, masts, flag poles, plant containers

Lighting Decide the type and location of lighting, level, columns, floodlighting, illuminated signs, bollards, ornamental and functional. Relate to safety and security. Discuss with local authority and electricity board

---

5. General design principles: Bibliography

Adie, Donald, *Marinas: Design and Construction*, Winston Churchill Travelling Fellowship Report, 1969.

Adie, Donald, 'Costa Brava comes to Brighton', *The Architects' Journal*, 30 January 1974.

Beazley, Elizabeth, *Design and Detail of the Space Between Building*, The Architectural Press Ltd., London 1960.

Beazley, Elizabeth, *Designed for Recreation. A Practical Handbook for all concerned with providing Leisure Facilities in the Countryside*, Faber & Faber Ltd., London 1969.

Bertlin, D. P., 'Marina Design and Construction'. Paper presented at the *Symposium on Marinas and Small Craft Harbours*, Department of Civil Engineering, University of Southampton, April 1972.

Briggs, R. T., *The Environment and Water Sports. Holme Pierrepont Watersports Complex, Nottinghamshire*. The Institution of Municipal Engineers Monograph No. 6, 1972.

Building Research Station, 'Airflow round Buildings', *Building Research Station Film* (6 minutes) 1967.

Building Research Station, 'Building Overseas in Warm Climates' (Shade, Hurricanes, Termites, Earthquakes, etc.), *B.R.S. Digest* 92 (Second Series), April 1968.

Building Research Station, 'Design and Appearance 1 and 2. The Effect of Design on Weathering', *B.R.S. Digest* 45 and 46 (Second Series) 1964.

Building Research Station, 'An Index of Exposure to Driving Rain', *B.R.S. Digest* 127, March 1971.

British Standards Institution, *Site Investigations*, C.P. 2001: 1957.

Cement and Concrete Association, *Launching Ramps for Boats*, The Association, 1971.

Chaney, Charles A., *Marinas: Recommendations for Design, Construction and Maintenance* (Second edition) National Association of Engine and Boat Manufacturers Inc., New York 1961.

Cohen, H. L., Hodges, D., Terrett, F. L., 'Brighton Marina: Planning and Design'. Paper presented at the *Symposium on Marinas and Small Craft Harbours*, Department of Civil Engineering, University of Southampton, April 1972.

Cullen, Gordon, *Townscape*, The Architectural Press Ltd., London 1961.

De Chiara, Joseph, and Koppleman, Lee, *Planning Design Criteria*, Van Nostrand-Reinhold Co., London 1969.

'Designing for leisure: the future of sports facilities,' *The Architects' Journal*, 14 October 1964.

Department of the Environment, 'Air flow around buildings', *D.O.E. Film* (6 minutes).

Duckett, Margaret, 'The Bloom on the Landscape', *The Daily Telegraph Magazine*, 19 May 1972.

Duffell, Roger J., 'Car Parking and Pleasure Travel in the Countryside', *Journal of the Institution of Municipal Engineers*, October 1973, Vol. 100.

Frazer, B., 'If Marinas there must be . . .,' *Boats and Sail*, June 1964.

Goldman, Charles R., McEvoy III, James and Richerson, Peter J. (eds.), *Environmental quality and water development*, W. H. Freeman & Co., San Francisco 1973.

Gruen, Victor Associates, *A Development Plan for Marina Del Rey Small Craft Harbour*, 1960.

Honeyborne, D. B., 'Changes in the Appearance of Concrete on Exposure', (Sea-water Damage etc.), *B.R.S. Digest* 126, February 1971.

Kaye, Sidney, 'Marina Architecture'. Paper presented at the *Symposium on Marinas and Small Craft Harbours*, Department of Civil Engineering, University of Southampton, April 1972.

National Yacht Harbour Association, *Yacht harbour guide*, Ship and Boat Builders' National Federation, London 1963.

Olgyay, Victor, *Design with Climate*, Princeton University Press, Princeton N.J. 1963.

Perrin, G., 'Building for recreation', *Journal of the Royal Institute of British Architects*, October 1968.

Roberts, K., 'Sport for the Disabled', *Physiotherapy* Vol. 60, No. 9, September 1974.

Roberts, K., 'Water Sports for Us All?' *Municipal and Public Services Journal*, 11 June 1976.

Royal Institute of British Architects, 'Design for Pleasure and Hard Wear in the Landscape. The Urban Scene.' (Reports of Symposium), The Institute, 24 May 1960.

Sillitoe, K. K., *Planning for Leisure*. Government Social Survey, H.M.S.O., London 1969.

Sports Development Bulletin 27. Sailing and Canoeing for Disabled People, Sports Council, London 1976.

Vian, R., *General Layout of Pleasure Ports*, Service Maritime et de Navigation du Languedoc —Roussillon, 1968.

'Water and planning', *Town and Country Planning* (special issue), June 1966.

Whitaker, Ben and Browne, Kenneth, *Parks for People*, Seeley Service & Cooper Ltd., London 1971.

Wise, A. F. E., 'Wind Effects due to Groups of Buildings', *B.R.S. Paper* C.P. 23/70, 1972.

6 Engineering

Coastal and inland sites

Whilst this chapter deals mainly with *coastal* engineering it should be noted that much of the information it contains is relevant to estuaries, lakes, reservoirs and other inland sites. Knowledge of the hydraulic characteristics will be just as necessary and examination of tidal data, salinity, siltation and flow will be equally essential, for in many ways some inland sites and estuaries present greater problems than their more obviously exposed counterparts on the coastline. The reason for this is that tidal range and current, siltation, pollution and instability of the bed conditions may well create greater problems in rivers and estuaries than in some sheltered and less variable coastal areas. The comparative table on page 158 sets out the benefits and drawbacks of coastal sites as against their inland counterparts. In estuaries and wide river mouths these distinctions become blurred and aspects from both categories may apply.

Glossary of terms

Coastal engineering is an involved and difficult subject and before describing the essential outlines a brief glossary of some common terms may be helpful.

TIDE The periodical rise and fall of the sea caused by the gravitational effects of the sun and moon. Only totally land-locked water is unaffected

SPRING TIDE A tide occurring immediately after a new or full moon. Maximum Spring tides occur at the equinoxes—in Britain during late March and late October

NEAP TIDE A tide occurring when the moon is at half phase

TIDAL RANGE The variable difference between the lowest and highest tide measured vertically

STORM SURGE The build up of coastal water to an abnormally high level by the coincidence of tide and weather. It is usually most severe at Spring tides

SEICHE The oscillation of mainly inland water caused by high winds blowing across the surface. This lowers the surface to windward and raises it to leeward. Oscillation may continue after the wind has dropped, due to changes in barometric pressure

WAVE A ridge created by wind action on the surface of a body of water which sets up a series of orbital movements. Although it may seem superfluous to define so common a phenomenon it should be remembered that shelter is the prime consideration in marina design: the stronger the wind the greater the height and length of waves

FETCH The distance over open water across which a wave has travelled.

6.A Inland and coastal sites compared

Inland situations

Advantages	Disadvantages
Generally shallower water—although lochs and similar formations can be very deep and steeply shelving	More prone to icing
More sheltered conditions likely	Higher density of craft per unit area of water
360° catchment area	Greater likelihood of pollution and algae growth
Sail training advantages	Future expansion may be more limited
Greater landscaping opportunities	Limited sailing conditions
Generally less costly particularly in engineering work	Greater sanitation problems on enclosed water
Cheaper statutory services likely	
Sites in or near population centres are possible	
Less handling equipment necessary	
Lower maintenance costs and overheads	
More sailing days and longer season likely	

Coastal situations

Advantages	Disadvantages
Greater appeal of sea and coast	Usually greater water depth
'Limitless' sailing space	Storm surge and other extreme conditions more likely
Variety of sailing conditions	'Semi circular' catchment area
Foreign and coastal sailing opportunities	Difficulties where tidal range is great
Larger sites, larger draft and higher charges possible	Greater salinity
	Higher cost of dredging, breakwaters, locks etc.
	Expensive handling equipment and storage often necessary
	General maintenance costs are high

6.1 Insufficient maintenance has allowed this valuable quayside berth to silt up

6.2 A 203 mm (8″) floating pipeline conveying silt to the reclamation area of Mercury Yacht Harbour, during construction in April 1972.

This is limited in enclosed waters but in great oceans it may be hundreds or even thousands of miles

BREAKER A heavy and violent ocean wave which occurs as a result of a change in the wave profile. This change consists of the leading part of the wave being slowed down more quickly than the back thus causing a steepening of the front and its eventual overturning

WAVE REFRACTION Waves change their direction as well as their shape upon approaching the shore. The leading end is slowed by sea bed friction, but the trailing end is not, thus causing a curve in the wave crest known as refraction

SWELL A heaving of the sea caused by long ocean waves which generally do not break, travelling into the region of shorter local waves

TIDAL CURRENT A directional movement of water, the strength of which is related to surge and tidal range

SHORE CURRENT A directional flow of water which is the result of wind and wave action across its surface

RIP CURRENT A particularly strong offshore flow of water. Wave refraction causes circulating currents along the shoreline which then return strongly seawards through the waves as rip currents. They can be very strong and dangerous attaining a speed of up to 2·43 m (8 ft) per second

SWASH AND BACKWASH Swash is the forward movement of a wave up the beach. Backwash is its retreat under the force of gravity

LONG SHORE DRIFTING The flow of coastal sediment produced by waves approaching at an angle to the shore. On plan each particle makes a saw-tooth progression

FLUME A long, transparent water tank in which waves are generated to enable the effects of various engineering structures to be studied

Some aspects of basic marina engineering have already been dealt with in the context of site selection (page 23) and off-shore considerations (page 31). Most engineering problems in marina construction shade off into areas of layout, materials, costs and architecture etc., but to facilitate reading and reference this essentially homogenous problem is divided here into the following separate elements:

Engineering data studies
Dredging
Dredging plant
Breakwaters
Piles
Bulkhead and quay walls
Siltation and erosion
Ice
Locks

Engineering data studies

The first question coastal engineers prefer to be asked is *where* to build rather than *how* to build the harbour. They also prefer to be consulted at an early stage, because to obtain information on the existing conditions may require recording equipment to be in position for up to a year. Whilst the prevailing conditions of the given locality are being recorded, meteorological records and data will be studied and perhaps hydraulic research work in the field and laboratory will begin. The engineer is used to organising his work in separate stages so that information needed for the feasibility study can be obtained in a much shorter time than that needed to obtain information for the detailed design.

The required data falls into the following main categories:

1   Historical data of site conditions

2   Data required to model the existing site conditions (this includes the dynamic forces as well as the topography)

3   Data needed to represent the engineering structures proposed

4   That required to interpret the results and to forecast the effectiveness of the various designs to combat the conditions predicted from 1

5   Data from physical samples—trial boreholes, soil and salinity analysis etc. The bed of the future harbour must also be examined. There is a joke amongst coastal engineers that the bed of any new harbour needs to be soft for dredging and simultaneously hard for retaining the piles

6.3 shows the sequence and relationship of studies that would be required from concept to final design where a major project is envisaged. With smaller developments or easier sites the programme may be shortened. The length of the study programme depends very much on what is required to be known. Taking the hydraulic model studies alone, the work would probably fall into one or more of the following categories:

1   Storm wave models

2   Flume studies

3   Long-period wave models

4   Littoral drift studies

5   River and estuary studies

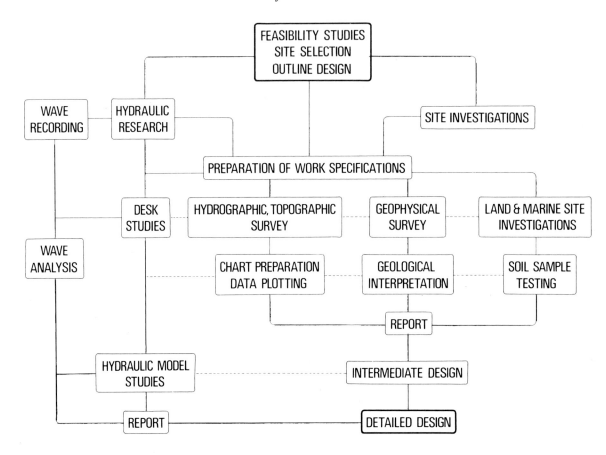

6.3 Flow chart to show the studies that may be necessary where a major project is envisaged

Not only will the results of those studies be needed to establish the most efficient harbour design but it will determine its performance over the years with regard to siltation and the effect of the new project upon the existing regime. This last question often has a strong influence upon the local authorities' attitude because if they can be assured that no adverse effect will be caused by the development upon the present conditions, there is a much improved chance of its acceptance.

Investigatory work and model studies may need to be more extensive with some sites than others but whatever the scope, *this aspect of marina work must not be skimped or hurried*. If the project is not feasible or economic the engineer can usually tell the developer quite quickly, but if the prospects are good the client must be patient and allow the engineers to assemble their recordings and surveys, analyse the data and work logically towards the detailed design.

It is important for the engineer to determine what information is *necessary*. The questions are as important as the answers for, whilst crucial data must not be omitted (and it is wasteful in time and money to have to return to the site) it is also costly to over-do the research. Location is a more important determinant than size and not every site needs to be studied by hydraulic model testing. Mathematical techniques have developed even since the first edition of this book was published and their application through computer technology has advanced so that mathematical simulation is often as accurate and trustworthy as the construction of physical models and of course much less expensive. The only ingredient missing from the computation method is that of *observation*. This is a disadvantage because it is important for an engineer to *see* his predictions. In this respect it is analogous to structural and even architectural models but like them can become an expensive indulgence unless really needed and unlike them is not transportable and cannot be used for explanation to a client or committee unless special visits are arranged.

Accurate site information is just as important to either the mathematical or the model system and sufficient time and money should be allowed. It is appreciated that the design problems of yacht marinas are much the same as for commercial harbours (in some ways they are *more* difficult) but the money available for data collection is usually much less. Any less expensive method brought about by new techniques are to be welcomed and some have quite recently appeared or been developed in the late 1970s.

Radio buoys

Wave height recording by radio buoy is now more common. These electric floats transmit wave height information by short wave radio to, usually, land-based receivers. These new models show a considerable improvement in continuity of data output. Designed to be operational for a complete year, they usually provide a 10-minute record every 3 hours. The 2,900 records thus provided then require processing and the data plotted, unless, like the 'waverider' buoy the wave recorder and punch tape output are a standard part of the installation.

Cine recordings

Deflections in structures such as floating breakwaters, timber jetties, cribs or anchorages may be registered by this method.

Calibrated turnbuckles

These may be used to measure maximum mooring forces on cables or structural stresses on bulkhead walls, breakwaters or other offshore protective measures.

Aerial survey

This new method of hydrographic survey uses the actual site rather as though it was a model. High-speed launches seed the nearby waters with cardboard floats at predetermined intervals. The aircraft makes parallel runs at 1,200 and 3,000 m (4,000 and 10,000 ft) taking vertical photographs from a timed camera. The speed and direction of each float is plotted relating time, direction and distance. Final surveys may be plotted at 1:3,000 and 1:10,500, the latter being the scale for local British Naval charts.

With the sharply increasing cost of construction work developers are constantly seeking methods to economise in the expensive field of marine structures whilst still providing long term protection. Adequate information is an essential ingredient of design efficiency and sufficient funds must be available early enough to benefit the engineers in this respect.

6.4 (a) The curving western breakwater of Brighton Marina, Sussex, England, photographed in July 1977. It is formed from a line of the caissons shown in 3.32, page 56 and 6.28, page 180

6.4 (b) The upper and lower-level access roads entering the site of Brighton Marina, Sussex, England, photographed in July 1977. The locked berths with their walkways, left, are being made ready beneath the cliffs and on the right is the outer harbour enclosed by the eastern breakwater

6.4 (c) Floating pontoons awaiting final assembly at Brighton Marina, Sussex, England in July 1977. The moulded concrete hulls have galvanised frames supporting treated timber decking

The aim of any harbour, in its protective role, is to reduce the effect of bad weather to an acceptable level for the vessels within. This is more difficult to achieve in a marina with its small boats than in a commercial harbour because the acceptable reduction is more or less inversely proportional to the size of craft.

Wave height

One of the engineer's great concerns will be to ensure that his eventual design does not, by some misfortune of size or shape, encourage the very conditions that he set out to abate. This is more easily done than the layman would suppose. It is possible for instance to choose a harbour size which is some exact multiple of the average wave length and thus become a victim of the phenomenon known as resonance, whereby energy increases as waves are reflected to and fro across the harbour. One way of ensuring against this is to include a 'spending beach' within the harbour to absorb wave forces before they intensify. Another method of reducing wave height is to construct an outer harbour in which entering waves will be diminished.

The positioning of the marina entrance will be an important decision and so will its orientation, width and shape, etc. The conflict the engineer has to resolve is to ensure that the entrance is constricted enough to minimise the ingress of waves, yet sufficiently wide for yachtsmen to enter safely in difficult weather conditions. During rough weather a badly positioned entrance will worsen the situation by creating reflected waves which give a steep, cross sea, notorious with sailors for causing them to 'broach to'. Similarly, the engineer must take the problem of 'overtopping' into consideration. Overtopping is an engineering term for water being thrown over the breakwater or quay by the force of its impact. In a free-standing breakwater this may not be very important but with a harbour's peripheral wall it is more serious, particularly if it is intended that moorings should lie on the other side.

Because the reduction of wave height is the marina's *raison d'être* it is essential to know as much as possible about off-shore wave characteristics under as wide a range of conditions as possible and, because it affects wave behaviour so much, the nature and profile of the sea bed. Offshore recordings and soundings will be necessary to obtain this information.

Wind control

The effects of wind must also be considered, particularly in very large projects where wind will act as a wave generator within the harbour as well as blowing directly upon the boats. This only becomes a problem where the water area is really large. The extent to which a marina projects out to sea is also important when calculating the effects of wind. Brighton Marina for example is three-quarters the size of Dover harbour and projects out to sea for half a mile.

Wind may be a greater problem in urban marinas with tall buildings nearby than in sites which are in fact more exposed. The funnelling turbulence caused by high buildings is well known, and in marinas this effect can make difficulties for manoeuvring craft. Certain breakwater and quay sections should also be avoided. In Florida a tall open C-sectioned quay wall profile was chosen running to windward in a tightening curve which built up such ferocious conditions that windscreen baffles were later added.

The coastal engineer's work requires a high degree of technical skill and stamina, and his will be the final decision in designing the engineering infrastructure. The architect however will have a contribution to make particularly where options on materials and design are available. Working together they should produce a functional and visually satisfying design.

Costs

It is not unusual for there to be difficulties in convincing the developer of the value of the studies to be undertaken, particularly at early stages of the project. There is very little to show physically and the release of early money is never easy. *The developer should however understand that it is at this time that his investment is at its most critical period. Money spent now is not wasted but invested.*

To ensure value for money any developer contemplating a large project will be advised to invite tenders for hydraulic research just as he would for any other component. Firms competing may to some extent write their own briefs and these may not be easily comparable in terms of value for money. It will however make clear in the itemised breakdown of costs what one is

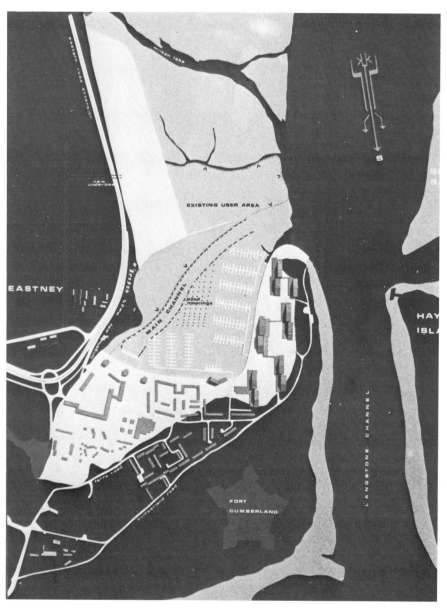

6.4 (d) Langstone Marina Development near Portsmouth involves dredging out a huge area adjacent to Langstone harbour. Some craft will be berthed and others accommodated at piled moorings. Architects: Bailey Piper Stevens Partnership. Consulting engineers: Archibald Shaw & Partners

getting for one's outlay. The number of laboratories in Britain capable of handling a sizeable hydraulic study programme is small. This however may emphasise the need for competition rather than invalidate it. Certain private laboratories, universities, contractors and national institutes specialise in hydraulic research work. An initial approach to them to ask for a draft of the work and an approximate estimate of its time scale and costs is a very necessary first step in any research programme.

Dredging

Marina construction nearly always involves dredging to some extent, and in many cases it accounts for a high proportion of the total money invested. Site conditions are variable, but certain questions are common to all developments, and these are:

1 Surface area
2 Depth of dredge
3 Method of operation and choice of equipment
4 Disposal of material
5 Drainage

Calculating the dredge area

The surface area may be computed when the finalised plans are compared with the existing site conditions. The advice of an expert in the early stages of design, however, may save a great deal of expense as a correct decision at this stage on the exact location of the quays, moorings, and river or sea walls will minimise the yardage to be removed. The same applies to 2, 3 and 4 for, as an example, cut and fill by a dragline over a short distance will be much less expensive than the use of a dredger, which has to carry material a long distance or transfer it into barges. Generally speaking, any land-based machines such as bulldozers and draglines are cheaper to operate than floating equipment and should be employed where possible, although they may only be working effectively at low tide.

An optimum position needs to be agreed for the positioning of the quay so that a balance may be struck between the amount of material to be dredged on the water side of the quay, and the yardage of land gained inland of it. Having constructed the quay as a freestanding wall, the unwanted material may be pumped hydraulically to the inland side to form the hardstanding perimeter of the marina. As there will probably be as much land area necessary to accommodate facilities as developed water area, the positioning of the quay wall will probably be near the low-water line in average river or sea-shore conditions. (For positioning the quay wall see also pages 193–196.)

The depth of dredge may vary within the same development, for it is often possible in the layout to separate the deep-keel from the shallow-draft vessels and the latter may then be positioned alongside the quay, thereby economising in the depth of the bulkhead wall. The formula for relating a vessel's draft to the depth of water may be summarised as follows:

Draft of vessel + sub-normal tide + 0·61 m (+ 2 ft).
Therefore, if in any one area of the marina the maximum draft is 1·37 m (4 ft 6 in) and the maximum sub-normal tide (i.e. below normal low-water level) is 0·45 m (1 ft 6 in), one then adds 0·30 m (1 ft) for clearance and another 0·30 m (1 ft) to account for siltation, thus giving:
1·37 m + 0·45 m + 0·30 m + 0·30 m = 2·42 m
(4 ft 6 in + 1 ft 6 in + 1 ft + 1 ft = 8 ft)
In a locked marina with a constant water level and no siltation, the 0·45 m (1 ft 6 in) sub-normal tide and the 0·30 m (1 ft) for deposits would, of course, be omitted giving a final figure of 1·65 m (5 ft 6 in).

6.5 sets out the draft of boats in a typical marina and such a graph may be computed from the type of vessels likely to be within the catchment of the new marina. By applying the above formula to the graph, a reasonable assessment may be produced of the necessary dredging depths and areas required and some guide obtained for selecting suitable boat-handling equipment. There is a school of thought which advocates deep (3 m +) dredging in virtually all circumstances. The theory is that anything shallower will merely limit the dredger unnecessarily and prevent it from dredging its moneysworth and secondly, that when dredging at 3 m (10 ft) or over one need not be unduly bothered about siltation until about 1 m (3 ft 4 in) of silt has formed which, under average conditions, might defer the need

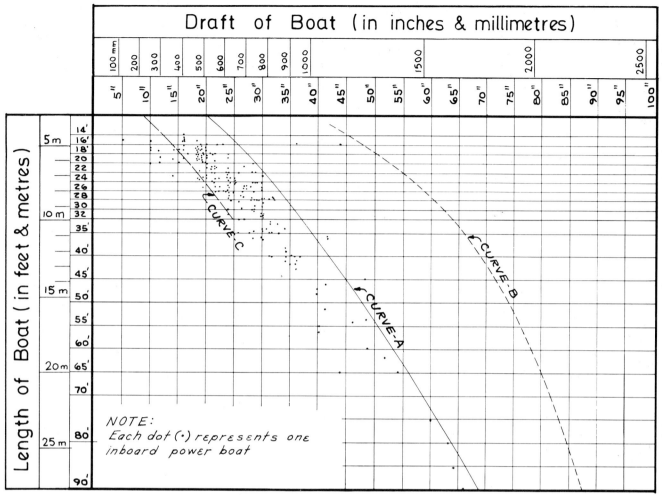

6.5 Typical pleasure boat basin: American figures showing drafts of boats for
establishing minimum dredging depths and the selection of boat-handling equipment
Curve A shows approximate maximum drafts of inboard boats
Curve B shows approximate maximum drafts of sailboats
Curve C shows approximate maximum drafts of hulls for outboard boats

6.6 Imported fill at Emsworth Yacht
Harbour. This may be a major item if
dredged material is not usable

6.7 Cabrillo Small Craft Harbour, San
Diego, California. The dredging completed,
the site awaits the basic engineering
construction

for maintenance dredging for many years. Given ample capital the theory is probably good provided the dredging agreement sufficiently reduces the cost per cubic metre as the amount to be removed increases.

6.B Mass densities of materials

Material	kg/m³	tonnes yd³
Aggregates		
Coarse	1,522	1·15
Fine	850	0·64
Chalk	1,602–2,723	1·20–2·05
Concrete: cement, plain:		
Aerated	961	0·72
Brick aggregate	1,840	1·38
Clinker	1,440	1·07
Stone ballast	2,240	1·70
Concrete: cement, reinforced:		
1 per cent steel	2,370	1·78
2 per cent steel	2,420	1·80
5 per cent steel	2,540	1·90
Earth		
Dry loose	1,280	0·96
Dry compact	1,550	1·16
Moist loose	1,440–1,600	1·07–1·20
Moist compact	1,760–1,840	1·33–1·37
Hoggin	1,762	1·33
Mud	1,760–1,922	1·33–1·45
Peat		
Dry stocked	561	0·43
Sandy compact	801	0·60
Wet compact	1,362	1·02
Portland stone	2,243	1·67
Sand		
Saturated	1,922	1·45
Undisturbed, dry	1,682	1·27
Shale	2,563	1·93
Stone		
Freestone	2,243–2,483	1·67–1·87
Granite	2,643	1·95
Portland	2,243	1·68
York	2,243	1·68
Water		
Fresh	1,001	0·75
Salt	1,009–1,201	0·75–0·90
Sea	1,009–1,041	0·75–0·78

Dredging material

The initial site investigation will have shown the types and amount of soil or rock to be removed and will have given some indication of the feasibility of the operation. Equally important is an assessment of whether this material is usable as fill and, if not, whether it can be deposited nearby on land or in water or will need to be dumped out to sea.

At this stage a number of questions will need answering, to determine how the dredged material may best be used:

1   What is the material's quality?

Good granular material compacts quickly and completely into workable

6C Safe bearing capacities of soils
from British Standard Code of Practice CP 101 : 1963 table 1

	FPS tonf/ft²	MT(1) tf/m²	MT(2) kgf/cm²	SI kN/m²
Rocks				
1 Igneous and gneissic rocks in sound conditions	100	1,090	109	10,700
2 Massively-bedded limestones and hard sandstones	40	440	44	4,300
3 Schists and slates	30	330	33	3,200
4 Hard shales, mudstones and soft sandstones	20	220	22	2,200
5 Clay shales	10	110	11	1,100
6 Hard solid chalk	6	66	6·6	650
7 Thinly-bedded limstones and sandstones				
8 Heavily shattered rocks and the softer chalks	To be assessed after inspection			
Non-cohesive soils				
9 Compact well graded sands and gravel-sand mixtures:				
dry	4 to 6	44 to 66	4·4 to 6·6	430 to 650
submerged	2 to 3	22 to 33	2·2 to 3·3	220 to 320
10 Loose well graded sands and gravel-sand mixtures:				
dry	2 to 4	22 to 44	2·2 to 4·4	220 to 430
submerged	1 to 2	11 to 22	1·1 to 2·2	110 to 220
11 Compact uniform sands:				
dry	2 to 4	22 to 44	2·2 to 4·4	220 to 430
submerged	1 to 2	11 to 22	1·1 to 2·2	110 to 220
12 Loose uniform sands:				
dry	1 to 2	11 to 22	1·1 to 2·2	110 to 220
submerged	$\frac{1}{2}$ to 1	5·5 to 11	0·55 to 1·1	55 to 110
Cohesive soils				
13 Very stiff boulder clays and hard clays with a shaly structure	4 to 6	44 to 66	4·4 to 6·6	430 to 650
14 Stiff clays and sandy clays	2 to 4	22 to 44	2·2 to 4·4	220 to 430
15 Firm clays and sandy clays	1 to 2	11 to 22	1·1 to 2·2	110 to 220
16 Soft clays and silts	$\frac{1}{2}$ to 1	5·5 to 11	0·55 to 1·1	55 to 110
17 Very soft clays and silts	$\frac{1}{2}$ to nil	5·5 to 0	0·55 to 0	55 to 0

6.D Conversion factors

Density

ton per cubic yard	kilogramme per cubic metre	kg/m³	1 ton/yd³	= 1328·9 kg/m³
	tonne per cubic metre	t/m³		= 1·3289 t/m³
pound per cubic yard	kilogramme per cubic metre	kg/m³	1 lb/yd³	= 0·5933 kg/m³
pound per cubic foot	kilogramme per cubic metre	kg/m³	1 lb/ft³	= 16·02 kg/m³
pound per cubic inch	gramme per cubic centimetre	g/cm²	1 lb/in³	= 27·68 g/cm³
	mogagramme per cubic metre	Mg/m³		= 27·68 Mg/m³

well-drained land and can be readily handled. If the material is light and easily pumped its cost per yd³ will be much less than harder stuff which is difficult to move and clogs the cutter blades

2   How much dredging and filling is necessary?

If these equate and the quality is good then no material need be imported or taken away

3 How much need be imported?

Good fill is expensive particularly if imported from afar

4 How much disposed of?

This too is expensive. Trucking inland to authorised sites is costly and dumping at sea may be regulated to certain grounds some miles from the site

5 Is the unwanted material good enough to sell?

If the answer to this last question is affirmative then further investigation may well be worthwhile.

About half the coastal and a quarter of the inland sites are likely to provide excavated material that is good enough for their own use and a proportion of these will yield a sufficient surfeit to make its sale a profitable proposition.

The most readily marketable materials are likely to be fine to coarse sand, medium and coarse gravel, pebbles (from 10 mm to cobbles), Hoggin (sand/gravel mixture) and various types and grades of rock. The tonnage to be won from some contracts can be considerable and constitute an important factor in the economics of the development.

The water area required for 100 craft is often about 1 ha (100×100 m) which, if reduced by a depth of 3 m produces 30,000 m^3. This would yield, excluding topsoil, 54,000 tonnes of hoggin, 51,000 tonnes of dry sand or 45,000 tonnes of coarse aggregate. Still assuming a depth of 3 m a yield of 400 to 500 tonnes per craft space would be averaged.

Clean and sorted material naturally fetches the best prices. It is not difficult to arrange for material to be washed and graded on site and this is very worthwhile if good material is there in quantity. Approval to extract material should be sought with the outline planning permission. A greater problem than the excavation of material, is likely to be its transportation elsewhere. This is a serious problem for many authorities which recognise the environmental disturbance caused by the transporting of such loads through residential areas A 5 tonne truck will make 100 trips full—and 100 empty—for each 3 m-deep berth space. Barging out the material would overcome this difficulty, but if double handling is involved this may make the operation uneconomic in what is a highly competitive market.

It is important not to overestimate the usable or saleable material. To assume an unchanging situation in depth on visual inspection or hand probings is unwise. A programme of investigation tailored to suit site conditions is necessary and according to individual circumstances this may involve:

1 Hand probe with pipe or bar

2 Hand augers

3 Rig with casing or piston sampler

4 Drilling for samples with split-barrel sampler

5 Wash boring to reveal strata

6 Test pits dug by hydraulic digger

Dredging plant

Dredging is a highly specialised skill and advice from experts is essential. Plant nowadays is varied and specialised and the right choice of equipment is vital to ensure economy. The type of plant chosen for the work will depend upon the following factors:

1 The accessibility of the site

2 The type and amount of materials

3 The average depth of water and tidal range
4 The season of the year
5 The closeness to boats, piers, buildings, etc.
6 The shape to be worked
7 The accuracy of finished levels
8 The distance to reclamation or dumping areas
9 Any local constraints on noisy machinery

1   Mechanical dredges

Due to their relative simplicity, mechanical dredges were the first to be developed. There are three basic types still in use today, all three resembling dry-land excavation machines mounted on floating structures.

The mechanical dredges have certain distinct advantages, in relation to the types of material to be moved, but they execute only one part of the two-phase operation of, first, excavation and, second, disposal. They therefore require the use of barges, tugs and sometimes trucks or trains to convey the excavated material to its disposal or spoil area.

1.1   Ladder bucket dredging

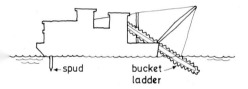

6.8 Ladder-bucket dredging

This is done by a vessel of the traditional dredging type with an 'escalator' of buckets lowered to the sea or river bed. The dredger is usually towed by tugs which also pull it into the material being removed. The vessels are shallow-drafted and have a variety of bucket sizes and shapes, making them adaptable and very suitable for marina work. They are however very noisy and will not be popular with any nearby residents. The dredged material is discharged into hoppers alongside. These can be either self-propelled or dumb and are usually bottom-opening.

1.2   Grab dredging

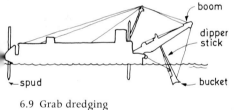

6.9 Grab dredging

This usually employs craft in which one or more of the familiar grabs or draglines are mounted onto pontoons. They are shallow-drafted and can work effectively in estuarine water. They either load into themselves or attendant hoppers from which the material is dumped or pumped ashore as fill according to its quality. Such dredges are advantageous for work in confined areas near docks and breakwaters, but are not particularly suited to removal of very soft material that is readily washed out of the buckets, or very hard material where penetration is insufficient to fill them.

1.3   Dipper dredging

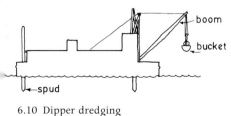

6.10 Dipper dredging

The dipper dredge is essentially a barge-mounted power shovel. Its main advantage is in the strong crowding action of the bucket, as the dipper stick forces it into the material to be moved. The best use of such dredges is for excavation of hard, compact materials—such as rock and other solid formations—after blasting operations have loosened them.

2   Hydraulic dredges

Hydraulic dredges use the water on which they float to increase the efficiency of their work. This is done by mixing the water with the material to be removed, and pumping it away as a sludge. Hydraulic dredges are more versatile, more efficient and economical to operate than mechanical dredges, because both the digging and disposing functions are performed by one self-contained unit in the majority of instances. However, they are not suited to hard rock removal, nor to use when disposal areas are small or very distant.

2.1   Cutterhead pipeline dredging

This dredge is the most versatile and widely-used marine excavating unit. The secret of a cutterhead dredge's remarkable efficiency lies in the rotating cutter that surrounds the intake end of its suction pipe, and without which it would be substantially identical to the plain suction dredge. The cutter loosens and breaks up the material to be removed, and mixes it with water into a liquid sludge that is drawn through the dredging pumps, and then conveyed by pipeline to a disposal area. All alluvial materials; most compacted deposits such as sand, gravel, clay and hard-pan, and even soft types of coral rock, sandstone, and limestone can be removed by this method.

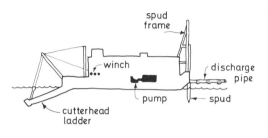

6.11 Cutterhead pipeline dredge

2.2   Plain suction dredging

Dredges of this type resemble giant vacuum cleaners in their operation. A mixture of water and material is drawn through a suction pipe lowered to the working face of the deposit to be removed. The sludge mixture is then discharged through a pipeline to a disposal area, or into barges or hoppers to be taken away. Dredges of this type can dig deeper than others, but are limited to soft, free-flowing alluvial-type material. They are often equipped with water-jet heads to loosen and agitate the material. Their primary applications are in the removal of silt from navigation channels and in land fill and beach erosion work.

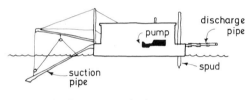

6.12 Plain suction dredge

2.3   Self-propelled hopper dredging

Resembling an ocean-going ship, a self-propelled hopper dredge functions in much the same manner as a plain suction dredge. The great difference between them is the hopper dredge's freedom of need for supplemental disposal equipment. In actual operation, a hopper dredge's suction pipe or pipes are dragged along the bottom, while the dredge moves ahead at slow speed, picking up a mixture of water and material that is pumped into hoppers or containers built into the dredge's hull. When those hoppers are full, the dredge moves to a deep-water dumping ground and releases the material through bottom dump doors, then returns to the work site to repeat the cycle.

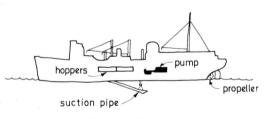

6.13 Self-propelled hopper dredge

3   Portable dredges

In addition to the equipment outlined above, very efficient portable dredgers of both the mechanical and hydraulic types are now working in the United States. These may be transported overland quite readily thus saving long sea journeys. They make construction work possible at otherwise inaccessible sites on lakes and similar areas.

4   Pumping-ashore plant

This equipment sucks material from the hoppers and deposits it onto the reclamation or dumping area either directly or through a discharge pipeline.

With most work in shallow waters it is important to know whether the dredger and hoppers will be able to float all the time or only at high tide, for equipment that is aground and idle increases the cost of the work.

Where dredging passes through good and poor material, a switch valve can be installed to direct material to the reclamation or dumping areas according to its quality (see 6.14).

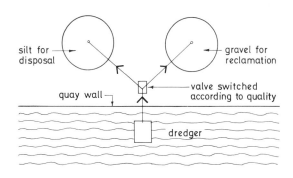

6.14 Pumping ashore

Maintenance dredging

If siltation is accepted as inevitable, a calculation of the amount of deposition and the cost of its removal should be entered as an allowable business overhead.

It is often a good idea to adopt a contractual system, whereby the original contractor for the development is required, at a fixed price per annum, to keep the marina at a minimum depth of water. This type of contract, often adopted in the United States, is an insurance by the owner against large and unexpected dredging bills.

Maintenance dredging involves the removal of all floating equipment in the area of operation but as it is not normally possible to extract piles these have to be dredged around, which is a delicate exercise. Incidentally it is worthwhile remembering that locked marinas cannot always accept dredgers and land-based draglines cannot usually reach the whole bed area and are often an unsatisfactory method of maintenance dredging.

In harbours which are inaccessible to standard or portable dredgers or where the total bed cannot be reached by land-based equipment, maintenance work may be done in the following ways:
1   By hydraulic pumping equipment, not unlike the pumping-ashore plant already described
2   By small floating draglines. A version of the grab dredge (1.2) above
3   By divers

Divers, or other specialist means may be necessary for maintenance work other than dredging. The repair of quay walls or breakwaters below the water surface, maintenance of lock gates or the removal of sunken vessels or other obstructions are some examples for which individual solutions or the services of specialists may be required.

Fortunately, the lock which may make access difficult also usually minimises the need for dredging by lessening the amount of silt entering the marina.

6.15, 16 Two views of a 1:00 natural scale storm wave hydraulic model of St. Helier Harbour, Jersey, showing the existing harbour and, in the distance to the south, a proposed extension at La Collette comprising land reclamation, berths and a breakwater. The wave recorder on the right of 6.16 is operated from the control room, where the wave profiles are simultaneously recorded on an X-Y plotter. Client: Land Reclamation Committee, States of Jersey. Consulting Engineers: Coode & Partners, London

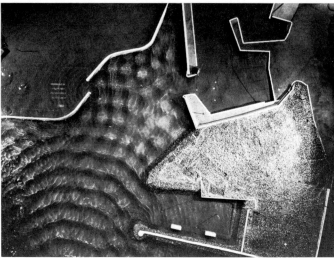

6.17, 18 Waves of 7 sec period approaching the model of St. Helier Harbour and the proposed extension. The waves are being refracted by the sea-bed and diffracted round the proposed new breakwater. The installation of the curved, vertical-sided, solid breakwater to the west 6.18, has caused the incoming waves to be reflected towards the harbour entrance. By further testing of the model it was possible to make modifications to abate these otherwise unacceptable conditions

6.19 Cross-section of a scale model of a proposed breakwater for Marsamxett Harbour, Malta, being subjected to waves of 8 sec period in a wave flume. Such tests examine the stability of the proposed design under storm wave attack and take into consideration the effect of such factors as run up and draw down, slope of cross section, size and type of primary armouring etc. Client: Director of Public Works, Malta. Consulting Engineers: G. Maunsell & Partners, London

6.20 General view of a proposed lay-out of 1:96 natural-scale, storm-wave hydraulic model of Brighton Marina Yacht Harbour, with waves of 6 sec period approaching from the south-east

6.21 View of 1:96 natural-scale, storm-wave hydraulic model of Brighton Marina showing waves from the south-east being reflected from the east breakwater. The reflections are reduced as the waves approach the entrance, by the absorbing action of rubble mound placed at the western end of the seaward face of the breakwater. Consulting Engineers for maritime works: Lewis & Duvivier

Breakwaters

The word 'breakwater' is a generic term which can describe many different structures or promontories whose purpose lessens the effect of the elements and provides more sheltered conditions for craft and marina facilities. Generally speaking they are constructions of a linear nature which have water on either side of them. Except for the floating types they rise from the bed of the sea or water and are visible above the surface at all times.

Jetties, wharves, piers and groynes may all act as breakwaters but not all breakwaters are jetties, wharves, etc. In a 'natural' harbour of course they are an inherent part of the topography and no further defence is needed. Where a marina is to be located on a small lake, reservoir or quiet stretch of river no breakwater is really necessary. However at the oceanside, estuaries and in other exposed waters, some form of general protection is often necessary and a survey prepared from weather stations, geographical maps, local coastguards, or from hydraulic models, will need to be prepared to establish the type, position and extent of the protection needed.

Siting

Positioning is a most important factor for it will determine the breakwater's physical form, effectiveness, cost and—where it forms the periphery—the marina's size and shape. In difficult locations a decision on breakwater siting can only be reached after the survey and recording work is complete and the data analysed.

The following list gives some of the serious consequences of breakwaters being wrongly positioned:

1   Encouragement of pollution
Correct positioning can create cleansing conditions whereby floating debris oil and silt are removed or minimised from within the harbour. The opposite effect can be a nuisance as at Sanford, Florida, where at certain times, large quantities of weed pour through the harbour entrance

2   Insufficient effect
Breakwaters are positioned to give an amount of protection for a given cost. If the effect is too little or the cost too great wrong siting is often the cause

3   Structural failure
Wrong positioning in respect of the direction of approach of damaging weather is not uncommon. Better to ward off the attacking wave than to attempt resistance by strength alone

4   Promotion of siltation and erosion
Unless an optimum position is chosen in relation to known currents, drifting and bed material then siltation or erosion or both can be a real problem not only at the marina site but a considerable distance from it

5   Harbour entrance problems
Breakwater positioning needs particular care at the entrance or lock for a tide-race can easily be created. Too narrow an entrance can have a funnelling effect: too wide an entrance gives insufficient protection from heavy seas

Length

The length necessary for a breakwater to do its job varies considerably although their effectiveness in differing conditions is in no proportion to their span. At Shilsole, Seattle, a breakwater 1,350 m (4,440 ft) long has been necessary to control the waters of Puget Sound. At Marina del Rey near San Diego the navigation channel was constructed well into the Pacific and no breakwater was considered necessary: however, the surge of water up the channel caused havoc among the moorings on several occasions and a short but effective breakwater was eventually placed across the channel entrance.

Types of breakwater

Breakwaters can consist of a wide variety of materials and forms, from an open row of timber piles to a man-made reef of thousands of old car bodies. Selection will depend upon suiting the practical and structural considerations of the site to the functional requirements expected of the breakwater.

The following is a guide to some of the factors in breakwater selection and the considerations stemming from them:

6.E Factors affecting breakwater selection

Factors	Related considerations
1 Depth of water 2 Bed conditions	Stability, settlement, scour
3 Tidal range	Relative water heights either side of breakwater
4 Wave height reduction	Type of structure, stress
5 Materials	Availability, durability, appearance
6 Construction methods	Available equipment and labour Likely weather/water conditions, in situ/precast
7 Permanence	Most are permanent but temporary, phased or moveable breakwaters are sometimes necessary (see floating breakwaters)
8 Height and form	Vehicular and pedestrian access Harbour side mooring, overtopping
9 Appearance	See Appearance and function, page 185
10 Cost	See Cost, page 186

A brief description of the more common types of breakwater is as follows:

1 Rip-rap or rubble mound

Loose heaps of large rocks or concrete blocks tipped from barges to form a (usually) continuous and rough-textured barrier of triangular section. A hopper is often used for laying rubble breakwaters in which rock loaded on deck is pushed overboard by hydraulic rams. This type is used almost exclusively in coastal areas with a small tidal range. The material used may be natural quarry or sea-shore rock or concrete blocks moulded to one of many different shapes

2 Single or double-row piling

Ordinary piles driven in line at a calculated distance apart: the second row being staggered in relation to the first. Sometimes shored on the leeward side

3 Timber crib or gabion

A series of cribs of 'log cabin' construction are built on land, floated into position and sunk by loading with rocks, stones or concrete blocks
Gabions are steel wire mesh boxes of various sizes filled with low grade or waste stones. These can be handled as building blocks to make a structure with highly flexible qualities. The principle has been in use since the 16th century when wicker baskets were filled with soil to protect military emplacements

4 Vertical face breakwater

This is used in deeper water or where, for reasons of availability of materials or constructional difficulty, the rubble-mound type is impracticable. The

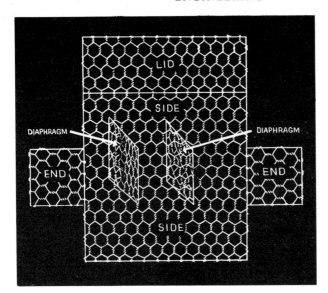

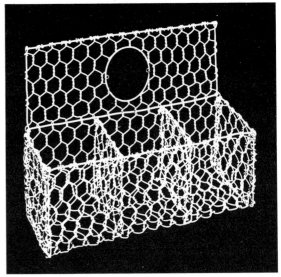

6.22 Modern gabions have evolved from basketwork. They are used as retaining structures, revetments and toe walls to embankments. As anti-erosion structures they form a strong and flexible means of constructing sea walls, river-bank defences, canal banks and protective structures for reservoirs and lakesides

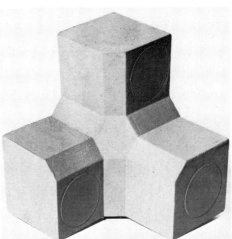

6.23–25 Models of registered design pre-cast armouring units for maritime structures such as breakwaters. These models which are accurately scaled for dimensions and weight are used in hydraulic laboratory tests for specific projects to assess stability under storm wave attack
6.23 Cob: Coode & Partners, London
6.24 Tripod: Sir William Halcrow & Partners, London
6.25 Stabit: Stabits Ltd., London

6.26 Stabit being placed in position on a harbour breakwater. These units which are the registered design of Stabits Ltd., London, have been used in large numbers as breakwater armouring in Libya and the Persian Gulf and in beach defence work in Shoreham, England. Consulting Engineers: Sir William Halcrow & Partners, London

materials and forms used vary from steel piling to reinforced concrete caissons. Such caissons may be of the Mulberry Harbour principle where purpose-made concrete caissons are floated into position and sunk, either filled or unfilled. Old barges are sometimes used or, off the coast of California stripped motor bodies where, incidentally, the protective conditions provided by such unorthodox structures have benefited fish-breeding quite dramatically

5   Precast cylinders
In this type of vertical-face breakwater large-diameter hollow piles are closely spaced. A continuous heavy beam or coping is often used to strengthen the system and enable the breakwater to be used as a pier or walkway

6   Hydraulic fill
When dredging by hydraulic methods the unwanted material is deposited offshore of the marina area to form a protective spit. This needs careful handling or the barrier will either be quickly washed away or will accumulate further material and create an ever growing bar causing navigational difficulties. It can sometimes be used as the core of a rubble mound or composite breakwater

7   Ice breaker
A breakwater with the additional function of cutting up ice. Often a series of bundles of piles triangular in plan with a sloping top

8   Composite types
These are often economic in deep water locations. The lower, trapesium section gives a broad and stable base from which the vertical breakwater may rise

9   Diaphragm breakwater
This is a special and comparatively recent adaptation of the caisson-type breakwater. The cylinders are not filled with sand but are instead hollow and slotted on their seaward side. Inside the cylinder is a permeable partition which, together with the curving outer wall has a remarkably dampening effect. The top of each cylinder is splayed off at an angle thus reducing the horizontal impact of the incoming sea.

   To compensate for the omission of sand fill each caisson has a thick concrete base and stability may be further improved by increasing the diameter of the cylinders. Each vertical column acts as a beam and the total calculated structure is more resilient than the sand-filled types which, because of their weight and rigidity are prone to tension cracks. A diaphragm breakwater is being constructed at Brighton, Sussex, after prolonged testing against mass concrete and rubble mound types. These breakwaters will be 1,800 m (5,900 ft) long enclosing an area of 50 ha (125 acres) and take up a substantial amount of the £7m estimated for the maritime works

10   Floating breakwaters
There are several types of floating breakwater with a considerable range of application

6.27 Breakwaters of concrete sheet piling with capping and buttresses

6.28 Lowering a 600 tonne caisson into place for the breakwaters of Brighton Marina

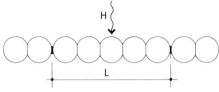

6.29 If long, caissons can be floated into position and placed on a prepared bed. The diaphragm type shown here is preferable

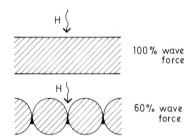

6.30 Variation of force with the horizontal profile of a vertical face

100% wave force

60% wave force

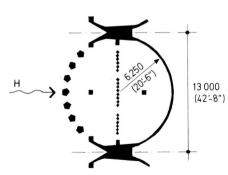

6.31 Slotted circular caisson breakwater with internal permeable wall

The first known example was constructed with heavy blankets on buoyant cylindrical tanks by a Captain Graves in 1842. A famous example was the Bombardon breakwater used at the Normandy landings. Designed to withstand 2·5 m (8 ft) waves, it was unfortunately destroyed shortly after installation by a storm generating waves estimated at 4·5 m (15 ft).

The main applications of breakwaters may be summarised as follows:

10.1   For permanent use—taking the sting out of a rough sea

10.2   For temporary use—to protect harbour and other coastal works during construction

10.3   For the protection of dredgers or pipe layers—mounted seawards of the vessel, perhaps with its own system of winches and anchors

10.4   Sea going types—capable of holding station without moorings. These devices can achieve remarkable efficiency by filtering out wave lengths according to a pre-determined design. The orbital motion of the waves is broken up usually by an open raft-like design

The great advantages of floating breakwaters over fixed varieties are:

10.1   Because they *are* floating they are always in an ideal position to deal with the attacking wave

10.2   They do not stop the natural water flow—most important with river authorities and areas with littoral drift and heavy siltation

Other advantages are:

10.3   Mobility

10.4   Low capital cost

10.5   Prefabrication

10.6   Deep-water uses

Against these points it is only fair to say that there may be mooring problems, maintenance may be expensive and waves may only be partially dampened.

Floating breakwaters are not a real alternative to fixed structures but can serve as a useful complement. In deep water areas, harbour entrances, on reservoirs, lakes and rivers where the hydraulic regime must not be altered, floating breakwaters may be the only practical answer.

The principal designs now either in use, patented or in an experimental stage are as follows:

10.1 The shallow draft breakwater (see 6.33). Mooring forces are kept very small, wave height is reduced by interfering with the orbital wave motion particularly in its vertical direction. Efficiency is proportional to the length (i.e., direction of wave advance) of the raft. Field trials have recently been undertaken by The National Physical Laboratory and the Department of Engineering at Southampton University in association with Archibald Shaw and Partners.

10.2 The sea-breaker (see 6.34, 35). Devised by Col. (Blondie) Hasler and sponsored by the National Research and Development Council, this is a shallow pontoon connected to an outrigger float to minimise rolling. The tubular superstructure limits flexing and torsion. The waves break on to the pontoon which destroys their motion and momentum. The dimensions are 40 m × 2·4 m (130 ft × 7 ft 9 in). A prototype has been moored in Stokes Bay, East Solent, since February 1971

10.3 Pneumatic and fluid-filled bags (see 6.36). New rubber and nylon fabrics have promoted further experiment with this type. They are light, easily transported and inexpensive but the fabric is vulnerable and mooring lines can be a problem

10.4 'A' frame and Winter types (see 6.37, 6.39, 40). This Canadian development has a central vertical plate supported by outrigger buoyancy pontoons on each side. Waves are suppressed by interference and turbulence. A variation by Winter incorporates a landing stage and portal frame.

10.5 The resa (see 4.13, page 80, 5.88, page 145). This is an adaptation of the 'A' Frame type designed for marina sites. It has a wave-return profile on the weather side and is available in prefabricated units

10.6 The reservoir marina type (see 6.38). Two pontoons separated by a perforated base and with a vertical barrier below the lee side

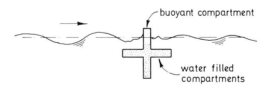

6.32 Bombardon breakwater

6.33 Floating break-water by Harris & Sutherland, Archibald Shaw & Partners

6.34, 35 Colonel Hasler's 'sea-breaker' under test in choppy conditions. The reduction in wave height is impressive

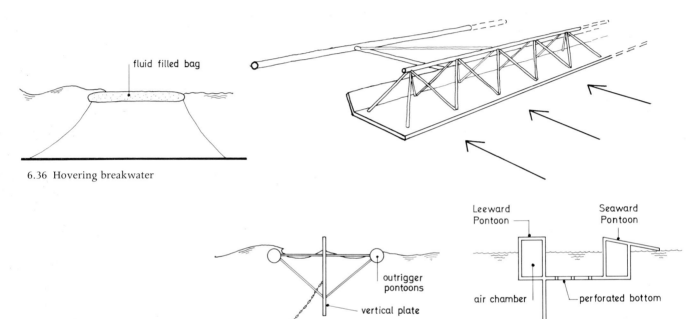

fluid filled bag

6.36 Hovering breakwater

outrigger pontoons

vertical plate

6.37 A-frame floating breakwater

Leeward Pontoon

Seaward Pontoon

air chamber

perforated bottom

vertical wall

6.38 Reservoir-marina floating breakwater

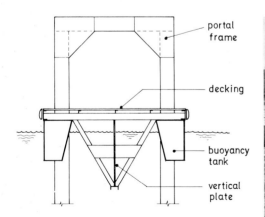

portal frame

decking

buoyancy tank

vertical plate

6.39 The 'Winter' dual-purpose floating breakwater

6.40 The 'Winter' pontoon breakwater unit under test at Cowes, Isle of Wight

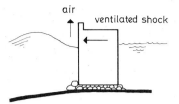

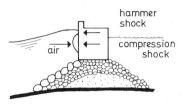

6.41 Types of shock force

Wave forces

Most structural and civil engineering deals with statics, gravity and dead loads: breakwater design is concerned with dynamics, variable forces and live loads. Many waves hitting the vertical face of a breakwater at right angles are reflected so that the next wave breaks at a critical impact distance from the face, resulting in a greater shock force on the structure than would have been imparted by an unreflected wave force.

Shock may be delivered in three principal ways:

1   Ventilated shock. Where air, upon impact, is forced through the wave-crest. The resulting explosion jets water high into the air, an effect known as tsunami

2   Compression shock. Where air remains trapped in the concave hollow of the breaking wave

3   Hammer shock. Where a plunging wave impacts with high local intensity

It is usually uneconomic to design breakwaters to withstand every eventuality irrespective of how likely or how frequently it may occur. Research will reveal the likelihood of really exceptional conditions and a calculated risk is often taken. To do otherwise will mean catering for a freak event which on average may occur once in a century and which may double the breakwater's cost.

The criteria for such decisions may be in the form of wave height within the marina, stress in the structure or wave forces on the breakwater.

With rubble mound types most of the energy is dissipated by wave breaking and percolation. The size of rocks or armour blocks used and their relationship with the angle of the seaward face will have an important effect upon the breakwaters' efficiency.

The forces countered by breakwaters are surprisingly high. For a 1·2 m (4 ft) wave the energy for each foot run of crest is 6 tons, generating between 1·9 kw (2½ hp) and 3·8 kw (5 hp) according to depth.

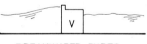

BREAKWATER TYPES

6.42 Breakwaters of types S and V: reflections from the outside
Most of the wave energy impinging on Type S is dissipated by wave breaking and percolation, but on Type V the major part is reflected
The configuration in plan of the breakwaters in Harbour (a) is favourable because the reflected waves (shown dashed) are dispersed
In the case of Harbour (b), wave conditions at the entrance would be unacceptable if the breakwaters are of Type V. During severe storms, craft may also be in serious trouble within a triangular area that extends far out to sea, because breakers occur among both incoming and reflected waves, which cross at an angle

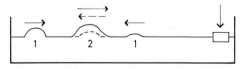

6.43 Superposition of waves

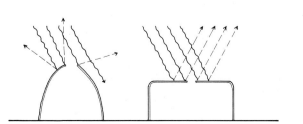

WAVE REFLECTIONS

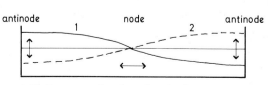

6.44 Geometry of node and antinode

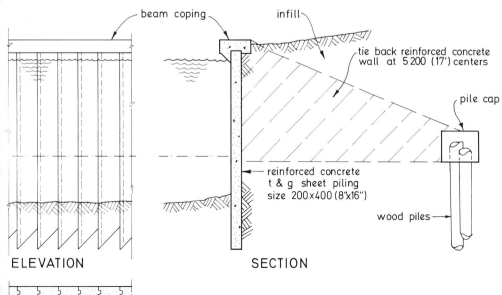

beam coping

infill

tie back reinforced concrete
wall at 5 200 (17') centers

pile cap

reinforced concrete
t & g sheet piling
size 200×400 (8"x16")

wood piles

ELEVATION   SECTION

PLAN

6.45, 46 Two stages in the development of
an American lake-shore development. The
lake-bed material is booster pumped from
the borrow area, through the pipeline
system to the reclamation site. The
eventual edge treatment is of concrete
sheet piling tied back through wing walls
to capped wooden deadmen piles and then
backfilled. See 6.47

6.47 Section through bulkhead

6.48 A new marina and wharf under construction at Prince George Wharf, Nassau, Bahamas

Appearance and function

Within the engineering constraints of breakwaters there should be plenty of scope for the architect to exercise his ability with materials and geometry. They are one of the most satisfying elements in the design of harbours and need to reflect their protective role. These are the fortifications, the ramparts, and strong modelling and stout materials are needed to match form with function. The sort of configuration given by 20 ton rocks and simple artless contours are the right sort of approach allowing the form to speak for itself. As to railings, steps, lighting and other 'furniture' this should be kept to the absolute minimum, and where unavoidable should be very simple. As an example of a jetty itself and the ironwork upon it, the cob at Lyme Regis is superb and a wonderful guide to those proposing a similar construction.

The shape on plan may be important if seen from nearby cliffs or buildings but unless this is so paper planning should be avoided. Much more important is the prospect from the shore, the sea and the marina itself. Remember that the breakwater will probably become the new horizon from everywhere at ground level within the harbour. Its profile will be silhouetted in a most unforgiving way from the sea and adjoining land but softened by masts and rigging from within the harbour.

If it is at all practicable or economic, the right to walk on the breakwater is enormously appreciated by the public; whether fishing or bathing is allowed will depend upon local circumstances.

One form of breakwater not covered in the definition of types is the large embankment breakwater providing sufficient width and stability to support buildings. Where the environment makes such a concept possible it should be seriously considered for the return may justify the expense. One or two storey residential units, clubhouses, customs, harbour master and light beacon are likely contenders. Housing will probably turn its back to the sea and face inland, with berths and terraces towards the harbour. Such an arrangement is a most attractive proposition and, with the careful handling of levels in cross-section, access for the owners, services and the public need not present a problem.

Cost

As these barriers are generally expensive, anything contributing towards lessening their cost should not be overlooked and any permanent offshore spit, shoal or shallows that can be used as a base from which to build, or any extension of a natural promontory may provide a means of minimising the amount of material needed and the depth of water thereby lowering the cost. Often, in areas where conditions require only moderate abatement, some form of protection analogous to a dotted rather than a continuous line may be considered, dissipating the water rather than providing total protection.

The rubble-mound or rip-rap types, described previously, where material is strategically positioned, are of course substantially cheaper than vertical-faced or precast constructions which require either on-site con-structure or are pre-fabricated elsewhere and afterwards positioned on site.

Piles

Foundations and materials

There are three main categories of piles in marinas:

1   Those supporting structures such as jetties, fixed moorings gantries or hoists
2   Those restraining floating moorings
3   Those supporting the perimeter bulkheads or quays

The piles in 1 may be part of a complex structure supporting piers or even buildings. Those in 2 are generally simpler, free-standing columns, although they may extend high enough to support a roof for covered moorings or be braced in pairs over walkways for extra stability. 3 are normally sheet piles retaining siding or sheeting. This type is dealt with under Bulkheads and Quay walls (see page 192). The three principal materials are timber, concrete and steel. Some pros and cons of material choice are listed in 6.C.

Selecting a suitable type

Although a useful guide, 6.C does not take into account the many different types of pile of each material, some of which are accepted forms and others proprietory designs. Many piles are combinations of these materials, particularly of course reinforced concrete which will display the good and bad properties of both materials. There are many types of piles and many ways of driving them in cross section. They may be round, square or octagonal and in length straight or tapered. Hollow sectional piles are usually filled with a concrete or reinforced concrete core after they are seated. Each material and shape has its uses and none may be said to be cheapest, as, dependent upon the quality chosen, their prices per foot when driven into position will overlap according to the grade and treatment of the timber, the

6.F Comparative choice of materials for piles

Timber		Concrete		Steel	
Advantages	Disadvantages	Advantages	Disadvantages	Advantages	Disadvantages
Light weight	Attacked by rot, crustacea, marine borers	Durable	Must be capped for driving	High strength	Rust
Easily worked bolted or spiked		Unaffected by rot etc.	Reinforcement	Usually durable	Costly surface protection
Will float	Fungi (in fresh water)		can rust	Unlimited length	
Withstands impact		Can be tapered, shaped etc.	Can fracture	Driven without damage	High maintenance
Can be impregnated	Limited length		Fabrication of rc		Possible delivery delays
	Vulnerable at water-line		Quality control necessary	Can be driven into hard bed	Surface coating worn by pile abrasion
Less damage to craft on impact			Hair cracks difficult to eliminate		
Sympathetic appearance					

mix and reinforcement of the concrete, or the type and surface treatment of the steel.

The specialisation of the contractor will also help to determine the choice. If, for example, considerable concreting work is to be carried out on bulk-head walls and breakwaters, then equipment may already be available on site to drive concrete piles and these would probably be chosen as a natural extension of the perimeter engineering work. The length and strength of piles will vary according to whether the marina has a variable tidal range or is a locked water area where the relationship between the water level and the walkways is constant.

Placing the pile

The number and positioning of piles will be largely determined by the layout of the marina, and their section, length and material will be decided bearing in mind the local conditions and money available. Fixing piles into the bed through a depth of water is naturally more difficult, specialised and expensive than its land-based equivalent. Equipment which is piling around the perimeter can work on land but free-standing piles in water must be positioned by plant that either walks on the bed or, more commonly, floats on the surface. The following are the more usual methods used:

1   Rotating by hand
This is sometimes possible. A weight is applied to the top, the pile being screwed in by lever arm and tackle

2   Augering
Drilling a hole into which the pile is fixed

3   Driving
3.1   By hammer blows from above to the pile top
3.2   By a steel rod or mandrel lowered vertically through a hollow pile. This bears onto the bottom of the pile and transmits hammer blows to it from above

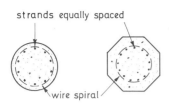

strands equally spaced

wire spiral

ROUND
PILE

OCTAGONAL
PILE

6.49 Pre-stressed pile sections

3.3   By pressure, where the pile is passed into the bed

4   Jetting
Where the passage of the pile into the bed is assisted by a pressure-jet of water

5   Blasting
In exceptional situations, usually in rock

6   Pressure injected footing
Used where there is little skin friction. A steel tube is driven to the required depth and 'no slump' concrete rammed to form a spread footing or bulge below it. The tube is then filled with concrete or reinforced concrete

7   Drilled-in caisson
A steel shell is driven to a solid bed and the material removed from within it. A socket is drilled into the rock and a heavy H column lowered into the steel tube which is then filled with concrete

6 and 7 above are primarily proprietary systems, one by Frankipile Ltd, the other by The Drilled-in Caisson Co.
   With the dugout system mentioned on 2.18, page 34 piles are driven dry into the excavated bed which is later flooded. This is a cheap and easy way of placing piles, but prior knowledge is needed of the soil's holding capacity when flooded.

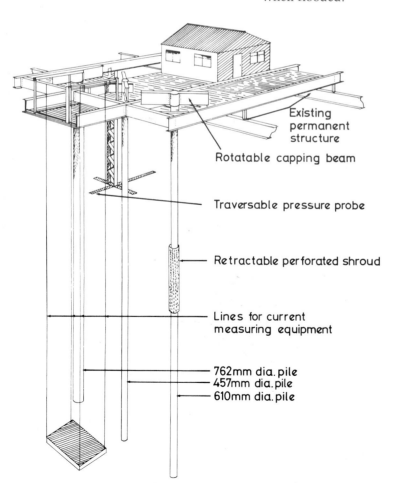

Existing permanent structure

Rotatable capping beam

Traversable pressure probe

Retractable perforated shroud

Lines for current measuring equipment

762mm dia. pile
457mm dia. pile
610mm dia. pile

6.50 Lay-out of pile test rig

6.51 View of the experimental rig from upstream

65.3 Covered wet moorings at Sanford, Florida. The columns are extended timber piles

6.52 Leonardo was right. The inscription translation by Professor E. H. Gombrich reads: 'Observe the motion of the surface of water, which resembles that of hair, which has two motions, one of which depends on the weight of the hair, the other on the direction of the curls: thus the water forms turning eddies, one of which follows the impetus of the main course, while the other follows that of incidence and reflection'

6.54 A simple fixed jetty at Fort Lauderdale. Except as mooring posts, and perhaps as markers at very high tides, the piles need not extend above deck level

6.55 Untreated ferrous metal as sheet piling in this marina at Miami has proved a failure even in such a mild climate

6.56 Piles need careful setting out. When driven out of line they cause the floatation system to jam

It is most important to ensure that piles in a pontoon system are accurately driven and do not afterwards lean out of true. Unless they are set out correctly and are perfectly vertical at best the system will creak and groan as stringer rubs against pile and at worst will jam and drown the piers and walkways. This danger increases with pile length and tidal range. It is better to have generous tolerances than a precise and tight-fitting arrangement.

In both the selection and placing of piles it is a good idea to find out how existing piles have behaved in similar condition nearby. Capping piles is often necessary particularly with timber and cylinder types. Copper, plastic or asphalted fabric are suitable materials.

Piles vulnerable to impact from craft may be protected by fenders for concrete and steel or by vertical timber battens spiked to timber piles.

Extended piles

Piles of any material may be extended in length to form supporting columns to the roof of covered berths. As these moorings often accommodate the larger type of keeled and masted boats, the length of pile-cum-column may be anything up to 18·28 m (60 ft). The handling and driving of such piles present special problems and costs are sometimes reduced by accepting a joint or splice, allowing the pile to be driven and extended later to form the roof-supporting column. Such splicing can also be used in maintenance rather than a complete replacement. Recessed-headed bolts through a long scarf joint can be satisfactory—tight metal sleeves or collars being added afterwards.

Suitability of bed material

Estuaries, tidal rivers and broken coastlines often provide good bed material for piling but conditions can vary from good to impossible within a short distance. Trial boreholes should be taken unless the engineer is absolutely sure of the sub strata. The following is a brief guide to some common conditions:

1 Fine light mud
Very little skin friction but often lies over a bed of better material. Choice depends on how far down the solid bed lies

2 Medium-grained silt and neutral clay
These hold soundly in most cases and accept driven piles very well taking all but the heaviest loads. Depends upon the cohesion of the material

3 Fine and medium sand
Usually good particularly if there is a solid bed below

4 Medium and coarse gravel
As above but driving may become difficult in some cases

5 Pebbles
Difficult if they continue too deeply but they usually give way to sand within a few metres

6 Rock
Varies from shale and soft stone to granite. Chalk rubble is quite treacherous. Generally welcomed as a bed at reasonable depth but often difficult and expensive to drill for piles and even more for dredging

7 Organic material
Found in inland lakes particularly. In its vegetable form is quite useless. If shallow may be driven through. May contain secret pockets of methane

Pile oscillation

One phenomenon worth particular attention is the oscillation of piles and piled structures due to currents or tidal flow. Very severe damage was

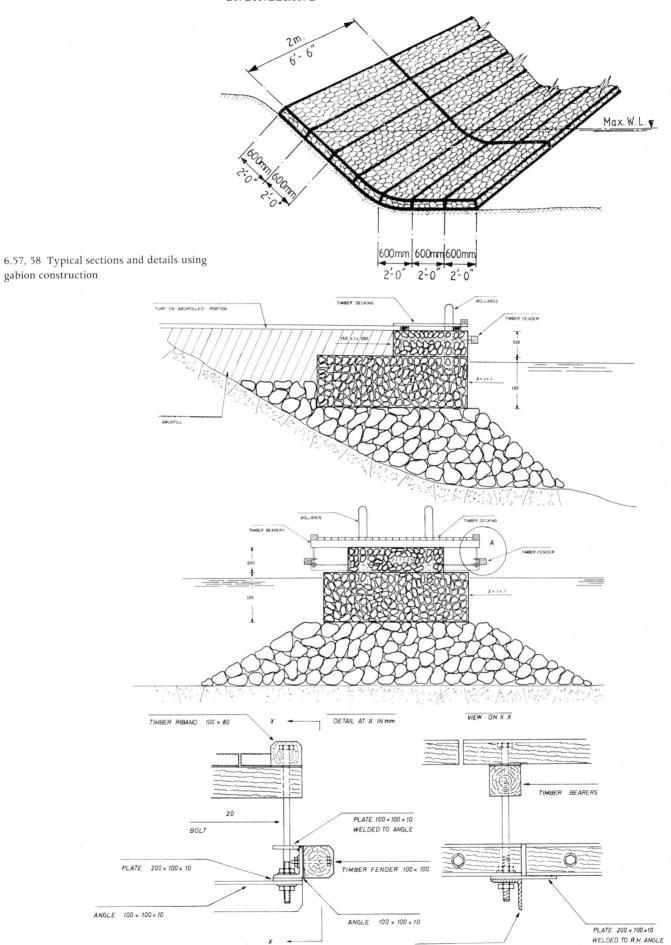

6.57, 58 Typical sections and details using
gabion construction

caused during the construction of the Immingham oil terminus on the River Humber in England. Following this, research by the Construction Industry Research and Information Association (CIRIA) established the hitherto unrecognised limits on the combination of pile diameter, pile length and current velocity which together determine vulnerability to in-line excitation.

Similar dangers due to *wind* have been well known since the Brighton Chain Pier and the Tacoma Narrows Bridge disasters, the two simplest characteristics being the oscillation of members at a particular speed (not necessarily a very high one) and the oscillation of the member in a plane at right angles to the wind direction.

The striking features of the Immingham research were that they were in water and that oscillation was *in line* with the flow. The effect is very similar to that which causes a flag to flap even in the steadiest wind, the flag-pole being analogous to the pile and the flag to the water currents.

The means by which the destructive effects of oscillation may be minimised are by adding perforated shrouds or offset dorsal fins to each pile. The exact methods are too complex to be dealt with here, and CIRIA Report No. 41 is recommended for those with a specialist interest.[1] In connection with the recent discovery of this phenomenon, it is fascinating to learn that Leonardo da Vinci—with incredible perception, grasped the problem, subsequently drawing and noting it with an exactitude which has taken 500 years to improve upon 6.52.

1. Construction Industry Research and Information Association, *Oscillation of piles in marine structures*, Report 41, The Association, London 1972.

Bulkhead and quay walls

Definitions

The terms bulkhead and quay are often interchangeable. Both usually refer to the perimeter of the water area where land and water meet after the harbour's completion. This, of course, may not coincide with the tide line or water's edge before development. Both infer some form of retention and may include a composite arrangement of vertical and sloping faces. Many of the observations on breakwaters made in a previous section will also apply to the quay or bulkhead walls.

Other than when a slope or revetment is used, the new interface of land and water will almost certainly involve a bulkhead. Whilst different in some respects both the breakwater and the bulkhead bring to equilibrium forces acting more strongly on one side than the other.

The treatment of slopes and shelving edges is again similar to the mound type of breakwater but the former will not encounter such great wave forces. Their use may necessitate some floating or fixed construction built out over the water to form a peripheral walkway.

Principal types

There are two principal methods of quay wall construction:

1   The sheet type
Driven into the ground or bed and braced back by piles, shoring or deadmen

2   The gravity type
Which retain the soil (or water) by their weight and shape

There are combinations of both and many forms of each and it will depend upon several factors as to which is the most suitable. Some of the more important of these factors are:

1 The relative heights of soil and water and the tidal range of the latter

2 The quality of the retained soil and sub soil with regard to its

2.1 water retention

2.2 angle of repose

2.3 consistency

2.4 ph factor

3 The relative pressures calculated from 1 and 2 above together with additional stresses exerted from wind and water forces, nearby buildings, vehicles etc.

4 The suitability of the construction method with regard to local conditions, available equipment etc.

5 The positioning of the bulkhead (see 6.63)

6 Appearance

7 Cost and maintenance

In some cases where the soil is of consistent clay, bedrock or hardpan a bank may be formed at the proper angle of repose. Thus the total difference in level between the water and the land is made up of a retained vertical face for the lower section and soil at its angle of repose for the remainder. This of course is only possible where water will never rise or be dashed above the vertical face, or where the slope is to be covered with impermeable material. Unprotected banks are generally only suitable for non-tidal or locked situations.

Gabions (see 6.22, 57, 58) make excellent bulkhead walls being both strong and flexible. The cost of materials may be more expensive unless suitable stone is available locally but ease of construction often overcomes this as the building method is a repetitive placement of rectangular building blocks and no excavations, foundations or formwork are required. During the early period of use, silt and vegetation combine with the rock filling to form a permanent structure by the time the mesh starts to corrode.

Positioning

The siting of a quay wall in relation to the original tide line is often a crucial factor in determining the type of bulkhead to use. Given a normal shelving coastline, riverbank or lake there are five principal locations for the bulkhead wall (see 6.63):

1 Inland
This 'pushes back' the land, allowing water to gain distance inshore

2 On the low-water line
This takes advantage of a medium position where fill and dredge are likely to equate, preserving the existing shoreline as far as possible but substituting a vertical for a sloping interface

3 Offshore
This gains land and pushes back the water

4 On the high water line but behind (inland of) a lock system
This, in itself, allows a minimum bulkhead construction but of course necessitates the expense of a lock to eliminate the tidal range

5 As a perimeter to a dug-out marina

Suitable edge treatments for vertical,
sloping and gravity walls

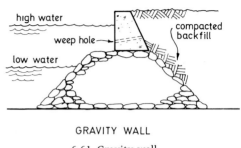

CONCRETE   TIMBER

GRAVITY WALL

6.61 Gravity wall

SAND BEACH

REVETTED SLOPE

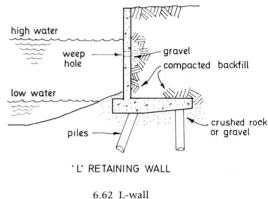

'L' RETAINING WALL

6.62 L-wall

CONCRETE SLAB REVETMENT

6.60 Typical beach and revetted slope profiles

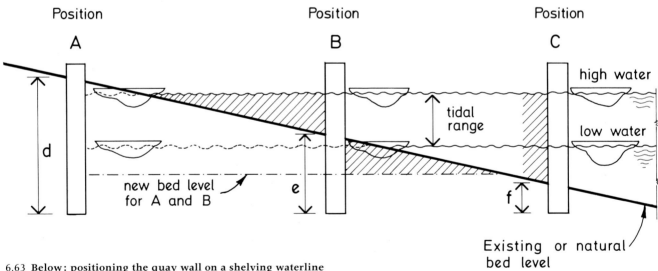

6.63 **Below: positioning the quay wall on a shelving waterline**

A

Quay wall can be built dry in trench (soil
  type permitting) or in open cut
Land area loss and water gain
Possible to dredge from shore (at low tides)
  with land-based equipment
Maximum depth of foundation into soil
  (seed)
Maximum overturning moment at low
  water

B

Fill equals dredge
Construction at low water line will be wet
  but no great water pressure problems
Advantage of suction dredging material (if
  fine) from offshore to new bulkhead wall
  onshore
Medium depth of foundations into bed
Minimum alteration to existing shoreline

C

Maximum land gain
Maximum imported fill
Minimum dredge
Water pumped from onshore side
Expensive wet construction
Minimum depth of foundation (dependent
  on bed conditions)
Greater wave height in deeper water

6.64 A river wall under construction showing gabion boxes in position ready for filling. A flexible gabion apron is provided to prevent undermining of the structure

6.65 This quay wall at Stavoren, Holland, changes levels with elegance and precision

The whole new water area is excavated dry in an inland site, bulkheads being constructed as earth-retaining walls. The reduced area is then flooded by breaching the existing shore-line.

The type of quay and its positioning will determine whether a coffer-dam is necessary. This is a (usually) temporary, watertight, linear structure which holds back the water and allows the permanent quay wall to be constructed in dry conditions. Whilst expendable and costly, coffer-dams are often necessary, particularly where sheet-type bulkheads are to be built off-shore on a shelving waterline.

Length

The length of bulkhead wall will not only depend upon the size and shape of the marina but also its basic type. 5.4–7, page 95 show how the percentage of bulkhead wall to the total perimeter alters with the land-to-water interface, giving approximately:

25 per cent for the offshore

50 per cent for the semi recessed

75 per cent for the built-in

100 per cent for the land-locked type

These relationships may be significant if the constructional difficulties or expense (which are closely related) of building a bulkhead wall is substantially greater than a perimeter breakwater. The situation could of course be the reverse where, for instance a land-locked, dug-out marina with its 100 per cent dry-built quay wall is much easier and less expensive than a deep water offshore type with its 75 per cent perimeter breakwater.

Costs

In short, the marina *type* determines the amount of bulkhead and, because the perimeter construction will probably be the largest single engineering expense this will play a large part in determining the harbours final cost and possibly its overall viability.

The difference in cost between the sheet and gravity bulkheads will depend upon factors 1–6 listed on pages 181–182 and will vary far more due to these than any intrinsic or constant differential.

Appearance

A concrete surface can be modelled by lining the formwork or textured by bush-hammering or grit-blasting but these operations are difficult when impeded by the presence of water. Steel and concrete can both be surfaced with a cladding material but this is expensive if only needed for appearance rather than protection.

In considering the appearance of bulkhead walls and other fixed structures in tidal waters the tide-line usually reads as a strong horizontal mark because of organic growth. Whilst it is not desirable (even were it possible) to prevent this, a strongly textured or profiled face may add definition and character to what are otherwise often rather dreary streaks.

Although timber is probably less durable it usually looks more acceptable as a material for quay and bulkhead walls than unfaced sheet piling. This is particularly so in rural settings.

Many materials laid in a variety of patterns are used for dampening waves on a sloping surface. Some of these are proprietary designs and others created by engineers for specific purposes. They vary in size from very large cyclopean stones and stabits or armouring pieces (shown in 6.23–6.26, page 178) to the small scale paving slabs of a river bank or spending beach.

A natural material with remarkable dampening properties is coral and in the Pacific, Australia and the Middle East—where it is known as ferroush—it is used as a facing slab. The hard sponge-like material traps and holds the water most effectively. A manufactured aerated concrete with similar qualities would find a receptive market.

Siltation and erosion

Siltation is the accretion and erosion, the removal of material from the bed, banks or shoreline. These phenomena are closely related and the action of one may be the cause of the other.

A close understanding of the existing regime is necessary before the effects of any new structures can be assessed with any degree of accuracy. This will require data collection, recording, survey, analysis and possibly an hydraulic model. Siltation and erosion occur in several different forms depending whether the site is on the coast, river, estuary or lake:

1 Coastal siltation

This is of three main types according to the distance from shore:

1.1 Offshore drifting usually runs parallel to the coastline, the direction and strength of transportation depending upon tide, wind, current and type of material

1.2 Littoral drifting takes place closer to the shore and is strongly influenced by the predominant currents and need bear no relation to the offshore movement. Breaking waves and surf encourage littoral drift and the coastline profile will effect and be effected by the amount of material transported

1.3 Coastal drifting occurs close inshore due to the to-and-fro motion of breaking waves. The amount and velocity depends upon the material and strength of current

6.66 Silting is a real problem in this old marina at Charleston

6.67 Small docks on colloidal rivers will soon silt up and dry out unless regularly dredged

6.68 A sad ending to a rigid structure unprotected against scour. The wall collapsed for three reasons: it was undermined; it was pushed over by the pressure of water trapped behind it and it was too rigid to withstand any ground movement

Siltation should be an important consideration in the location of a marina and if other factors outweigh this aspect, and the marina is positioned where the deposition of silt is high, then methods to minimise the nuisance and expense deserve careful thought. In some cases protective measures which have not been located correctly have worsened the situation rather than improving it.

Indiscriminate dredging can also increase the deposition of silt in the near vicinity. At a marina near Los Angeles the developer, having purchased a large area of mudflats eventually discovered that the dredging for his first phase development caused heavy siltation on land safeguarded for expansion, and within two years this material had to be removed at great expense before further engineering work could commence.

2 Inland siltation

An important phenomena is the heavy siltation that occurs in rivers where the fresh water coming down stream meets the salt water coming up stream with the tide. These do not readily mix, but remain separated for some distance in the form of a plane laying at an oblique angle to the river bed (see 6.69). This plane travels up and down with the tide. The fresh water which will be carrying silt in suspension, will quickly deposit this upon contact with the salt water. It is along these stretches of the river that deposits will form. Marine clay is very hard and intractable stuff and resists erosion much more than the ordinary sand or gravel of a river bed, so that a shoal area along these lengths of the river will form. If a marina is contemplated along these stretches, then a lock with a reservoir system may well prove worthwhile to overcome heavy and continuous maintenance dredging.

The harbour itself and its entrance is where siltation is likely to be most troublesome. Recessed, unlocked marinas are in most danger particularly on a colloidal river or estuary. Silt-laden water will often enter a harbour-

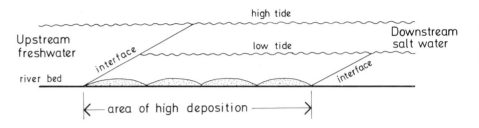

6.69 The likely extent of high deposition in tidal rivers

mouth near the bed, circle the enclosed water area depositing its material en route and exit again from the mouth as relatively clear water at surface level.

3 Erosion

Although erosion is less likely to occur in a marina which has been developed inland from the bank of a river or the sea, its effects can be even more serious than siltation. The scouring of the bed tends to loosen piles and undermine jetties and quays, and is often expensive or even impossible to amend.

Deposition and scouring, being often caused by the same hydraulic conditions may be found very close together, the scoured material being deposited within a short distance. Erosion is most serious where it undercuts breakwaters, quay walls or embankments which are subject to pressure loading from the opposite side. As these conditions are difficult to reverse, overturning and failure may occur.

The effects of the natural hydraulic conditions may therefore manifest themselves in the form of a continual drain upon the economy of the whole project or by a sudden physical collapse. The best insurance policy against such adversity is the really early employment of hydraulic engineers who will discover the existing conditions of siltation and erosion and pronounce upon the project's viability in engineering terms. Because the basic engineering costs are likely to absorb 50–80 per cent of the total outgoings this must be easily the most important advice that will be received and money spent on getting it should not be begrudged.

Ice

Damage to boats and marinas in northern waters is sometimes caused by the formation of ice. This can put the marina out of commission, by causing damage to the piers and by pulling the piles out of the harbour bed. If a substantial thickness of ice forms and this is followed by a rise in the water level, the ice, having gripped the piles will rise with the tide, uprooting or loosening them so that subsequent winds or currents will knock them sideways.

Ice is less likely to form in salt or moving water or water of considerable depth. The slow-moving fresh waters of lakes, canals and quarries however are quite likely candidates for ice conditions particularly on exposed sites or where there are large expanses of relatively shallow water suitable for dinghy sailing.

The majority of boats will be out of the water in the winter months where ice is an annual hazard. When warning of severe conditions is given all boats remaining are best removed from wet mooring as part of the marina service. Whether the floatation system is taken out depends upon individual circumstances. Pontoon maintenance is a winter-time task best undertaken at the time of worst weather, but annual removal would be a laborious and expensive business.

If ice formation does look like being a problem then it is best to sound out local knowledge and records to establish how serious and frequent these conditions are likely to be and discover what the traditional methods are of overcoming them. Marina operators in areas which experience hard winters often remark on the strange fact that quite thin ice within a short spell can cause great damage one year and yet the same marina may suffer extreme conditions with heavy icing on another occasion and escape with little or no harm to the marina or craft.

De-icing methods

De-icing methods include permanent, mobile and traditional methods. Which to choose will be determined by:
1   The likely amount and frequency of damage
2   The number of craft remaining in the water
Whilst there is a growing need to encourage winter sailing by creating open conditions it would be unwise to install expensive equipment where severe conditions are likely to be infrequent. Some of the more common ice prevention methods are as follows:

1   A permanent aeration system
This is best installed during the construction stage of the marina development and consists of perforated piping laid on the bed, usually around the piers and connected to an air compressor. This activates and brings the warmer water to the surface. Siltation, dredging and anchors may cause blockage or damage however. A better and cheaper method may be:

2   A mobile unit
A compressor is either placed in a boat or taken on wheels along the piers and a flexible weighted (often plastic) perforated hose, not unlike the vibrating pipe used in concreting, is moved along the bed of the marina. 1 and 2 will melt existing ice as well as preventing further formations

3   Floating electric blankets
These have been developed in Canada to overcome very severe conditions. They are only partly successful because they are quite expensive to buy and operate and present problems in the space occupied, the labour involved, and deterioration during storage

4   Manual methods
These simple means include the use of a timber pole to break the ice when still thin and running a stout-hulled craft through the ice along the clearways and piers

5   Grease coating
A band of thick grease put on the piles at low water helps to prevent them lifting up. This is however a lengthy and unpleasant job

6   Pile clusters
In rivers and estuaries in particular, permanent clusters of piles, often with sloping tops, are placed in the path of damaging ice floes to break them up and divert them from the harbour

7   Operating the lock
If a lock is opened at regular intervals a current of water can be created sufficient to prevent or minimise icing

---

Locks

The conditions for locking

A lock reduces or eliminates the natural tidal range and therefore maintains a more constant water level. Marinas in tideless or virtually tideless areas such as the Mediterranean, or in inland situations and localities where there is only a moderate range such as Glasgow or Southampton, have special advantages for they can dispense with the costly constructions necessary to impound the water and enjoy uninterrupted passage at all times. At the other extreme are locations where all other conditions for a marina are admirable except that of tidal range. Here, the acceptance of a lock is the means of creating a harbour from an otherwise difficult or impossible situation.

Between these extremes lies a borderland where the advantages and disadvantages of locking are evenly balanced. It is here that skill and experience are needed in deciding to lock or not and, if so how, where and at what cost.

Whether a lock is necessary in any marina will depend upon the height and frequency of tides and the seriousness of siltation and erosion in the area. Wherever a locked marina is contemplated, however, one essential aspect has to be carefully considered: this is the 'turnover' of the lock in relation to the number of boats berthed. This must be carefully computed so as to ensure that the lock has sufficient capacity to deal with the quantity of boats likely to be going in and out of the marina at peak periods to avoid a build-up of waiting craft. If this cannot be assured by an increase in lock size or speed of operation, then a second lock will have to be considered or the constraints of a tidal basin accepted.

Modern locks can be very fast-acting. The lock size in terms of the length and width of water area will need relating, as far as this is possible, to the range in length and beam of boat sizes. Whilst several permutations of boat

6.70 The Government locks at Seattle, set in parkland, are a popular attraction and draw 100,000 visitors every year

size may be accepted to fill the lock, it is frustrating and inefficient to find that the lock is operating at low capacity because water areas 'left over' after accommodating a couple of average vessels are just too small to take one or two more smaller craft.

Conditions vary, but in any area where the rise and fall is more than 3·04 m (10 ft) with an average tide, a locked marina may prove worthwhile. It is worth mentioning that locks, as well as being an engineering necessity, in many locations prove to be a popular spectacle for the public. At Seattle the area around the large complex known as the Government Locks has been landscaped as a park by the State of Washington, and draws many thousands of visitors each season. No charge is made.

Design features

Even small locks, however, should be made accessible where possible, as an additional source of interest to the marina itself. The incorporation of a footbridge into the lock design is often well worthwhile. The wide harbour entrance necessary navigationally for most tidal basins creates a permanent obstacle to any coastal walkway so that the public are forced either around the perimeter of the water, which may not be acceptable for security reasons, or made to divert further still to avoid the marina altogether. A lock, being narrow enough to be bridged, provides both a pedestrian link when the lock is closed and a feature of public interest. The lock-keeper's office is often best located at high level to give good visibility in all directions. This, together with the lock itself, the machinery housing, the footpaths, noticeboards and so on creates a centre of interest for the architect to handle in a bold and simple manner, retaining the best in waterside traditions whilst integrating these elements into the design of the marina as a whole.

Both erosion and siltation may be minimised by locking the marina, although silt may still be carried through the lock to some extent. However, various systems are now proposed in locked marinas whereby the amount of water necessary to fill the lock is drawn either from the marina water itself or from a storage reservoir or reservoirs positioned beside the lock. This is then re-used on each action of the lock, thus effecting an almost entirely closed system and maintaining a constant level of water in the marina. Water is filtered, controlled and kept free from any erosion or siltation problems as well as being cleaner, sweeter and hence more acceptable aesthetically. The extra initial cost of such a locked system can be worthwhile, for the eventual savings in maintenance, especially in colloidal areas, may be considerable. Such a locked system, ensuring a constant level of pure water, enables fixed piers to be adopted throughout: these being more rigid and not articulated, are more lasting and require less maintenance.

Advantages and disadvantages

The factors listed below identify some of the pros and cons of locking. These of course are general points applicable to a broad range of conditions. The individual geographic, hydraulic and climatic circumstances of the site will, naturally need careful study and the more equally balanced the case is, the greater will be the need for thorough research to ensure the correct decision. If this decision *is* to include a lock then it is bound to bring with it several

6.G Locking

Advantages	Disadvantages

Creates stable water level	Expense
	  capital
	  maintenance
	  staff

Minimises siltation and erosion	
Cleaner, clearer water	Limits free entry and exit
Allows fixed mooring system	Creates bottlenecks
Almost fresh water is possible in coastal areas if lock is fed by stream or spring	Limits 'turnover'
Flushing out of engines is possible in salt water areas where harbour water is fresh	Most troublesome at peak periods
	Difficulty of manoeuvring craft

Creates closed and safer conditions because of
  calmer water
  greater security
	Possible damage to craft

	Major problem if:
	  gates jam
	  machinery fails
	  maintenance or repair periods
	  staffing troubles

May be only means of using site for marina at all	

Public attraction	May limit craft type, e.g.
	  length
	  draught
	  beam
	  catamarans
	  dredger
	  historic vessels

Footbridge access across closed lock is possible, giving uninterrupted waterside footpath for public, staff and owners	
Protection from major oil pollution	
	Likely to condition size to over 500 berths

factors from both sides of the list below. Choosing those which apply to a particular site and adding to them the local factors should produce an equation from which a judgement may be made. There will be an element of cancelling out to simplify the issues. One point of special importance is that of size. Locking usually means a large marina of at least 500 or more berths for it is a costly item which adds to the final cost-per-berth figure. The converse view is that the larger the number of moorings the less the proportional cost of the lock will be when related to the total expenditure.

Siltation problems

Difficulties are encountered where impounded water is of a different density or salinity from the water outside. When two waters of different densities come into contact they do not readily mix because the saline (denser) water flows under the fresh water, spreading over the bed remarkably quickly, with a velocity which can be as high as 1 m in 3 seconds. Such flows are known as density currents and they always present problems where a locked entrance separates fresh from saline water.

There are 3 methods of overcoming the problem:

1 The salt-water catch pit or sump lock
This involves trapping the salt-water current in a recess at the inner lock gate. This is then pumped back into the river or sea as water flows out of the lock

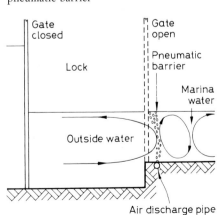

6.71 Two methods of preventing lock water from mixing with marina water. Above: salt-water catch pit. Below: pneumatic barrier

2   The pneumatic barrier or air curtain
This principle creates circulating currents at the interface between salt and fresh water to discourage interflow

3   The complete interchange of lock water
Studied and perfected in Holland by the Delft Hydraulics Laboratory this system can only be adopted where such quantities of fresh water are available that there is no objection to losing a lockful of fresh water at each operation. Water is expelled at one level and replaced by water of a different density, brought in at a different level so that the layers do not mix. Pumping must be controlled to prevent turbulence (and hence mixing). Locking times are usually lengthened and capacity thus reduced.

Siting the lock
In locating the lock within the marina boundary certain factors will need careful consideration. These, apart from special problems of a local nature are likely to be the following:
Accessibility: approach channels, outside clearances, inside clearances
Entrance conditions: tides (range and duration), currents, shelter, wind, depth of water, erosion, siltation
Overall marina type: recessed, offshore etc.
Mooring pattern
Orientation: sun, wind, visibility
Pedestrian or vehicular access to or across the lock
Turnover: capacity, speed of lock cycle

Generally speaking a straight, well-defined approach of ample depth from both sides is essential. Entrance conditions should provide as much shelter as possible whilst at the same time discouraging the penetration of heavy seas into the lock. This last condition may require the placing of permanent or floating breakwaters or the positioning of the perimeter walls in such a manner as to give protection to the lock and its entrances.

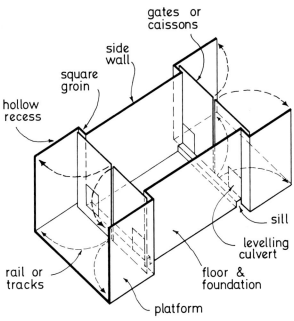

6.72 The main components of a lock

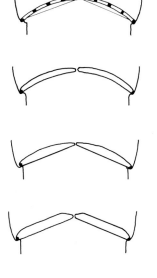

Early mitre gate in timber.

Early wrought iron gates copying the timber pattern. Expensive and not bouyant at low water level.

Modern segmental mitre gate. Economic and stiff.

Special profile of modern mitre gate. Simpler to set out and construct - minimises the depth of recess necessary to house gates when open.

6.73 Examples of mitre gates

Main design features

Apart from the floor of the lock, its foundations and the gates themselves the principal components of most locks are as follows:

1 Sills

These are the thresholds of the lock gates and are often curved in plan and sometimes in section. The accuracy of their construction is important as a close fit with the gates will minimise leakage. They guard against the build up of silt or other debris. Weight and homogeneity are important to counter hydrostatic pressure

2 Platforms

These are the concrete beds or floors over which the gates or caissons pass as they open and close. They may house metal rails or tracks upon which the gate rollers will run. Bedding and adjustment is important. A platform bed may need to withstand extremes of unequal loading where the load free condition of an empty lock is only a gate-width from the full water pressure of a high tide plus the weight of the gates themselves

3 Aprons

These are simple areas of concrete or masonry laid at bed level on the seaward or upstream side of the outer sill. They prevent scour and ease the removal of silt or debris

4 Walls

These must be parallel and vertical to prevent craft locking outwards or 'nipping'. They are similar to quay walls, except that they derive considerable support from the lock floor, but probably receive harder wear and greater pressure variations. They are usually of fair-faced concrete, masonry or sheet piling and tough enough to resist impact and abrasion without the need for rubbing strips of fenders, the onus then being upon owners to protect their craft

5 Side recesses

These house the open gates. They may be straight or curved depending upon gate design. They are usually deeper than the gate width to minimise impact from craft. At one end of each recess is the hollow quoin which houses the heel post of the gate, at the other end is the square quoin

6 Levelling culverts

These are openings which are operable independently of the gates. They allow water to pass through either to fill the lock, to balance any leakage or to maintain levels. They are usually at gate platform level behind the hollow quoin. In this position the flush of water keeps the platforms and wheel tracks clear of silt and debris. For the same reason where caissons or sector gates are used the culverts open into the caissons chamber

7 Foundations

This is the sub structure of the side walls and floor which are usually linked together and work as one unit. The function of most lock foundations and floors is not so much to withstand downward pressures as to prevent uplifting from the effect of the downward thrust of the walls and the upwards thrust of water pressure below the floor

The basic types

All marinas are of two types, namely, tidal or enclosed. The latter type are again divided into semi-tidal or impounded. One variety of the semi-tidal basin is the half-tide marina where water is impounded so as not to fall below a minimum level and where 'open' conditions occur at the top of the tide allowing boats free entry and exit 'over the bar'. This is an interesting and economic solution to some conditions.

Another compromise is the partially-locked marina where some boats (usually the larger) berth in an outer basin, protected but unlocked, whilst the remainder pass through a lock into an 'inner sanctum' usually having fixed piers and walkways. Brighton is the largest example of this type.

In marina design most locks and their component parts are light and small compared with those found in even moderate sized commercial docks and harbours. However they will nevertheless represent a substantial proportion of the total development cost and their efficiency will determine the success of the project.

Marina locks can be of a variety of lengths, depths and types but except for these variations and those of floor section the differences between locks is largely the result of differences in gate design.

N.B. The following is a brief description of the basic types of closure, but individual and hybrid forms abound. One pattern cannot be recommended being better than another for all will depend upon individual circumstances. Most locks take their name from a description of their moving components —those parts which open and close. These are broadly speaking divided into two principal categories or families: gates and caissons. The difference between them and the relative merits of each may be summarised as follows:

1   Gates

These are single or double flaps which may be either vertically or horizont-ally pivoted. Those with vertical axes need recesses to house them when open. This is expensive and adds to the length of the side walls. Gates are usually smaller and less expensive. Being more two-dimensional than caissons they are more subject to racking and being fouled by obstructions. They can accommodate pedestrian but not usually vehicular traffic across their top transoms

2   Caissons

Whereas the gate family may be likened to double or single *hinged* doors, caissons may be likened to *sliding* doors. When operated they close off the water channel but when opened they slide, roll or float back into deep slots or housings in the lock wall. They are generally more costly than gates if compared size for size, but can more readily accommodate vehicular traffic across them when closed. Most types slide or roll but larger examples float or are partly buoyant. Some are known as Ship Caissons for they take on a section not unlike the hull of a vessel. Caissons can be made to act more easily in both directions than gates and they are less likely than gates to be obstructed by debris

3   Two-leaved gates

Much the most common general method with gates is to use two symmetrical leaves, shutting on a centre-line which, when closed, form a blunt arrow pointing towards the impounded water. There are two main types:

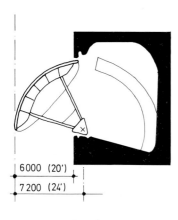

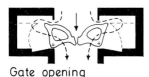

6 000 (20')
7 200 (24')

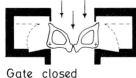

Gate closed

Gate opening

Gate opening

Gate opened

6.74 Method of working sector gates

6.75 Chichester Yacht Basin: the inner pair of lock gates. The marina could not have been developed economically without a lock as the excavation required for a tidal basin would have been very large and siltation a constant problem and expense. The sector-gate lock is very fast-acting and has proved highly successful. The free flow period over high water when both gates are open lasts for up to 5 hours

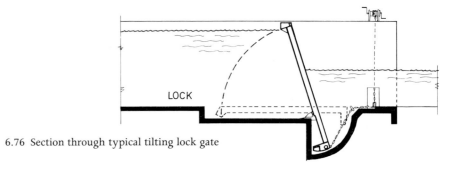

LOCK

6.76 Section through typical tilting lock gate

3.1 Horizontally ribbed. With this type, horizontal members join the heel and mitre posts with, perhaps other intermediate verticals, the gates acting on the principle of the arch

3.2 Vertically ribbed. Here, all the members are vertical except the top and sill members. The principle is different too, the forces being transmitted from the verticals to the upper and lower transoms which act as beams

4 Mitre gates
This term describes any type of two-leaved gate which has a curved front face designed as an arch to meet the unequal acting forces it is countering as opposed to the simple, earlier flat-faced 'doors'

5 Single gate locks
These locks simply impound water and are found in some commercial docks, but their severe limitation of access to an hour or so at each tide prevents their adoption for pleasure craft

6 Storm or flood gates
These are not part of the entrance lock but a safety exit allowing impounded

water out, thus reducing the water level in time of flood or danger. They are mostly used in areas which are subject to cyclones, tropical storms, exceptional tides or rapid flooding

7 Sector gates

These double-acting or reversible gates are equally effective whichever side the head of water may be. First adopted in Sweden in the 1920s the gates are struck to the arc of a circle, the resultant forces passing through the vertical axis about which the gates revolve. The concentration of thrust upon the 'hinge' allows the gates to move with the water rather than strain against it. The usual water pressures being absent, other devices are used to provide a watertight interface at sills and mitres

8 Falling or tilting gates

These, sometimes called Flat Gates, are usually a single leaf pivoting at sill level. They are more common for canal use than harbours. There are several varieties including the Box Falling Gate, named after Edward Box, a Tyneside engineer, and the Cherre System commonly used on French canals. One drawback is the rather complicated floor profile necessary for the gate recess and counterweight pit

9 Swinging caisson

This type is midway between the Caisson and Gate families and in design is not unlike a broad-beamed, single leaf gate. Within it is an air chamber

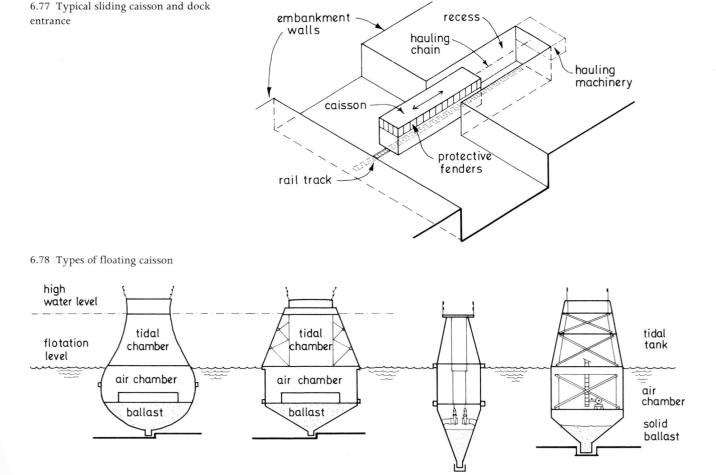

6.77 Typical sliding caisson and dock entrance

6.78 Types of floating caisson

giving bouyancy and its width will allow traffic to pass at quay level. As with the gate type the caisson swings back into a niche-like recess

10   Sliding caisson

These have keels or rubbing plates upon which they are pulled or pushed from their housings across the lock into the closed position above the sill. Air is often pumped into bouyancy chambers within the caisson to lessen friction and reduce the power necessary to operate the lock. The method is very simple and can be used effectively from quite small openings to massive structures such as the Gladstone Graving Dock, Liverpool, whose 40·23 × 15·85 × 7·62 m (132 × 52 × 25 ft)-caisson is of this design

11   Rolling caisson

With this type the sliding way is replaced by a track and rollers which are either on the underside of the caisson or on the bed of the housing and the sill. A groove in base of the caisson may straddle the sill to act as a grid and to make the junction watertight

12   Floating caissons

Nearly all caissons are made more bouyant to lessen their dead weight and help them move more easily. Floating caissons rely entirely upon floatation and have no guides or rollers. There are two main types, box and ship caissons. The former are rectilinear, shaped like a book and have sufficiently large air cavities within them to allow complete floatation. Being a simple shape their construction and their housing is simple too and therefore the least expensive of this category. The ship caisson has a more complicated profile not unlike a navigable vessel. Their bouyancy is controlled by floatation chambers and their 'keel' runs along a groove, step or sill. They are usually designed for large openings where their weight and size overcomes their tendency to instability in bad weather conditions

Lock management

Any lock, by its nature and function restricts the free passage of craft into and out of the marina. It does this by concentrating vessels to a particular place and to particular times of the tide. Whilst this is generally held to be a nuisance there are certain advantages of control and safety which often remain unexploited. Where there is a lock, marina managers can ensure that information about tides and weather conditions are made available to all who leave the harbour and that marina regulations, conditions, dues, tide times, free berth positions and other local information are passed to incoming visitors.

The ability to keep a check on craft in and out is also a great help in ensuring that craft are not removed without the owners consent. At St. Katharine's Yacht Haven in London a printed card unique to each craft, acts as an information booklet and pass. Without it no vessel can pass through the lock.

Lock safety

Locks can be dangerous both for crews and the general public. Staff must be vigilant, lifebelts and boathooks must be easily seen and quickly available and handrailing adequate and properly maintained.

6. Engineering: Check list

Coastal engineering

Examine the overall hydraulic and coastal characteristics
Assemble available data regarding off-shore conditions, generally

        tidal range
        siltation and erosion
        storm frequency and severity
        stability of bed conditions
        flow of tidal and shore currents
        long-shore drifting

Where facts are inadequate arrange for survey and engineering data studies
Decide upon type of recording equipment on site and length of trial(s) (see pages 159–162)
Form harbour shape to discourage resonance
Consider the value of a spending beach and/or outer protective harbour
Investigate options of harbour shape for maximum protection

        for economic berthing plan
        to maximise wave-height reduction
        to minimise costs

Allow design to grow from coastal topography and sea-bed conditions
Determine position and form of entrance and its effect on craft in all weathers
Consider overtopping problem re seaward profile
Consider retention of specialists for survey and hydraulic data studies
Explain value of engineering studies to client

Dredging

Ascertain type of dredge material from initial investigation report
Calculate
  surface area
  depth of dredge
  total cube removal
Select equipment to suit
  material
  shape of dredge
  method of disposal
(See principal dredger types pages 171–173)
Will working at low tide
  prolong contract?
  increase costs?
Consider deep-dredge theory in light of future maintenance dredging costs
Is dredged material suitable for fill?
What is the distance to the nearest dumping area?
Answer the 5 questions under dredging on pages 168–170
What is the dredge^3/reclamation^3 equation?
Contact dredging firms and get their general advice
Select short list of firms for spot quotations
Consider the options on plant for pumping ashore
Approach dredging firms about initial plus maintenance dredging deal

Breakwaters

Physical conditions
Determine necessary wave height reduction
Examine tide and meteorological tables, storm frequency, etc.
Establish water depth, type of bed material and load capacity
Consider effects of siltation, erosion, drifting, etc.
Calculate weight/deadload against likely forces
Design seaward profile to minimise overtopping

Position
Relate length and configuration to total enclosed water area
Consider basic harbour type
  off-shore
  recessed etc.
Calculate total length, cross-section, quantity of material.
Use any permanent shoals, spits or reefs where possible

Form

From calculations determine basic classification

Select sub-type

Will breakwater carry vehicles, pedestrians?

Decide for or against public access, fishing, etc.

Could breakwater support buildings?

Consider appearance from off-shore, marina, coastline, etc.

Determine section, profile, geometry

Will breakwater obscure horizon line? Consider silhouette

Consider appearance

       function

       cost

Establish position re local materials

              transport

              construction methods

              equipment

Will any land area be needed for fabrication

                  equipment

                  materials

                  railway etc.?

Relate base spread and sectional geometry to bearing pressure of bed and overturning moment

Consider temporary (perhaps floating) breakwater during permanent construction

Miscellaneous

Remember services: lighting, columns, ducts, handrailing and trim generally.

Piles

Consult piling specialists

Arrange for survey of bed material and organise trial boreholes if necessary

(see 'bed materials', page 190)

Find out which type are successful locally

Is erosion or scouring a problem?

Will tidal range necessitate overlong piles?

Consider splicing for long piles

Select material for piles in light of strength and length

               durability

               method of placement (see pages 187–188 for methods)

               holding capacity of bed

               cost

Can extended piles support roof to covered moorings?

Select pile system in light of, locked or tidal basin

             fixed or floating moorings

             need for noise control during placement

Determine acceptable tolerance of placement—verticality and accuracy of grid, particularly with floating systems

Is ice likely to lift piles or bed material pull them down?

Decide on fenders or vertical battens where impact from craft is likely

Are piles vulnerable to rot, borers, algae? If so decide treatment

Consider pile capping—either permanent or temporary (during driving)

Is pile oscillation or similar problems likely to be encountered?

Bulkhead and Quay Walls

Type

Determine the relative heights of soil and water

              the tidal range

              the quality of retained soil

              all static and dynamic forces acting on the wall

Decide between gravity and sheet types

Consider gabion type in poor bed conditions

Calculate need

       size

       type

       position of drainage through wall

Is a coffer-dam necessary?

Will sheet piling stability be by tie rods

                 anchorage

                 deadman?

Position
Consider need for land gain
Calculate cube of fill and/or dredge
How will moorings be located in relation to quay wall?
Will a permanent walkway be necessary?

Length
Calculate relative bulkhead/breakwater lengths and costs for each viable alternative

Appearance
Select materials for suitability
                              availability
                              colour
                              texture
                              strength
                              maintenance
                              planting possibilities
Should sheet piling be clad or unclad
Consider lighting
            handrails
            edge profile
            bollards
            rings

Costs
Work out cost options for gravity and sheet types
Calculate what proportion breakwater/quays are of total engineering costs for each viable
      alternative

Siltation and erosion
What data is available about site conditions? Tap local knowledge
If data is inadequate organise survey
                                    data collection
                                    recording
analysis: consider hydraulic model
Will development worsen conditions within the harbour
                                    nearby?
What volume, frequency, cost of maintenance dredging is acceptable?
Determine type of deposit: mud, marine clay, etc.
Contact dredging firms about maintenance dredging contract

Ice
Examine effect on canals
                  lakes
                  quarries
                  reservoirs
                  rivers
Determine nuisance value from local knowledge
Will piles be uplifted? Is special edge treatment called for?
Consider aeration system, permanent or mobile.
Could conditions warrant ice breakers? Consider type
(See ice prevention methods, page 198)

Locks
Can initial final harbour size warrant cost of lock?
Calculate lock expenditure in cost-per-berth terms
Determine basic type:
                        half-tide
                        impounded etc.
Is siltation a problem and will locking lessen or worsen this?

Calculate locks probable throughput and speed of lock cycle
Consider entrance conditions
    channels
    lighting
    marking etc.
Can access to the public be given for viewing
                  for continuity of waterside path?
Investigate possible gate types
Consider lock operation
    control
    keeper's office
    management
    maintenance
    notice boards
    weather board

6. Engineering: Bibliography

Aggregates—The Way Ahead. Report of the Advisory Committee on Aggregates, H.M.S.O., London 1976.

American Society of Civil Engineers, *Report on Small Craft Harbours*, Manuals and Reports on Engineering Practice No. 50, The Society, New York 1969.

Atkinson, Ian, 'Brighton Sets the Trend in Breakwaters', *Contract Journal*, Vol. 254, 12 July 1973.

Block, J., 'A Pile Driving Rig for Marine Structures, *Proceedings of the Institution of Civil Engineers*, November 1972.

British Standards Institution:
Foundations C.P. 2004: 1972.
Foundations and substructures for non-industrial buildings of not more than four storeys C.P. 101: 1972.
Protection of buildings against water from the ground C.P. 102: 1963.

Building Research Station:
'The Assessment of Winds Loads,' *B.R.S. Digest* No. 119, July 1970.
'Choosing a Type of Pile,' *B.R.S. Digest* No. 95 (Second Series) July 1968.
'Damp-proofing Solid Floors,' *B.R.S. Digest* No. 119, July 1970.
'Fill and Hardcore,' *B.R.S. Digest* No. 142 (Second Series) H.M.S.O., London 1972.
'Materials for Concrete' (Waterproof cements etc.) *B.R.S. Digest* No. 5 (Second Series) 1960.
Overtopping of Earth Flood Banks, B.R.S. Film (ten minutes) 1961.

Canadian National Organising Committee/Coastal Engineering Research Council, *Abstracts of Papers Presented at the Thirteenth International Conference on Coastal Engineering, Vancouver B.C.*, National Research Council, Ottawa 1972.

Chaney, Charles A., *Marinas: Recommendations for Design, Construction and Maintenance* (Second Edition), National Association of Engine and Boat Manufacturers Inc., New York 1961.

Clarke, R. H., 'Environmental Conservation and the Municipal Engineer', *Journal of the Institution of Municipal Engineers*, August 1972.

'Coastal Engineering': Data Sheets Nos. 10.03 alternate to 10.0, *Civil Engineering and Public Works Review*, November 1970 and alternate months onwards.

Construction Industry Research and Information Association, *Oscillation of Piles in Marine Structures*, Report 41, The Association, London 1972.

Cornick, Henry F., *Dock and Harbour Engineering*, Charles Griffin & Co. Ltd., London 1958. (Four Vols.) See: Vol. I, *The Design of Docks*.

Cross, Ralph H. and Sollitt, C. K., 'Wave Transmission by Overtopping', *Proceedings of the American Society of Civil Engineers* (Waterways and Harbour Division) Vol. 98, Paper No. 9137, August 1972.

'Dampness in Buildings', *Building Trades Journal*, 8 September 1972.

Department of the Environment, *Accuracy in Setting Out*, Film (seven minutes).

Department of the Environment, *Hydrography as applied to Dredging,* Film (ten minutes).

Department of the Environment, *Overtopping of Earth Flood Banks*, Film (ten minutes).

Dredging and Construction Co. Ltd., The, *I Drink the Earth*, The Dredging and Construction Co. Ltd., Kings' Lynn, Norfolk, England—A Subsidiary of Royal Dutch Dredging Company 'Adriaan Volker' Ltd., Rotterdam, Holland.

Fairweather, L. and Sliwa, J. A., *A.J. Metric Handbook*, The Architectural Press Ltd., London 1970.

Ganly, P., 'Design for Brighton Marina in Sussex', *Civil Engineering and Public Works Review,* March 1971.

Ippen, A. T. (ed.), *Estuary and Coastline Hydrodynamics*, McGraw-Hill, New York, 1966.

Institution of Structural Engineers, The, *Earth Retaining Structures*, Civil Engineering Code of Practice No. 2.

Kind, C. A. M., *Beaches and Coasts*, Edward Arnold Ltd., London 1972.

Lee, C. E., 'On Wave Damping in Harbours', *Proceedings of the American Society of Civil Engineers* (Waterways and Harbours Division) Paper No. 6241, November 1968.

Lundgren, H., 'Coastal Engineering Considerations'. Paper presented at the *Symposium on Marinas and Small Craft Harbours*, Department of Civil Engineering, University of Southampton, April 1972.

Minikin, R. R., *Wind, Waves and Maritime Structures* Charles Griffin & Co. Ltd., London 1950.

Oliver, A. C., *Timber for Marine and Fresh Water Construction*, Timber Research and Development Association, London 1974.

Painter, R. B., 'Hydrological Research and the Planner', *Surveyor*, 11 December 1972, Vol. 140.

Perfrement, Denis, 'Piling Techniques for Maritime Works, *Civil Engineering and Public Works Review*, March 1971.

Sargent, J. H., 'Dredging Equipment and Techniques'. Paper presented at the *Symposium on Marinas and Small Craft Harbours*, Department of Civil Engineering, University of Southampton, April 1972.

Seiffert, K., *Damp Diffusion and Buildings: Prevention of Damp Diffusion Damage in Building Design*, Elsevier Publishing Co. Ltd., London 1970.

Stevens, Sir Roger (Chairman), *Planning Control Over Mineral Working: Report of The Committee Under The Chairmanship of Sir Roger Stevens*, H.M.S.O., London 1976.

Stickland, I. W., 'Site Investigation and Hydraulic Model Studies'. Paper presented at the *Symposium on Marinas and Small Craft Harbours*, Department of Civil Engineering, University of Southampton, April 1972.

Vollmer, Ernst, *Encyclopaedia of Hydraulics, Soil and Foundation Engineers*, Elsevier Publishing Co. Ltd., London 1967.

Webber, N. B. and Harris, R. J. S., 'Floating Breakwaters'. Paper presented at the *Symposium on Marinas and Small Craft Harbours*, Department of Civil Engineering, University of Southampton, April 1972.

Weigel, R. L. *Oceanographical Engineering*, Prentice-Hall International Inc., Hemel Hempstead 1964.

Wyatt, T. A., 'A Review of Wind Loading Specifications', *Structural Engineer*, Vol. 49, May 1971.

7 Landscaping

In recent years standards of many built forms such as housing, industry and public buildings have been progressively upgraded. Whilst most professionals would agree that this has not gone far enough and while external space-treatment standards have not remotely reached those for interiors, there has nevertheless been improvement in some areas and a general recognition that external layout is significant. Meanwhile, demand for recreational land is still increasing and this has influenced landscape opportunity, for the more valuable the commodity more respect it is treated with.

Landscape expertise should be employed nationally as well as locally in the recreation scene and the strategic location of water-side sites on the coast, in National Parks and other places of natural beauty should not be left entirely to the planners, for the services of landscape designers can be as important strategically as they are in the local environment.

The re-use of derelict areas for water recreation and the consequent need to reclaim land and find ways to create a new, man-made but attractive landscape setting will undoubtedly demand the skills of an expert within the marina development team. Probably the greatest pressure on land today is the demand for recreational use. As a result much land is coming out of agriculture and must often be reallocated for dual or multiple purposes; forestry interspersed with picnic sites; water catchment areas crossed with trecking routes and nature trails; water and its adjacent land being shared between hitherto competing sports. Furthermore, public pressures on land for recreation cause much greater wear and tear than before, so that a form of reinforcement needs to be planned.

The growth of the conservation movement and the increased interest in environmental issues will undoubtedly, and quite rightly, demand a high level of design in the majority of areas where marinas are proposed. Any developer, public or private, who attempts to get away with a sub-standard development is not likely to be saving money and more likely to be embarking upon an expensive, embarrassing and time-consuming exercise even assuming that such a proposal overcomes the initial hurdles of an outline planning approval and a successful Public Inquiry.

The role of the landscape architect

Private developers, local authorities and others concerned with the siting and design of marinas should be fully aware of the value of the professional landscape architect within the design team. His or her rôle has changed considerably during this century, yet otherwise knowledgeable people still confuse them with gardeners, landscape contractors or external decorators.

7.1 Taking advantage of local species can give sensational results. This poinciana thrives close to the sea at Key West, Florida

They are of course none of these, but qualified professionals who rely entirely upon being commissioned to provide a service to their client on a fee-earning basis. This service has widened greatly ever since the Second World War. Those working for large commercial organisations, government departments or county councils may be concerned with environmental, conservation and development proposals at a national or regional level. Their expertise in the development of water recreation is particularly valuable and the range of their expertise cannot adequately be covered by the architect, planner or engineer.

If the hard-headed marina developer is not convinced of the necessity for professional guidance in landscaping during the physical development of the project he may be persuaded of such a need at the Public inquiry stage. No aspect of development impresses the public more than the knowledge that the eventual scheme will harmonise well with the existing environment. If any prospective developer can display expertly prepared evidence backed by professional testimony he is much more likely to win the confidence of the public and the inspector.

To get the best from the landscape architect he must be commissioned at the very beginning of the planning period. He is commonly considered as the per-

7.2 The attractions of coastal vegetation exemplified near Nassau, Bahamas

7.3 Shadow pattern adds to the effect of
these young holm oaks in New Orleans
Yacht harbour

7.4 Grass and trees taken almost to the
water's edge at Coral Harbour Club and
Marina, Nassau, Bahamas

son to tidy up the site and do some planting before the opening ceremony,
which is absolutely the wrong approach. Retained in the early stages he can
influence the positioning and mounding of dredged material, incorporate
surface drainage and sprinkler systems, advise on the retention or removal
of existing trees and shrubs and start planting in areas unaffected by construction—the shelter belt perhaps. Most important, he can work with the
team during the formative design period. Too often he suffers a last-minute
commission with an exhausted budget, during the wrong season, with the
plan already fixed and topsoil buried under several metres of clay.

General landscape design principles

Handling the marina site

Whether they are sited on the coast or inland, new marina projects will be
imposing themselves upon an environment which, in the majority of cases,
will be predominantly rural. In the interests of both good public relations
and good design the careful integration of the project will greatly influence
the scheme's success in terms of design and acceptability.

Whilst the siting of a new marina in coastal or rural settings may produce
the greatest impact, development in urban situations often proves equally
challenging. Landscaping, in the broadest sense will assist in fitting the
project successfully into its surroundings whether they are a low-lying and
remote estuary or the hard-edged grandeur of a Georgian dock.

7.5 Landscaped parkland adjacent to the Municipal Marina, New Orleans, Louisiana

It is sometimes forgotten that landscaping is not necessarily connected with planting but may deal almost solely with hard surfaces, changing levels, the selection of materials and the skilful manipulation of the spaces between buildings. The water surface itself may offer many qualities—choppy, reflective, opaque, transparent—each capable of exploitation by a creative designer.

The modelling of the land and the choice of species require particularly careful handling in low-lying and coastal districts. Landscaping, itself a fusion of skills, must combine with the buildings, equipment, boats, water, roads and car parks, knitting them together and enhancing them individually. The appearance of large buildings may be softened by climbing plants or their impact lessened by a tree screen.

The basic elements

The objectives of good landscaping are simple and change little in time or place. The creation of space, setting shelter, colour, contrast and amenity are constant, but the means of achieving them vary according to climate, soil and finance.

Like the architect, the landscape consultant must bear in mind several factors at the same time, some of which will be complementary and others competitive. If there is one issue in landscape work which dominates others in importance it is that of scale. If the scale is well chosen in terms of space, plant selection and the relationship of the principal elements then it will be most unlikely that the design will fail. Conversely most schemes that do not succeed are guilty of shortcomings in this respect.

The relationship between landscaping and the water's edge will alter with the location and type of marina. It seems natural in warm climates for luxurious planting to border and even overhang the water, whereas it is better in northern climates to keep it back and provide a more hard-wearing treatment. Inland marinas and built-in coastal types allow the water beyond the harbour to be seen through trees and planting, giving a very different effect from the off-shore marina where planting stops behind the land/water interface to give a hard-edged immediacy between sheltered and open water.

At its best planting should look natural rather than imposed and this applies to location as well as the choice of subject. At its worst it can look foolish: such features as a tree on a breakwater, bedding plants in quayside tubs or hanging baskets on floating walkways are entirely inappropriate.

Landscaping should not be confined to the more obvious areas of garden and lawn. Car parks for example can look a lot worse empty than full when they are unrelieved asphalt, but they can with advantage be reduced in level, surfaced with grassed blocks, enlivened with groups of trees, screened with planting or walls and treated in all manner of imaginative and effective ways.

Landscape types

A general review of the kinds of site normally suitable for marina projects has been dealt with in Chapter 2. The landscaping opportunities may be divided into various types of spaces in much the same manner as different types of buildings, that is to say, by function, relationship between component parts and a regard to the scale and character of the natural and built environment influencing the site and being influenced by it.

Some of the more important areas needing consideration are as follows:

1 Climatic category
Whether temperate, tropical, Mediterranean etc.
2 Climatic characteristics
Prevailing winds, temperatures and rainfall
3 Geology
Soil and subsoil type, permeability, drainage water table, height above sea level
4 Water category
Whether coast, river, estuary, lake, quarry, reservoir or canal
5 Water type
Whether salt, fresh, brackish, wave height, tidal range and depth etc.
6 Character of site
Aesthetic considerations, pollution, noise
7 Topography
Levels existing and proposed
8 Soil characteristics
PH value, particle size, drainage texture
9 Visual aspects
Orientation, aspect, vistas and views from site/into site/within site
10 Planting
Suitability of types, shelter belts, screen clumps, groups etc.
11 Existing form
Design potential, masses and voids, existing structures, siting, contours, scale, external spaces
12 Zoning
Planning category of site and surroundings now and proposed
13 Function
User requirement, client's brief

The landscape architect does not, generally speaking, choose his site, he inherits it when appointed, complete with the existing characteristics outlined above. His task, where possible, is to conserve and exploit its good attributes and to eliminate or minimise the bad. The quality of site in

7.6 Lake Monroe, Florida. Inland sites are much easier to landscape as the perils of salt and wind are absent

appearance and potential is likely to vary considerably. To some extent the less promising the initial material, the more rewarding the final treatment: conversely, very attractive sites pose problems within their boundaries and the impact upon their surroundings. As demand for recreation grows, certain sites of outstanding value are being opened for leisure activities. Inland areas such as Lake Windermere in the Lake District and Loch Lomond in Scotland are examples of National Trust property in Britain where marina development has recently been allowed. Stretches of 'heritage coastline' are controlled by the local planning authority and the Countryside Commission and permission for leisure harbours is rarely given. Within such areas planning control is naturally strict and accompanied by restrictions regarding overall size, class of coast, height and form of buildings and type of materials. Landscaping in such areas is largely a question of retaining the existing ambience by controls on topography, planting, scale and materials.

Planting for shelter

On exposed sites, particularly on the coast, there will be a need both to provide shelter and to take advantage of it. It is not unlike a battle, the enemies being wind, salt and sand. The landscape architect provides cover by mounding, screening and utilising existing buildings. Then he deploys his planting—toughest troops first in full blast of the enemy, then the second-line species and so on. The sacrifice line of planting thus protects the more tender species. Although prevailing winds may stunt growth on the windward side by reducing new leaves and shoots, if a suitable species of tree has been chosen (see 7.A, 7.B) this should do no more harm than checking the windward growth. Windward facing slopes rising direct from water level are the most difficult to plant satisfactorily, particularly in coastal areas, and may be best avoided for it is rarely found in nature. Lessening the impact of wind is everything with most marina sites. The effect of the wind off the sea cannot be overestimated for the combination of wind and salt can burn the hardiest plants a kilometre inland. Artificial screening of lath or wattle may be necessary, particularly for young plants. Planting should be closer together than for sheltered sites. Nothing is gained by planting large specimens as young ones outstrip them in a few years.

Trees as shelter belts should be planted at right angles to the prevailing winds. It should be borne in mind that open textured planting filters wind, reducing its force, and that the broader the planting belt, the more effective the filtering effect—solid planting creates turbulence on the leeward side. Shelter belts give wind reduction for a distance of 10 times the height of the trees.

The function of all planting should be clearly understood and relate closely to the whole design concept for the site if both the visual and the climatic effects are to be achieved. Screens and hedges can be used as follows:

1 To relate buildings to the site and to each other, and to link external spaces

2 To demarcate boundaries and areas

3 To accommodate changes in level and ground modelling

4 To give privacy, screening and visual barrier and security

5 To shelter from wind, dust, strong sunshine and—to some extent—from noise

6 To form spaces by enclosing or dividing areas

7 To direct pedestrian circulation

8 To channel views to or away from buildings or objects

9 To provide contrast in form, texture or colour, with buildings, pavings or water

10 To contrast with or complement sculpture

Although temperatures in Britain are rarely very low, wind-chill (wind rapidly removing heat from the body surface) makes the weather seem colder. The need for shelter sometimes conflicts with that for sunlight, particularly in winter when the sun is very low. The balance of these needs will vary from region to region. As wind control is important, an understanding of wind behaviour is necessary. The source of wind greatly influences its hostility and as east and north winds are harsher than the

7.7 Comparative effect of solid and perforated screens as wind breaks

(a) Solid screen gives shelter for a limited distance as shown

(b) Trees act as wind screen. Air passing below foliage gives local high velocity but the general effect of this arrangement is to give shelter further downwind than would be the case with a solid screen

(c) Perforated screen. Effect of air passing through perforations is to increase the downwind distance for which wind velocity is reduced

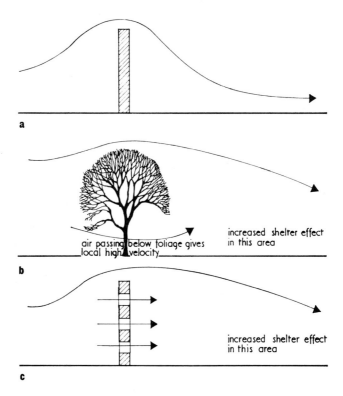

more frequent south-west winds, their exclusion may prove more necessary.

With urban marina sites, buildings will often divert the wind's natural course, funnelling and increasing its force until intolerable conditions are sometimes created. Wind-tunnel testing of a scale model will provide the most accurate prediction of future conditions on site. Long, low structures such as high walls or terraces of buildings provide sheltered areas with some alleviation on the windward side and ten times their height on the leeward side (see 7.7). However the ends of such structures are particularly turbulent, the wind reducing towards their centre. If the wind direction is parallel to the structures no shelter is provided and funnelling may occur. Solid structures always induce turbulence and openings in solid barriers act like weirs. Screens are sometimes beneficial in breaking the force at troublesome points. Shelter from trees and structures can be very effectively combined to help to overcome wind exposure. A decision to use trees, however, requires allocating space at the design stage as they need space in plan and section, sufficient for their ultimate size.

For quick screening effects existing features should be reinforced wherever possible—ground slopes, hollows, hedges, tree groups or woods. Graded earth banks can be used as a start, perhaps by employing dredged material. Their lees can shelter quick-growing hedges shrubs and trees (see 7.A, 7.B). Short fencing such as wattles and timber poles, give contemporary screening for visual purposes and plant protection. Decorous planting can be effective where summer activities are the main concern and there is nothing to screen in winter. Boat parks may need slower growing evergreens and other more permanent screens.

Landscaping in urban marinas need not try to be too natural and soft. Hard landscaping, screen walls and fences with adjacent trees or planting will probably look better in high density urban sites than grass and flowerbeds. Screening can take many forms depending upon function and cost and any of the following could be considered:

1   Walls
Open, solid or permeable
2   Screens
Canvas, tarpaulin, sailcloth etc.
3   Glassfibre or polycarbonate fibre
Resistant to vandalism
4   Asbestos cement
Inexpensive and quite attractive if framed
5   Woven fences
Wattle, lath or hurdles

Screening can, of course, take the form of changes of level, undulations, mounding and lowered levels for car parking etc. The larger the object to be screened the greater the problem. Boat stackers and other large structures may require their impact to be lessened. Nothing much can be done by way of reduced levels because the boat transporters will need a level surface from the quay side to the floor of the stacker. In these cases screening by tall trees or softening the structure's surface by creepers may be alternative methods.

Plant selection

One of the difficulties of exposed and coastal sites is that some species survive the harsh conditions more because of their prolific growth than their natural adaptation and may cause more problems than they solve— elder is an example. Similarly, laburnum survives well but is poisonous to cattle, box and cherry laurel to horses and sea buckthorn berries to birds. Branches in water can foul propellers and damage hulls, overhanging branches may catch masts and roots can erode banks and spoil edging. Nevertheless there are compensations. The climatic influence of the sea is usually good and humidity and daylight are seldom problems. Inland and estuarial sites often provide above average soil conditions.

It is not possible to be too precise about the geographic suitability of species, for conditions can vary so much within a few miles. The orientation of the shoreline to the coldest winter winds will affect both choice and location. The depth, quality and drainage of soil and subsoil also determine the choice of subjects.

Marram and sea lyme grasses are excellent for holding and reclaiming shifting sand. Evergreen trees and shrubs are most useful as winter shelter although deciduous trees branch to the ground better. The evergreen or holm oak does well near the sea, growing from 12–18 m (40–60 ft) if it is not too cold. It can also be grown as a windbreak hedge to almost any height. Austrian, cluster and other maritime pines are coastal subjects and elms are often seen almost down to the sea although they do not hold well into really wet soil. Sweetbay, arbutus, willow and sycamore all survive fairly rigorous coasts and the mountain ash is particularly good. Gorse and tamarisk grow right down to watermark and some junipers and brooms do well. Most bamboos prosper and can be sensationally good on some inland lakes and rivers and tea trees (Lycium) will even accept being swamped at high tide by brackish water. For covering and climbing the common ivy, Virginia creeper and honeysuckles are good and Russian ivy, being so quick and rampant is often too good. Escallonia is excellent and can be trimmed or grown as a hedge and rosemary will grow to a good size.

Before choosing any species, however, it is always valuable to get advice from a good nurseryman, familiar with the vagaries of local soil conditions and climate and already experienced in the cultivation of plants for his particular area. *The degree of maintenance which different species require should also be considered at the design stage* (see pages 223–224). Needless for to say, before planting begins all ground should be well prepared. Ensure that appropriate top soil is used because *plants should not have to battle against adverse soil conditions as well as climate.*

Suitable species
The following are lists of plants *generally* considered suitable for the conditions described in their respective headings. Soil conditions, temperature or even cost may eliminate some and there may well be others which, for some areas, may be even better than those listed.

Common names vary from region to region. The Latin names used here in general follow those recommended in C. Chicheley Plowden's manual of plant names.[1]

1. Plowden, C. Chicheley, *A Manual of Plant Names*, Allen & Unwin Ltd., London 1968.

7.A Guide to the selection of species for coastal sites

Coastal: Trees capable of taking the first brunt

Common name
  Botanical name
douglas fir
  Pseudotsuga taxifolia
cluster pine
  Pinus Pinaster
corsican pine
  Pinus laricio var nigricans
austrian pine
  Pinus rigra austriaca
holm oak
  Quercus ilex
goat willow
  Salix caprea
maple
  Acer platanoides
grey poplar
  Populus canescens
white elm
  Ulmus americana
red thorn
  Crataegus rubens
cypress
  Cupressocyparis leylandii

Coastal: Trees for semi-exposed conditions
alder
  Alnus glutinosus
basket willow
  Salix viminalis
bat willow
  Salix alba
ash
  Fraxinus
monterey cypress
  Cupressus macrocarpa
white poplar
  Populus alba
turkey oak
  Quercus cerris
arbor vitae
  Thuja plicata

Coastal: Trees for slightly more sheltered areas
scots pine
  Pinus sylvestris
aleppo pine
  Pinus halepensis
Canadian jack pine
  Pinus banksiana
atlantic cedar
  Cedrus atlantica
black poplar
  Populus nigra
aspen poplar
  Populus tremula
lawson cypress
  Cupressus lawsoniana
hornbeam
  Carpinus betulus pyramidalis
juniper
  Juniperus
chestnut
  Castanea dentata
fir
  Abies (many varieties)

Coastal: Shrubs capable of taking the first brunt
tamarisk
  Tamarix
common elder
  Sambucus
buckthorn
  Rhamnus frangula
dwarf willow
  Salix herbacea
hazelnut
  Corylus avellana
common juniper
  Juniperus communis
cotoneaster
  Cotoneaster
spindle tree
  Euonymus europaeus
Spanish broom
  Cytisus

Coastal: Shrubs for semi-exposed conditions
buffalo berry
  Shepherdia argentea
strawberry tree
  Arbutus unedo
salt bush
  Atriplex
bladder senna
  Colutea arborescens
flowering currant
  Ribes sanguineum
saxifrage
  Saxifraga
calophaca
  Calophaca

Coastal: Shrubs for slightly more sheltered areas
berberis
  Berberidaceae
hydrangea
  Hydrangea
common dogwood
  Cornus sanguinea
viburnum
  Viburnum
bay tree
  Laurus nobilis
butcher's broom
  Ruscus aruleatus
furze
  Ulex
buddleia
  Buddleia
ceanothus
  Ceanothus
comfrey
  Symphytum
laurel
  Laurus

7.B Guide to the selection of species for inland sites

Inland: Trees for river and lakeside planting
Japanese cedar
  Cryptomeria japonica
gingko
  Gingkoaceae biloba
metasequoia
  Metasequoia glyptostroboides
swamp cypress
  Chamaecyparis taxodium distichum
beech
  Fagus (many varieties)
tupelo
  Nyssa aquatica

stag's horn sumach
  Rhus typhina
catalpa
  Bignoniaceae
mountain ash
  Sorbus americana

Inland: Shrubs for river and lakeside planting
buddleia
  Buddleia
dwarf buckeye
  Aesculus humilis

bamboo
  Bambusa
hydrangea
  Hydrangea
rhododendron
  Rhododendron
azalea
  Rhododendron
viburnum
  Viburnum
auspice
  Pimenta officinalis
celastrus
  Celastrus

Inland: Plants for hedges and screens
Tall screen: height 12·5 m (40′)
Common name
    Botanical name
norway maple
    Acer platanoides
wheatley elm
    Ulmus stricta wheatleyi
dawyck's beech
    Fagus sylvatica fastigiata
hornbeam
    Carpinus betulus pyramidalis
lime
    Tilia euchlora
lawson's cypress
    Chamaecyparis lawsoniana
abor-vitae
    Thuya plicata
spruce
    Picea excelsa
corsican pine
    Pinus laricio var nigricans
japanese larch
    Larix leptolepis
lombardy poplar
    Populus nigra italica
poplar
    P. robusta (short lived) for
    temporary use in shallow soil

Medium screen: height 7·5 m
willow
    Salix alba
whitebeam
    Sorbus aria
box elder
    Acer negundo
field maple
    A. campestre
wildpear
    Pyrus communis
thorn
    Crataegus grignoniensis

Rustic hedge: height 4·5 m
blackthorn
    Prunus spinosa
hawthorn
    Crataegus monogyna
dogwood
    Cornus sanguinae
hazel
    Corylus avellana
elder
    Sambucus nigra
buckthorn
    Rhamnus cathartica
spindle tree
    Euonymus europaeus

Tall (boundary) hedge: height 4·5 m
laurel
    Prunus laurocerasus
portugal laurel
    P. lusitanica
plum
    P. cerasifera
thorn
    Crataegus monogyna
hornbeam
    Carpinus betulus
tamarisk
    Tamarix vars
rhododendron
    R. ponticum
cypress (gold)
    Chamaecyparis pisifera plumosa aurea
cypress (blue)
    C.p.p. squarrosa
cypress (green)
    C.p.p.

Tall (formal) hedge: height 4·5 m
holly
    Ilex aquifolium
holm oak
    Quercus ilex
yew
    Taxus baccata
beech
    Fagus sylvatica
purple beech
    F.s. purpurea
copper beech
    F.s. cuprea
bay
    Laurus nobilis
firethorn
    Pyracantha atalantoides
privet
    Ligustrum ovalifolium
golden privet
    L. o. aureo variegatum
lawson's cypress
    Chamaecyparis lawsoniana
cypress (blue)
    C. I. Triomphe de Boskoop, C. I. allumii
cypress (gold)
    C. I. lutea

Medium (informal) hedge:
height 1·2–2·5 m
barberry
    Berberis darwinii
barberry
    B. stenophylla
cotoneaster
    C. lactea
escallonia
    Escallonia macrantha
escallonia
    E. Donard Seedling
briar rose
    Rosa rugosa

elaeagnus
    Elaeagnus pungens
daisy bush
    Olearia macrodonta

Medium (formal-clipped) hedge:
height 1·2–2·5 m
holly
    Ilex aquifolium
yew
    Taxus baccata
box
    Buxus sempervirens
box
    B. s. handsworthensis
lonicera
    Lonica nitida fertilis
beech
    Fagus sylvatica
purple beech
    F. s. purpurea
copper beech
    F. s. cuprea
privet
    Ligustrum ovalifolium
golden privet
    L. o. aureo variegatum
firethorn
    Pyracantha watereri
cotoneaster
    Cotoneaster franchettii

Dwarf (informal) hedge: height 1 m
barberry
    Berberis verruculosa
barberry
    B. candidula
St John's wort
    Hypericum patulum
lavender
    Lavandula nana Hidcote
Jerusalem sage
    Phlomis fruticosa
rosemary
    Rosmarinus officianalis
pernettya
    Pernettya mucronata

Dwarf (formal-clipped) hedge:
height 1 m
box
    Buxus sempervirens
box
    B. s. suffruticosa (edging)
lavender
    Lavandula nana hidcote

7.8 Grass may be unsuccessful in hot climates unless costly watering is accepted. A hard surface and shrubs would be better

7.9 Planting provides screening and shade to the car park at Long Beach, California

Marina landscaping in tropical climates

It is only possible here to give an outline of factors influencing the landscaping of marinas in hot climates. It is of course a large and complicated subject and those involved in development in these regions will be wise to seek expert advice, preferably from those working in the same area or specialising in this type of commission. Local knowledge is of particular value in tropical regions as areas with sharp climatic changes may be only a mile or two apart—so close that the difference may not be readily discernible on a map, although very important to the plants and people who live there. The dictum 'Landscaping should be done on site' applies particularly to tropical regions where drawing-board designs and plants selected without local knowledge may prove valueless.

Whilst certain basic principles pertain wherever landscape design is practised, each area of the globe affords its own opportunities and presents its own hazards. It is a great mistake to copy European traditions in regions where the climate, species and environment are entirely different.

Climate

Climate is the principal factor, controlling the growth of plants. The term 'tropics' generally refers to climate where the average temperature is near or more than 19°C (70°F) (see 7.10). Tropical climate should not be confused with the climate found in regions on the equator where it is neither very hot nor very cold except at high altitudes—the weather here being covered by the term equatorial. The seasons in tropical regions vary very greatly, some seasons being hot, some cold, some are wet and others dry.

The most likely locations for marinas will, of course, be coastal. The

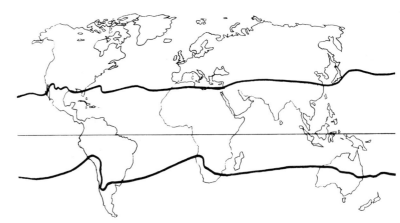

7.10 The areas between bold lines
experience an average temperature of
21·1°C (70°F) or above

average temperature for days and nights throughout the year is 6°C (10°F)
higher at sea level near the equator than in the tropics, but during summer
months there are many more hotter days in tropical areas than ever occur
near the equator. Equatorial nights at sea level are often nearly as warm as
the days, whilst in the tropics a 22°C (40°F) difference in diurnal temperature is not uncommon.

Topography and soil

In many tropical and equatorial regions the coastal and estuarial conditions
follow the same pattern—long lines of shallow waves advancing towards a
beach backed by palms—ideally suited to the offshore or semi-recessed

7.11 Holm oak and mesembryanthemum at
Nassau, Bahamas

7.12 Private moorings seen through palm
fronds near Hurricane Hole, Bahamas

7.13 A row of palms contains the sea view at the Coral Harbour Marina, Nassau

The following trees are subjects which grow well in coastal areas in the tropics

Common name
　Botanical name
cashew nut
　Anacardium occidentale
mudilla
　Bassingtonia speciosa
she-oak
　Casuarina equisetifolia
seaside grape
　Coccoloba uvifera
Indian almond
　Terminalia catappa
tulip tree
　Thespesia populnea
wild tamarind
　Lencaena glanca

The following shrubs are subjects which grow well in coastal areas in the tropics

Common name
　Botanical name
tamarisk
　Tamarix gallica
American aloe
　Agave americana variegata
peacock flower
　Caesalpinia pulcherrima
shoe flower-hibisars
　Hibiscus (many varieties)
Indian lilac
　Lagerstroemia indica
oleander
　Oleanda

7C. Suitable species for tropical marinas.

marina. Coral soils are quite common and need special treatment as they are frequently deficient in magnesium and boron, trace elements which almost all plants need, albeit in small quantities.

As a general rule sandy soils are more in evidence in these regions than either loams or clays and reinforcement with humus or organic fertilisers will be needed, particularly where heavy rains leach out these vital constituents.

Landscape treatment

Shelter is even more necessary in tropical marina environments than in temperate climates. Damage by salt air and strong winds will damage plants severely during storms and hurricanes, but even considerable devastation will be concealed by quick recovery and growth provided plants have not been destroyed or damaged beyond repair. In tropical regions there is more opportunity but less need for colour. Beds of small flowers and banks of very bright colour look inappropriate against the ocean scale and the hues found in most natural vegetation in coastal areas and much exotic planting anyway suffers badly from scorching by strong salt winds.

In many respects the purpose of landscape will differ in hot climates from those in temperate zones. The provision of shade is much more important and species of trees should be chosen with this in mind not only for human comfort but to allow the opportunity of planting subjects which only survive where some shade is available.

Fountains, brooks and lakes are cooling and decorative as well as being a possible source of water for plants and lawns. An irrigation system may not be necessary but adequate water points and distribution by hose or sprinkler should be considered.

Very heavy rainfall over short periods will require an extensive surface water drainage system. Outlets, soakaways, drain sizes and falls may need to be designed with monsoon conditions in mind.

To avoid unpleasant glare large surfaces of white or reflective materials should be avoided, so also should black, heat-retaining paths or sitting areas as these are uncomfortable to bare feet. Where the climate permits or where irrigation is adequate lawns are an ideal surface.

Choice of planting

The principles of planting in tropical marina sites are similar to those in temperate climates but opportunities for exotic effects are greater. It is advisable particularly in coastal sites to incline towards shelter, texture and greenery rather than very colourful displays. Planting should contain or frame views of the water and boats and not screen them off. Suitable species tropical marinas are given on page 227.

Maintaining the landscape

It is essential that maintenance is carefully considered *at the planning stage*. Only then can the correct design and detailing be incorporated so that maintenance costs are minimised. The total landscape design concept, the complexity or otherwise of the hard landscape, planting and ground modelling will all determine a large part of a marina's overheads for many years. A wild or natural appearance for instance will require little attention whilst formal flower beds, rockeries and ponds will make much heavier demands. Materials of high initial cost, such as paving and brickwork, require little attention whereas soft areas of lawn make heavier maintenance costs.

Generally speaking simple shapes, careful detailing of edges and avoidance of exotic plants and elaborate styles are the most practical ways of designing to lessen future costs. Labour costs in landscape maintenance can be reduced as follows:

1 Choose species with maintenance in mind, i.e. those which require minimum staking, pruning, tying etc.
2 Choose plants suited to local soil and climate
3 Avoid annuals, herbaceous borders and exotics
4 Avoid bare earth. Instead, suppress weeds with ground cover
5 Remember that large plants, such as trees and big shrubs, make more impact whilst requiring less maintenance
6 Do not skimp on the basic routine of maintenance particularly during the initial period. Young plants and trees, for example, will require regular watering even in conditions where water abounds, for until their root system taps the water table they will be vulnerable to drought

Equipment

Simple landscape requires little equipment. However where it is necessary buy the best. Remember that equipment itself demands maintenance.

Equipment should be housed in robust, dry buildings preferably with concrete flooring. Allow plenty of room for cleaning and maintenance. It is best to keep equipment for the grounds separate from boat-handling machines and marina or transport workshops.

One member of the marina staff should be responsible for the maintenance and safekeeping of equipment.

Staff

In terms of the marina in general and the landscaping in particular labour accounts for the higher on-cost of all. Simple design needs less labour and less *skilled* labour. Remember accommodation for the staff—even if they are not resident they will require a decent place to change, rest, wash and make tea—not just a corner in the tool shed.

Vandalism

This is a problem of growing importance. Marinas as a whole suffer quite

2. Ward, Colin (ed.) *Vandalism*, The Architectural Press Ltd., London 1973; Van Nostrand Reinhold Co., New York 1973.

badly and planting, garden furniture and tool sheds are frequently a target. Simplicity in design is again helpful whilst security (see pages 329–332) is the overall answer. If the grounds should suffer malicious damage, do not be discouraged but replant as quickly as possible. With regard to soft landscaping, Peter Shepheard's well-known 'Law of Diminishing Vandalism' suggests that persistence in replanting and replacing destroyed plants 'until the children get tired of destruction or learn better' is rapidly repaid.[2]

Aquatic weed control

It would be wrong to make too much of this problem, for although many, if not all, countries are faced with serious aquatic weed problems, most marinas are protected or enclosed and, having a limited water area, are unlikely to suffer unduly. Weeds can affect a marina, however, spoiling the quality of boating by limiting and hampering marina-based craft.

Except for japweed which threatens from time to time to establish itself in Northern Europe, the problem is much more serious in warmer climates. Water weed (and algae) problems tend to be accentuated in highly developed countries by the extra plant nutrients reaching the water both from the high levels of fertiliser used on agricultural land and from sewage effluent. New reservoirs and impounded water suffer greatly, as nutrients are released when fertile land is submerged and water weeds exploit and infest such habitats.

The elimination or control of aquatic weeds has importance beyond the nuisance created in established marinas, for it may be a factor in determining the siting and development of new marinas in particular and boating as a recreation in general. Millions of acres of potentially admirable water for recreation are now sterilised by weeds and algae.

Identification of species

Identification of aquatic weeds is rarely a problem, except perhaps with some algae (see pages 291–292). The following tables 7.D, 7.E and illustrations 7.15, give ten species of weeds most likely to be encountered as a nuisance at harbour sites throughout the world. The first six are primarily

7.14 Fronds of japweed growing in the harbour at Portsmouth, Hants. The weed has attached itself to pontoons: air bladders can be seen glistening in the sunshine

Latin name	Common name	Habitat	Dispersal	Control
Sargassum muticum	japweed	Rocks in low water lagoons; along and just below water line on floating pontoons	Plants have air bladders and can float about in the sea. Possibly introduced from Pacific with imported oysters or on boat hulls. Reproduction by minute eggs liberated from fertile plants	Removal of entire plants by hand-picking
Ascophyllum nodusum	knotted wrack	Rocky shores with little wave action; mid-tide level. A great nuisance to fishing lines and wrapping round the propellers of small craft	Plants have air bladders but do not float far if torn off rock. Otherwise spread by reproductive cells liberated into sea	Usually none but plants can be cut by hand for industrial use
Chorda filum	bootlace	At and just below low-water. Plants can be up to 6 m (20 ft) long, floating to the surface. Even worse than *Ascophyllum* for clogging engines	No air bladders. Reproduction by minute eggs	Too prevalent to attempt except in limited areas
Ulva lactuca	sea-lettuce	Grows luxuriantly in sheltered bays and harbours forming free-floating masses Forms a green mat on mudflats along the south coast of Britain	Reproduction is by minute spores which can travel short distances	Handpicking. Chemicals may help to control heavy infestations but are not normally very effective
Enteromorpha (species)	'grass'	Usually grows with *Ulva* and causes considerable problems. Free-floating masses foul small engines and large quantities of cast-up driftweed will rot unpleasantly on beaches	Plants will fragment and travel fair distances. Reproduction is by spores	Usually none, but chemicals have been tried (herbicides and lime) but are not very effective and kill off too many other things
Fucus (species)	wrack	On rocks from mid- to low-water level. Wrack smells badly and large quantities are cast up onto beaches and into harbours throughout Europe	The illustrated species is bladdered but wrack does not float for long if broken off rocks. Reproduction by eggs as in *Chorda, Sargassum* and *Ascophyllum*	Nothing is very effective

7.D European aquatic weeds

Latin name	Common name	Habitat	Dispersal	Control
Azolla filiculoides	azolla	Tropical waterways, ricefields	Spores and vegetatively	Chemical methods best. Very invasive and difficult to control
Eichhornia crassipes	water hyacinth	Tropical waterways, ricefields	Vegetatively and seeds	Chemical control; handraking on a small scale. Very invasive and difficult to control
Pistia stratiotes	water lettuce	Tropical waterways, ricefields	Vegetatively and seeds	Handweeded in ricefields. Prohibited in many countries, very invasive
Salvinia natans	Salvinia, kariba weed	Tropical waterways	Vegetatively and spores	Chemical control. Invasive and may quickly reach epidemic proportions

7.E Tropical aquatic weeds

7.15 **Below: varieties of aquatic weed**

an annoyance in Europe whereas the last four are mostly confined to tropical waterways.

The story of the genus *Sargassum* (japweed) is particularly interesting and raises issues of the long-range transplantation of marine organisms. Early in 1973 some 30 plants were found attached on a sheltered shore in Hampshire, England. The specimens were well grown, about 1 m long and experts were convinced that they constituted a recent introduction. It is thought that the species may have been accidentally introduced with the spat of the Japanese oyster. The species grows very quickly—up to 10 mm a day and could create not only a threat to European coasts, fishing nets, and the propellers of small craft, but also to species of wild life such as the Brent Goose whose staple diet (Zostera) would be endangered by a species as fast growing as *Sargassum muticum*.

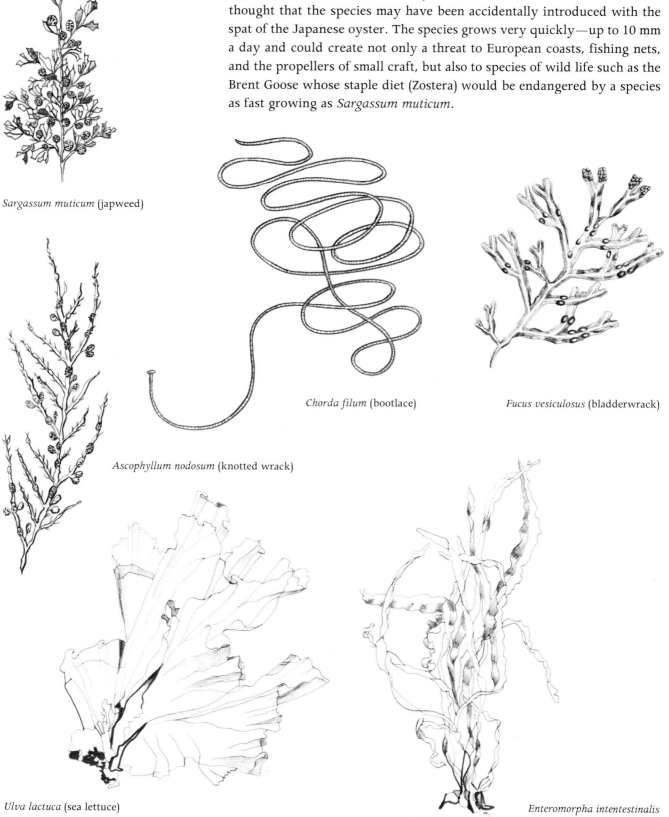

Sargassum muticum (japweed)

Chorda filum (bootlace)

Fucus vesiculosus (bladderwrack)

Ascophyllum nodosum (knotted wrack)

Ulva lactuca (sea lettuce)

Enteromorpha intentestinalis

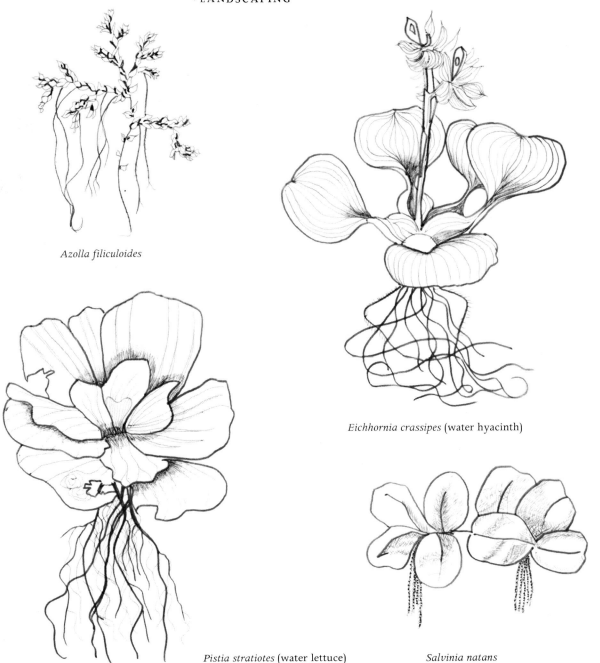

Azolla filiculoides

Eichhornia crassipes (water hyacinth)

Pistia stratiotes (water lettuce)   *Salvinia natans*

Methods of control

The principal methods may be grouped as follows:

1   Manual: cutting, digging out, netting (for floating species)

2   Mechanical: cutting, macerating, dragging out, harvesting for use

3   Chemical spraying: from boats (mist spraying), from bank, aerial (mist or pellets), additives (pellets, emulsions)

4   Biological: mammals, birds, fish, snails, insects, fungi

5   Preventive: division of floating varieties by nets etc.

Manual methods can be simple and effective if small areas are involved. Machines have been developed for channel clearing and in Britain, weed cutting boats, one designed in New Zealand, are in use. In Germany a machine for cutting and macerating water plants and then pressing out 95 per cent of the water content has been designed.

Submerged weeds can be controlled where one can temporarily lower (draw down) the water level. Spraying from the ground, from boats and from the air with mist-blowers has been successful and air propeller driven boats of shallow draft have been developed in America. Aerial operations are swift and effective but drifting is a hazard. Aerial spraying of granules has been successful in Australia and New Zealand.

Biological control is of growing interest. Manatees did not really succeed when introduced into Guiana but cattle keep down water hyacinth in the Sudan. Coypus have been tried in the Cameroons and water fowl can eat large amounts of water plants but are already a pest in Britain. The possibility of ducks and swans has been suggested in Pakistan and Rhodesia. Herbivorous fish can be useful if protected from predators and the common carp and the grass carp have been introduced. Snails, principally *Salvinia*, have been used in America and Puerto Rico, where its ecology has been studied, and the possibility of using this snail in conjunction with herbicide has been considered. A beetle, agasicles, has been used with promising results against alligator weed in America and a grasshopper, *Paulinia*, against kariba weed in northern South America.

The utilisation of the weeds themselves has received attention and 'a curse or a crop' is a phrase sometimes used. When dehydrated, many species can be used as a valuable mulch.

7. Landscaping : Check list

Procedure

Establish client's brief
Obtain clear instructions and requirements
Establish: available finance
                proposed numbers of maintenance staff
                proposed security arrangements

Establish user requirements
Size
Number of marina clients
Seasons of operation
Parking allocation
Likely aesthetic preferences

Planning
Check: use zoning
       planning approvals
       listed buildings
       building and improvement lines
       advertising control
       adjoining land use now/in the future

Legal

Check: easements

rights of way

building and improvement lines

party walls and fences

restrictive covenants

tree preservation orders

drainage

connections to water courses, water supply

listed buildings

firepaths

adequacy of site roads for future traffic

safety precautions

insurance cover for landscape/people

Contract

Establish: form of contract

form, scale and number of copies of drawings

specification

schedule of rates

defects liability interim certificates

retention fund

maintenance agreement/period

Invite quotes

Arrange for clerk of works

Job control

Prepare planting plan

Prepare landscape programme within overall contract

Set up regular site/landscape meetings

Start site diary and record site visits

Order, inspect and approve materials samples

Agree final programme

Agree measurement and daywork procedures

Agree progress reports

Arrange progress photographs

Arrange interim certificates to client

contractor

quantity surveyor

file

Completion

Organise maintenance drawings, inspections and reports

Convene pre-handover meeting

Prepare defects report

Prepare for handover

Prepare final account

Issue final certificates with copies to client

contractor

quantity surveyor

file

Landscape principles

Context and setting

Survey surroundings

Contact neighbouring owners and local people

Establish long/short distance views of site/from site

Take photographs and buy aerial views with photomontage in mind

Establish skyline elements

Consider boundary treatment

include off-shore boundary

remember views from the water

remember security requirements

Consider future development within site

beyond boundary

Establish topography and contours

Consider treatment of high/bulky building(s) with
                                        screen planting
                                        camouflage
                                        creepers
Check entire concept so far for scale and relationship
                                                within/beyond site
                                                now/in future

Design philosophy

Prepare material and report for possible public enquiry
Consider creation of
            enclosure
            shelter
            space
            setting
            colour
            materials
            contrast
            amenity
Integrate main design elements
                        buildings
                        equipment
                        car parks, etc.

Consider existing and proposed levels, contours, use of dredged material
Check orientation, views, aspect, avoiding glare from low sun, water, etc.
Establish percentage of hard/soft areas
Envisage interim/mature stages
Establish pedestrian/vehicular segregation
        circulation patterns
        desire lines
        path treatment
        serial vision
Keep overall budget and maintenance problems in mind

Influences and constraints
Macroclimate/microclimate
Climate: temperature, rainfall, sun, wind, shelter, length of season
Aspect, orientation, views
Topography, terrain, gradients, contours, sightlines, drainage
Local traditions in building, landscape, materials and their availability
Density: off-site, on-site, average, peaks
Wear and tear on site
Plant availability
Night-time conditions: lighting of grounds, ambient light from buildings, moon and reflec-
    tions in water, safety and security
Soil and sub-soil types: ph factor

Planting principles

Physical conditions
Contact local meteorological station, nurseries, parks departments
Identify prevailing/winter winds
Establish degree and type of atmospheric pollution
Establish soil and sub-soil data
        soil types
        ph factor
        particle size
        drainage
        water table level
Establish or exploit sheltered and sunny areas
Encourage soil/plant affinity
Plot existing/future drains, services, and foundations to avoid root damage
Avoid encroaching on road site-lines
Consider orientation in relation to windward slopes, frost pockets and the water's edge
Establish salinity, purity, etc. of water
Consider likelihood and effect of flooding of grounds

Plant selection
Survey and record position, species, size, condition of existing plants
Note particularly successful local species
Decide on retention or removal of local flora
Favour natural, hardy, low-maintenance species
Consider ground cover beneath trees as weed suppressant
Consider deciduous/evergreen balance
Select boundary plant carefully in relation to neighbouring land, remembering future
    expansion possibilities
Avoid shallow-rooted species in very dry/waterlogged conditions
Consider purchase of trees bearing in mind suitability of species, cost and size:
    nursery
    advanced nursery
    mature, etc.

Planting function
Screening from wind
Concealment of poor views, buildings, etc.
Consider views from grounds, particularly clubhouse, terraces, balconies
Consider: visual linking
        enclosure
        space division
        focal points
        screening from traffic
        noise
        dust, etc.
Decide form, density, colour and texture for each season
Balance planting with hard areas

Technical Information

Arrange site survey
Record existing services
      plants
      waterline, etc.
Calculate amounts of soil removal
       relocation
       import
Decide areas of cut and fill and disposition of dredged material
Establish need for and amount of fertilisers, top soil, etc.
Consider drainage of ground
        channels
        ditches
        gulleys
        soakaways
        pits
Decide whether sprinkler system is necessary
Assess need for firepaths and their location and surface
Consider extent, levels and materials of car-parking areas
List necessary maintenance equipment, tools, etc.
   provide for its storage and maintenance
Consider furniture and trim
        lighting of grounds
        bollards
        seats
        fountains
        tubs and planters
Consider children's play areas
        equipment
        sand pits, etc.
Remember shelters and mess room for maintenance staff

Maintenance

Remember natural balance between soil and plants helps to minimise maintenance
Remember low maintenance may mean higher initial costs
Carefully consider all edge treatments and allow for mowing margins against walls, beds and water
Avoid future pruning and staking if possible
Suppress weeds by ground cover or hard areas
Simplify all bed shapes
        paths
        boundaries
        design of water's edge
Ensure that paths take sensible routes
Avoid exotic plants, layouts and detailing
Design against vandalism by avoiding vulnerable plants and furniture
Discuss security and lighting arrangements
Provide litter bins
Remember that maintenance of small areas is relatively more expensive than that of large ones
Remember that undulations are better than terracing
Consider at design stage provision for
        watering
        weed control
        mulching
        feeding
        inspection
Advise purchase of good-quality equipment and employment of experienced, reliable staff

7. Landscaping: Bibliography

Air Ministry, *Climatological Atlas of the British Isles*, H.M.S.O., London 1952.

Bean, W. J., *Trees and Shrubs: hardy in the British Isles* (Eighth Edition), John Murray Ltd., London.

Beazley, Elizabeth, *Design and Detail of the Space Between Buildings*, The Architectural Press Ltd., London 1960.

British Standards Institution:
Bitumen emulsion for roads B.S. 2542: 1960.
Bitumen macadam with gravel aggregate.
Building drainage C.P. 301: 1971 B.S. 2040: 1953.
Cast stone B.S. 1217: 1945.
Clay tile for flooring B.S. 1286: 1945
Cold asphalt B.S. 1690: 1962.
Coping units B.S. 3798: 1964.
Granite and whinstone kerbs, channels, quadrants and sets B.S. 435: 1931.
Hand rollers for road and constructional engineering B.S. 1623: 1950.
Mastic asphalt for roads and footways B.S. 1446: 1962.
Pre-cast concrete flags B.S. 368: 1971.
Pre-cast concrete kerbs, channels, edgings and quadrants B.S. 340: 1963.
Sandstone kerbs, channels, quadrants and sets B.S. 706: 1936.
Specification for pre-cast concrete kerbs, channels, edgings and quadrants B.S. 340: 1963.
Tarmacadam with crushed rock or slag aggregate B.S. 802: 1967.
Tarmacadam and tar carpets (gravel aggregate) B.S. 1241: 1959.
Tarmacadam 'tarpaving' for footpaths, playgrounds and similar works B.S. 1242: 1960.
Tars for road purposes B.S. 76: 1964.

Caborn, J. M., *Shelterbelts and windbreaks*, Faber & Faber Ltd., London 1965.

Cement and Concrete Association, 'Paving patterns', *Concrete Quarterly*, Vol. 43, 1959.

Central Electricity Generating Board, *Design memorandum on the use of fences*, The Board, London 1966.

Chaplin, M., *Riverside Gardening*, Collingride Books Ltd., London 1964.

Colvin, B., *Land and landscape* (Second Edition), John Murray Ltd., London 1970.

Conover, H. S., *Grounds maintenance handbook* (Second Edition), Dodge Corporation, New York 1958.

Forestry Commission, The, *Shelter Belts and Microclimate*, Bulletin 29, H.M.S.O., London 1957.

Gaut, Alfred, *Seaside planting of trees and shrubs*, Country Life, London 1907.

Greenfield, I., *Turf culture*, Leonard Hill Books, London 1962.

Greenshill, T. M., *Gardening in the tropics*, Evans Brothers Ltd., London 1964.

Hackett, B., 'Maintenance costs and landscape design', *Municipal Journal* 6 March 1953.

Institute of Landscape Architects, *Lancaster conference on land and water*, The Institute, London 1969. *Available only in the institute's library.*

Institute of Landscape Architects, *Landscape maintenance, report of symposium held at the R.I.B.A.*, The Institute, London 1963.

Jellicoe, Susan and Geoffrey, *Water: the use of water in Landscape architecture*, A. & C. Black, London 1971.

Jones, Gareth and Farnham, William, 'Japweed: New threat to British coasts', *New Scientist and Science Journal* 8 November 1973.

Kelway, Christine, *Gardening on the coast*, David & Charles Ltd. 1971.

Le Seur, A. D. C., *Hedges, shelterbelts and screens*, Country Life Books, London. *Out of print.*

Lederman, A. and Trächsel, A. (eds.) *Playgrounds and recreation spaces* (Second edition), The Architectural Press, London 1970.

Little, E. C. S., 'The control of water weeds', *Weed Research*, June 1968, Vol. 8, No. 2.

MacMillan, H. F., *Tropical planting and gardening*, MacMillan & Co. Ltd., London 1956.

McMillan, R. C. 'Problems of maintenance', *Journal of the Institute of Landscape Architects*, April 1962.

Manley, G., *Climate and the British Scene*, Fontana Books, London 1962.

Menninger, Edwin A., *Planting and maintaining salt-resistant gardens*, Hearthside Press Inc., New York 1964.

Menninger, Edwin A., *Seaside plants of the world*, Hearthside Press Inc., New York 19

Ministry of Agriculture, Fisheries and Food, *Control of aquatic plants*, Bulletin 194, H.M.S.O., London 1973.

Ministry of Agriculture, Fisheries and Food, *Identification of water weeds* Bulletin 183, H.M.S.O., London 1968.

Ministry of Agriculture and Fisheries for Scotland, *Administrative leaflet 5: Shelterwoods or belts*, H.M.S.O., Edinburgh 1959.

Mono Concrete Ltd., *Paved areas* (Current Edition), 1974.

National Playing Fields Association, *Hard porous all-weather surfaces for outdoor recreation*, The Association, London 1970.

Newton, Norman T., *Design on the land: the development of landscape architecture*, Harvard University Press, Cambridge, Mass. 1971.

Perry, F., *Water gardening*, Country Life Books, London. *Out of print.*

Road Note 25: Sources of white and coloured aggregates in Great Britain, H.M.S.O., London 1959.

Russel, E., *The World of the Soil*, Fontana Books, London 1961.

Stodola, Jiri, *Encyclopaedia of water plants*, T. F. H. Publications Ltd., London 1967.

Sudell, R. and Waters, J., *Sports buildings and playing fields*, B. T. Batsford Ltd., London 1957.

Tandy, Cliff (ed.), *Handbook of Urban Landscape*, The Architectural Press Ltd., London 1972.

Thomas, Arthur, *Gardening in hot countries*, Faber & Faber Ltd., London 1965.

Thomas, G. S., *Plants for ground cover*, J. M. Dent & Sons Ltd., London 1970.

Ward, Colin (ed.) *Vandalism*, The Architectural Press Ltd., London 1973; Van Nostrand Reinhold Co., New York 1973.

Webster, A. D., *Seaside planting for shelter, ornament and profit*, T. Fisher Unwin Ltd., London 1918.

Weddle, A. E. (ed.) *Techniques of landscape architecture*, William Heinemann Ltd., London 1967.

8 Statutory services

The provision of public utility services both to the marina and within the marina frequently brings problems. Marina sites are often in remote places and even in suburban areas, being of necessity at water-side or coastal locations, are commonly bypassed despite the mains not being very far distant.

In cases where one or more essential service is not available, the potential developer will need to give serious thought to the site's suitability. The absence of main drainage or town gas can possibly be overcome by chemical closets and bottled gas, but a site denied mains water or electricity would find itself badly handicapped both in operation and in its construction. Where services are entirely absent or quite far distant the utility company may consider that providing a connection is a poor proposition when the estimate of cost is set against predicted consumption.

Even within the same utility most countries experience variations in policy from region to region. It is dangerous to assume that action taken by a company in one area will apply elsewhere. Agreements must be obtained in writing at local and regional level from each company, board or, with main drainage, the local authority. A clear unambiguous contract spelling out the date by which the service will commence must be arranged where utilities are required to be connected. *It is very important to establish the exact situation with the authority responsible.* Never rely on promises that sites without a particular service will be provided with it 'in a couple of years'. If a decision *is* made to forego a service (even if only temporarily) *check that the bye-law and planning authorities are agreeable to the marina operating without it or with an alternative.*

The services network must be conceived within the total framework of the development, both in design and installation. However, by separating individual services and looking at their relationship it is possible to clarify the normal sequence of operation and present them in a logical order. (See 8.1.)

Ground conditions

Ground conditions at marina sites, both before and after development are often poor. The degree of difficulty in providing for most public utility services is likely to be well above average and the allocation of time and money should reflect this. Existing land is frequently marshy, saline or vulnerable to flooding, and even after the basic site engineering has been completed the new levels will have been created by tipping, the disposal of dredged material or some other form of made-up land. Developers can rarely wait for natural consolidation and accelerated methods seldom achieve thorough compaction. Piped supplies are at greater risk in poor ground conditions than underground cables. Overground systems may not be a valid alternative aesthetically or because of wind conditions.

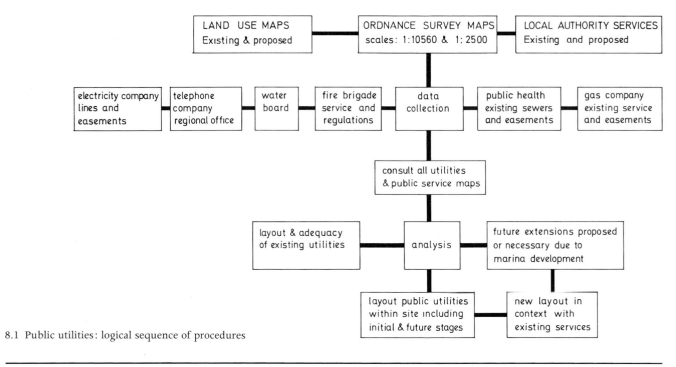

8.1 Public utilities: logical sequence of procedures

Survey

An appreciation of the overall topography and contours is often necessary. This can be obtained from 1-in and $2\frac{1}{2}$-in Ordnance Survey Sheets, (1:63360) and (1:25000) respectively, which are useful for rapid appraisals of local significance. More detailed information on boundaries and spot heights is shown on 6-in (1:10560) and 25-in (1:2500) sheets although these are not available for all rural areas. From these four sources the overall situation regarding services may be assessed in terms of contours and ground modelling. River, coastal and harbour authorities will provide information on river courses, liability to flooding and water pollution. Where detailed tests of soil are necessary (possibly for reasons such as bearing strength, suitability of building materials or landscaping) this information will be useful in establishing the conditions likely to be encountered by underground services. Such sampling is usually done by hand auger and the soil tested to find its engineering classification. Soil sampling also gives an opportunity to measure water-table levels, although these will only be indicative of the time of sampling. Several water-table tests will be needed to cover all conditions of rainfall and seasons. Some information about existing services such as mains, and invert levels is included on larger scale Ordnance Survey maps. More detailed maps are usually available from the Utility Boards and local authorities. This will include pipe diameters, material, depth of service, junctions, manholes, cable sizes and capacities, type pressures and calorific values of gas mains and possibly the dates of laying and subsequent repair work. Where services are known to exist, but their location is unknown or inaccurate, a 'Pipe Finder' will be useful. These instruments will find underground metal pipes either by magnetic attraction or electronic detection. The foregoing information on ground conditions and existing services will provide a valuable background on which to make decisions regarding future service runs, their depth, materials and layout pattern.

General layout

A marina usually presents many more problems in the layout of services than would a housing estate, school or shopping centre. As well as the difficult ground conditions already mentioned the often awkward shape makes dead ends difficult to avoid and the need to run services along breakwaters, quay walls and to the berths themselves is more than usually difficult, especially where floating moorings are to be provided. The overall layout of all services should be as economic as possible. Services should not be run too close to quay walls (2 m minimum) either when running parallel to them or changing direction.

Services are usually placed beneath footways or verges rather than carriageways:
1 to avoid interruption of traffic during emergency repairs, routine maintenance or provision of additional services
2 because paving materials can be easily-removable, small-scale units and access covers are accessible and readily accommodated into the paving units

Where they are required to serve both sides of the road, duplicate services will prevent crossing beneath the carriageway. Where this is unavoidable the service can be laid within a duct or conduit for easy withdrawal. In any event it should be protected from undue pressure or stone slabs above. Normally the order of laying services between the back line of pavement and the kerb will be: water, gas, electricity, telecommunications.

8.2 shows typical depths for various services below a footway and 8.3 shows how services at recommended depths may share a common trench in similar circumstances. Utility companies are notoriously bad at rationalising their separate services within communal trenches, channels or ducts and if this is thought to be desirable it will have to be proposed and detailed by the architect. Even so, the agreement of the authorities will be needed as statutory companies hold strong views about the compatibility of their service with others nearby. A joint committee of the Institution of Civic Engineers and the Institution of Municipal Engineers prepared a report recommending standard practice for locating public utility services under paved areas.[1] Another useful book gives guidance in the co-ordination of services and the provision of common trenches and ducts.[2] Not all services can necessarily share a single trench. District heating and oil pipes are excluded because of the excessive trench width and organisational problems. Instead a separate trench is recommended.

Service trenches

Where a considerable length or volume of trenchwork for services is necessary, mechanical equipment will be essential. The soil classification and depth of dig will determine whether formwork is needed and if so, what type. Back fill should be well compacted every 230–300 mm (9–12 in) preferably by power-driven rammers. The backfill material should be at its natural moisture content. With large contracts or where the soil is very unreliable it may be preferable to cart away material removed from trenches and replace imported granular fill in the form of course sand, fine gravel or other stable material. Flooding during trenching, laying or backfilling should be avoided if possible because this being a common cause of paving failure later on—insufficient compaction being another. Trenches should not be too near to trees, although the fear of roots damaging services is often exaggerated except perhaps in the case of Poplar, Elm or Willow. In most

1. Institutions of Civil Engineers and Municipal Engineers, Joint Committee of the, *Location of Underground Services*, The Joint Committee, London 1963.
2. Ministry of Public Building and Works/ Directorate of Building Development, *Co-ordination of underground services on building sites: 1 The common trench*, H.M.S.O., London.

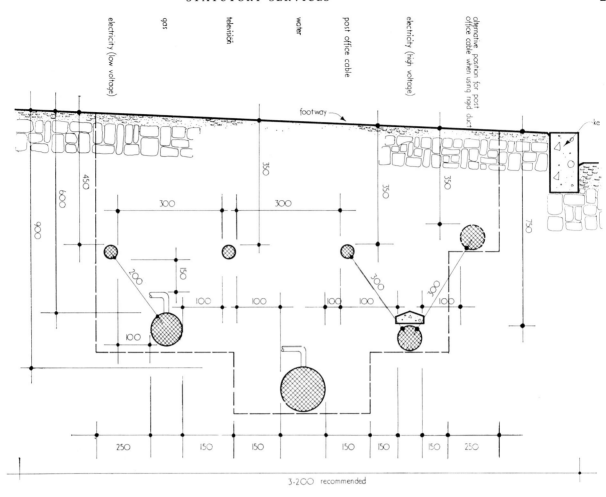

8.2 Cross-section through paved footway showing arrangement of public utility services and minimum depths and spacings between services recommended in MPBW R & D bulletin

8.3 Cross-section through common trench for public utility services recommended in MPBW bulletin. The trench would be back-filled in three stages as indicated. Depths of cover for sewers will vary and the local authority will need to be consulted. For land drains cover will vary between 0·700 m to 1·200 m depending on soil type

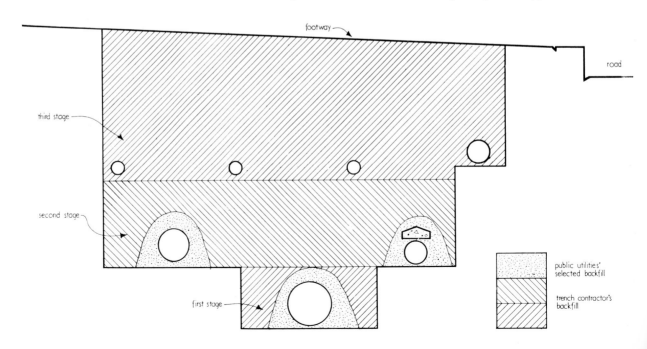

cases the tree is likely to suffer more than the service, particularly from savage root cutting or from leaking gas mains.

Water

An adequate and convenient supply of water is necessary not only for the specific use of the vessels and buildings, but for fire-fighting and sanitation. The supply will generally be from existing mains or will be brought to the marina area by the local water board, which is empowered to make By-laws to prevent misuse, undue consumption, contamination and waste of water under the provisions of the 1945 Water Act. Water bye-laws are quite separate from Building Regulations. This authority together with the local fire service will advise upon the bye-laws governing materials, layout, pipe sizes and metering.

Almost every facility within the marina (see the Check List on page 149) will need serving with water. It will probably be necessary to work through such a list to determine the total demand which in turn will establish the necessary sizes for distribution pipes and mains.

Numerous other facilities within the grounds may require outlets. The following check list covers the most likely needs:

Offshore	End of breakwater	Chemical closet emptying points
	The berths themselves	Drinking water points
	Police boat mooring	Drying rooms
	Fire boat mooring	Clothes washing
	Fuelling vessel	Public lavatories and wash rooms
	Bunkering facility (land based)	Boat hardstanding
	Harbour Master's office	Gatekeeper and security stations
	Refuse points	Groundsmen's mess and equipment store
	Play areas, sand pits	Watering points for grounds
	Mobile sewage suction unit	Car washing points

The quality of equipment is generally expected to be high, as exposure to salt water and heavy use of taps during the season will demand this. The material chosen for water piping will depend upon factors such as:

1 Speed of laying
2 Overall budget
3 Degree of mechanisation
4 Soil conditions
    chemical
    bearing pressure
5 Exposure to salt, air or water
6 Range of temperature
7 Use of granular fill in trenches
8 Experience of labour force

The following is an assessment of the properties and attributes of the principal materials now available:

8.A Materials for piping

Material	Advantages	Disadvantages
Cast Iron	Very strong Very durable under most conditions	Liable to chemical attack in some conditions Very heavy Special fittings required Caulking necessary
PVC e.g. Alkathene Telcothene Vulcathene	Corrosion resistant Easily and speedily jointed Withstands temperature changes Very light and easily handled Joints and pipe allow some movement Chemically inert Smooth surface which stands frost well	Can distort under ground pressure Relies to some extent upon internal pressure
Asbestos	Chemically inert Comparatively light Relatively inexpensive	Very rigid Rather brittle
Lead	Very expensive, its use now largely confined to repair work	
Nylon	Nylon is a generic term embracing many forms some of which, whilst made into tube, are unsuitable for water supply pipes having a water absorption of 6–7%. To overcome confusion in nomenclature numbers have been allocated, Nos. 11 and 12 being suitable for water supply piping. Only black should be used, being unaffected by ultraviolet light	

Pipe sizes

These will vary according to local bye-laws. Factors such as average and peak demands, water pressures and type of outlet are interpreted differently from one authority to another. On *average* the internal diameter of water supply mains and distribution pipes would probably be as follows:

	Largest	Smallest
Mains	914 mm (36″)	101·6 mm (4″)
Distributors	25·4 mm (1″)	12·7 mm ($\frac{1}{2}$″)

8.B Pipe sizes

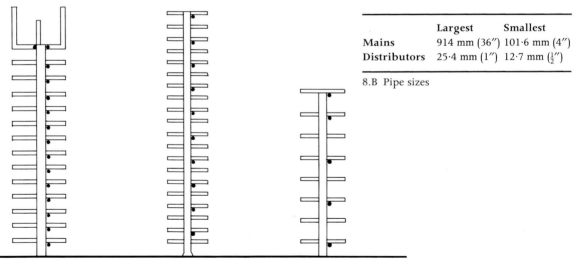

Frequency	:	2 berths per outlet	4 berths per outlet	6 berths per outlet
Berths served	:	30	40	30
No. of outlets	:	15	10	5
Berth size	:	to take 9m (30') oal	to take 6-8m (20'-30') oal	to take 6m (20') oal
Pipe size	:	25mm (1") diam.	19mm ($^3/_4$") diam.	12.5mm ($^1/_2$") diam.

The critical factor in determining sizes and location of mains can often be the demands of the fire fighting service rather than normal use or the number of outlets, 101·6 mm (4 in) main usually being the minimum require- (see Fire Fighting Equipment, pages 247–250).

In terms of outlets to moorings, distribution-pipe diameters would be expected to serve outlets as follows:

12·7 mm ($\frac{1}{2}$ in) diameter—3–5 outlets
19·0 mm ($\frac{3}{4}$ in) diameter— 10 outlets
25·4 mm (1 in) diameter— 30 outlets

Water supply to moorings

8.B illustrates the relationship between pipe sizes and the number of berths to each outlet. Supply to the moored boats is generally by taps of the lost-key type, positioned at the junction of the finger piers and the walkway. No more than eight boats should be expected to share one outlet, and preferably only two or three.[3] Easier maintenance and greater protection are provided if the service is run in ducts at the edge of the walkway. This also prevents weakening any support timbers by notching pipes in them. Fixed piers present no difficulty but laying pipes along floating walkways brings the problem of flexibility. If the pontoons are to move differentially at each junction (either from wave movement or from the weight of persons or gear) then the service pipes must do the same. Pipes of pliant material may need no special joints but the maximum movement between the pontoons must be less than the minimum flexibility of the pipe (i.e. in very cold weather). Unless this can be established then sections or joints of more ductile material must be introduced at each pivot point. Special joints for water pipes have been designed for similar purposes in the past but they are little used now.[4] With most proprietary floatation systems this question will no doubt be solved within the design. This means that the make of pontoon will often determine the material for the water service. The method used must be satisfactory and permanent for anything less will cause fatigue in the material or failure at the joints and be a continual trouble and expense.

A further problem in tidal waters is the means of overcoming the change in water level as it relates to a fixed outlet on the quay.[5] Where this is small —up to 1·5 m (5 ft)—then a slack, flexible hose of plastic material will suffice, preferably with an outer protective sleeve of pliant metal or nylon similar to those on hand-shower fittings. For larger tidal falls an incrtia reel may be fixed to the quay wall above the highest tide level and the flexible water pipe wound round it sufficiently to descend to low tide. A third method is to run the pipe below the walkway access ramp, using flexible joints at either end.

Frost damage

Water pipes laid at the recommended minimum depth of 762 mm–914 mm (2 ft 6 in–3 ft) will in most climates have sufficient ground cover to prevent freezing. As they surface above this level and travel along the walkways they can be protected by lagging although full protection is expensive and bulky. Low-voltage electric wiring is not recommended in off-shore or wet conditions. P.V.C. pipes that are able to expand slightly without bursting are an advantage. Drain cocks positioned at the lowest point of each leg will allow drainage during severe weather and this is not too inconvenient for

3. It is important to ensure (and to warn owners) that no connection must be made between the distribution pipe and any appliance aboard as this is not permitted by any water authority except via an approved tank or cistern. The walkway tap must be of the 'free-fall' type (i.e. where the water flows to the open air).

4. This is a ball-type joint with a packing gland and is still available from water authorities and specialist merchants.

5. The problem is most commonly encountered by water authorities where a mains supply is required aboard ships in harbours and docks.

8.4 Badly positioned taps are easy to trip over and how in this instance do you fill a bucket? Better to raise it a metre and paint it white

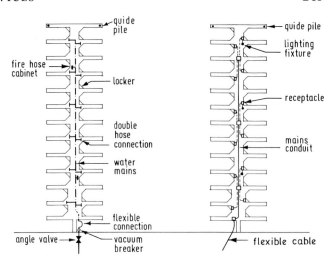

8.5 Lockers, water, electricity, fire equipment and telephones at Bahia Mar, Florida

FIRE & WATER SERVICES

ELECTRICAL LAYOUT

8.6 Services lay-outs

8.8 Lighting, telephone plug, electric power point and meter in one unit: a less than elegant attempt at unification. The free-standing foam extinguisher seems vulnerable to carelessness, vandalism or theft

8.7 Water terminals at Fort Lauderdale

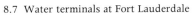

8.9 Service duct on walk-way

8.10 Rather inelegant service arrangements

8.11 A simple services duct. This example at Denia, Valencia, Spain, is unfinished

most marinas in their quiet season. Water from taps along piers can be unpleasantly warm in summer and, whilst lagging and ducting can be a help, the best way of overcoming this is to have ice available from the clubhouse, chandler or dispensing machines.

Lagging

There are several forms and various means of installing lagging. Powder mixed with water and plastered on to the piping is a messy process which may require warmth and time to assist in drying out. Preformed lagging is more popular. Rigid sections of mineral wool, glass fibre or expanded polystyrene are available ready moulded in two halves to the correct pipe size and after wiring to the pipe, are often protected with hard-setting cement, wrapped in hessian and painted. Metal clad pre-formed lagging is the easiest to maintain, for if properly designed and installed, it can easily be replaced. Others will require stripping for maintenance or repairs, a long and awkward task.

The maintenance engineer should have proper instructions for all trap drains. These items should have regular and frequent maintenance checks, for pipe-scale, dirt and corrosion are the worst enemies of drain traps.

Fire fighting

This section deals with the provision of fire-fighting services. (Fire *prevention* is covered in Chapter 13, pages 331–332.)

It is essential to contact the fire authority at an early stage when considering services in general. This is not only for public health and humanitarian reasons but because the fire authority's needs will be an important determinant in such factors as size and location of mains and the layout of grounds and fire paths. New marinas and the conversion of existing docks and harbours present many difficulties for the fire services because they are often in remote places, are awkward in shape and contain many substances by which fires may be ignited and spread. (See Chapter 13, page 331.) Even when suction hoses are available to draw water from the harbour the fire authority usually requires a 102 mm (4 in) main with couplings at strategic positions.

The location of hydrants, hose reels and couplings will need the consent

8.14 Defective electric wiring was the initial cause of this fire at the Sandusky Marina, Ohio on 30 May 1958; 14 boat-houses, 8 boats and 3 cars were destroyed.

Petrol in the boats, paints and thinners stored in the sheds and the wood-framed construction added to the fire spread. Damage was estimated at £132,000

8.12 Lighting with power plug and telephone at Huntington, Long Island

8.15 9 boats, 5 partially-completed hulls, a truck and the entire storage structure were destroyed in this fire at the Warwick Marina, Rhode Island on 15 February 1964.

The marina, left unguarded at night, was entered by children who started a fire below deck in one of the vessels. Damage was estimated at over £100,000

8.13 Lighting columns at Javea Harbour, Valencia, Spain

of the fire authority. Usually the position will be:
1   Along the requisite size of main
2   Where appliances can reach them without hindrance from overhead or ground obstructions and without bogging down
3   Where men, appliances and hoses have the maximum coverage of moorings and buildings
Appliances are usually pump escapes or turntable escapes, both weigh from 8·5–10·0 tons, have a track of 1·83 m–2·0 m (6 ft–6 ft 6 in) and a sweep circle of 18·3 m and 22 m (60 ft and 72 ft) respectively, giving a minimum carriageway width of 3·66 m (12 ft).

In situ or precast concrete pots, usually hexagonal or pyramidal in shape, make an effective fire path. They are usually approximately 100 mm (4 in) thick, the hollows being filled with a mixture of soil and grass seed which eventually grows to provide an unobtrusive surface sufficiently strong for fire appliances.

Nearly half a million buildings come under the Fire Precautions Act 1971.

Each building has to be inspected by the Fire Authority which makes recommendations and inspects the completed work.

Continuous vigilance is important because more than half of all major fires (costing more than £10,000, 1970 figures) start during the night when buildings are empty and the occupants asleep. The provision of manual alarm systems is mandatory in factories with more than 20 people (Factories Act 1961).

Whilst the type and quantity of fire-fighting equipment provided will depend upon the size of the harbour and the nature of the materials at risk, it is likely to be drawn from the following list:

1   Mains water
Serving couplings, hydrants, hose reels and fire brackets. The fire authority will advise on the location, number, and type of equipment as well as the size and location of the mains supply

2   Harbour water
Hardly 'equipment' as such, but an ever-present source of extinguishment drawn from the harbour by the fire authority's pump appliances

3   Fire boat
Such craft are standard equipment in large American marinas

4   Fire paths
The fire authority will determine the width, turning circle, bearing pressure and location

5   Fire extinguishers
Containing all-purpose foams or special chemicals or powders depending upon the likely fire source. Invaluable for containing an outbreak. Very necessary aboard as well as standard marina equipment

6   Fire points
Like water points these should be central and easily accessible. They may

Features of portable extinguishers

extinguishing medium	typical sizes	jet range (m)	characteristics and application	class of fire*	pressurisation
water	5–10 litres	10	good cooling agent, good penetration, prevents reignition disadvantages: electrically conducting, spreads liquid fires	A	chemical, CO_2 compressed air or N_2
foam	10 litres	8	combined smothering and wetting action: prevents reignition disadvantages: electrically conducting, not effective on free flowing liquids, some liquids break down the foam	A–B	chemical
powder	1–14 kg	3–8	nonconducting, acts faster than foam on flammable liquids disadvantages: no cooling action	B–C	N_2 or CO_2
CO_2	1–7 kg	1·5–3	nondamaging, acts faster than foam, effective on escaping liquids, nonconducting disadvantages: no cooling action	B–C	high pressure container
vaporising liquid	1·5–5·5 kg	2·5–6	nonconducting, rapid action disadvantages: only for small fires	B–C	N_2

*A: most ordinary materials—wood, paper, textiles   B: flammable liquids—petrol, oil, paint   C: fires in electrical equipment

include water and sand buckets, extinguishers, special blankets and besoms or fire beaters—useful in rural areas for undergrowth and heath fires

7   Hose reels
These are coupled to the mains supply and are usually swivel-mounted. When strategically placed, they provide a quick and effective coverage to large areas. They have the advantage of being simple to use and are usually more reliable than extinguishers, provided that they are regularly maintained. They can be misused but, unlike extinguishers, are not exhaustable

8   Portable fire pumps
The types used in marinas are usually big-wheeled, hand-manoeuvred cannons with or without a hose-extension. They are useful within the marina grounds and, possibly, the fixed piers, but are no good for floating moorings

Mains gas

Although a normal service installation from a mains supply would be expected to the on-shore buildings, this is not normally provided to the piers, because of the technical difficulties and hazards involved. Two gas *service* pipes are 25·4 mm (1 in) in diameter and are usually laid approximately 500 mm beneath foot paths although access is rarely needed to them, and then only to valves or for pumping pipes to remove condensate.

Mains are usually cast iron or steel, 100 mm (4 in) minimum diameter and have 600 mm–700 mm (2 ft–2 ft 4 in) ground cover.

Liquefied petroleum gas

Bottled gas is popular aboard and a contract to supply this to the clubhouse or chandlers is desirable and profitable. Boat owners should be acquainted with the dangers, particularly of such gas escaping into bilges, as this fuel is heavier than air. The provision of gas bottles to the boats and the collection of empties can be a popular and remunerative service. It can also be supplied in bulk and distributed and metered in a similar way to a mains gas supply.

Liquefied petroleum gas has a calorific value higher than both normal mains or natural gas, and a quality that is defined by *Commercial butane and propane* B.S. 4250:1968. The low volume to high power ratio makes it easy to transport and store. The two principal types of liquefied petroleum gas are Propane and Butane. The basic difference between these two gases is that Propane has a much higher rate of vaporisation from its liquid form—especially at low temperatures, making it suitable for storage in large outdoor tanks for bulk industrial use.

Propane and butane are produced where crude oil is being refined for petrol and its allied products. A number of processes are used, but in each case propane and butane are driven off as volatile gases.

Bulk tank storage
Specially constructed tanks are supplied at a small rental and come as complete units, including supports, vapour outlets and pressure relief valves. Propane for bulk users is stored in special tanks of various sizes up to 12 tonnes (12 tons). These are sited above ground and topped up by

8.16 Security gates at San Francisco Marina. Only boat owners and authorised persons are issued with keys. The design seems influenced by near-by Alcatraz

8.17 The police may require a mooring for coastal patrol: a welcome aid to the marina's security

tankers on a regular basis. The usual site requirement for a bulk tank is open air compound with a level concrete plinth—with reasonable access for tankers. The actual size of the area and tank plinth details will be given by the supplier when the installation is first considered and is based on the size or the number of tanks required. This is calculated on estimated consumption, the maximum rate of that consumption and the refilling schedule to be adopted.

From one or more bulk storage tanks, gas is piped directly to any part of the site for whatever application is required. Tanks are situated in any convenient spot which allows access to tankers. With the range of tank sizes available this kind of installation is adaptable to provide power for a wide variety of marina sites where mains gas is unavailable.

Installation and all necessary pipe work is normally undertaken by the supplier. Pipework and installation beyond that point can also be carried out by the supplier who will usually offer a quotation.

Alternatively, the contractor may use his own staff. If this course is taken, advice from the supplier is usually available regarding pipe sizes, jointing compounds, pipe runs and protection. On completion of the installation the supplier will test the system and instruct on the correct operation of valves and controls.

Cylinder storage

To ensure convenience and safety, advice on storage and handling should be sought from the supplier and the fire authority. It is usual for the marina staff to handle the cylinders and to deliver them to the craft. Trolleys or trailers will be necessary as well as a space for storing them. The following 10 points are the principal recommendations of the draft code of practice:

1 Cylinder storage areas should be accessible and marked with warning notices

2 LPG cylinders can be stored with like materials but not with oxygen or toxic materials, e.g. chlorine

3 Buildings for LPG storage should be purpose designed (or modified) to have high fire resistance and a high rate of ventilation

4 All cylinders should be kept upright and not be stored below ground or adjacent to cellars or drains

5 Storage areas should be enclosed to prevent access by the public and smoking prohibited

6 Where more than 5,000 kg (5 tons) of gas are stored the fire authority should be consulted with respect to fire-fighting arrangements

7 Electrical fittings should be flameproof

8 Within shops or office premises with residential accommodation not more than 15 kg (34 lb) may be stored

9 In industrial premises with a specially designed storage area within the building not more than 1,000 kg (1 ton) may be stored

10 The minimum separation distance from any building or boundary is determined by the total quantity of gas stored and the size of the largest stock in the area

Marinas consuming less than 3,000 kg (3 tons) per year pay a cylinder refill charge and receive deliveries at retail price through suppliers. Liquid petroleum gas can be used to operate vehicles, construction site equipment, fork lift trucks, industrial engines and generators.

Design of stacks and gangways

1   The maximum size of any stack should not exceed 30,000 kg (30 tons)

2   The gangway between palletised stacks should be not less than 2·5 m (8 ft 3 in) and unpalletised stacks 1·5 m (4 ft 9 in)

3   The maximum height of any stack should not exceed 2·5 m (8 ft 3 in), nor be such that the amount of LPG in any column of the stack exceeds 110 kg (243 lb) when palletised and 55 kg (122 lb) when unpalletised

Drainage

Main drainage systems

The two principal systems of main drainage are:

1   The combined system in which soil and rainwater share the same pipe

2   The separate system in which soil and rainwater are in individual but (usually) parallel pipes

Soil pipes should have as few contortions as possible. The layout should aim to collect short, straight branches into one main pipe. This is achieved within a manhole so that the lifting of one or two covers should reveal any obstruction. Manholes are necessary at every change of direction and are, therefore, frequently positioned at the corners of buildings. The gradient or fall varies with the pipes diameter and, to some extent with its material and method of jointing. Generally it would be approximately as follows:

101 mm (4 in)—1:40 to 1:70

152 mm (6 in)—1:60 to 1:100

No drain should have less than 762 mm (2 ft 6 in) ground cover. Ventilation of the system is usually by a vertical extension of the WC soil pipe and by fresh air inlets (FAIs) at the interceptor, which is usually placed near the site boundary.

Materials

1   Salt glaze. Usually 101 mm–152 mm (4–6 in) in diameter. Normally laid on a benched-up concrete bed but when passing beneath buildings must be surrounded by 152 mm (6 in) of concrete

2   Pitch fibre. Suitable for soil and surface water drains. High resistance to chemicals. A concrete bed may not be required. Usually 3 m (10 ft) lengths with standard couplers which do not need a jointing compound

3   Cast iron. Often used beneath building in preference to surrounding glazed pipes with concrete. The flexible telescopic joints which are available are particularly suitable for marina sites. Pipe length is normally 2·7 m (9 ft). All cast iron pipes should be protected by a bituminous solution.

4   Spun concrete. All joints must be flexible—usually by using rubber rings—otherwise fracturing of pipes or joints is a danger due to the material's brittle nature, particularly in conditions susceptible to settlement or thermal movement

For the design and construction of sewers the report of Working Party on the Design and Construction of Underground Pipe Sewers is particularly recommended.[6]

Soakaways, cesspits and septic tanks

In areas not served by a main drainage system the marina developer will be obliged to provide for sewage processing within the site boundary and arranging for the removal of sludge (usually by the local authority) at

6. Department of the Environment, *Third Report of the Working Party on the Design and Construction of Underground Pipe Sewers,* H.M.S.O., London 1971.

Section B is useful to those building abroad as it deals with overseas practice. The Appendix also gives reports on sewer design throughout the world.

regular intervals. Modern methods of treatment are often very clean and efficient and, provided that the capital cost of plant is acceptable, the absence of main drainage need not be so great a disadvantage as to preclude the consideration of such sites for marina development. Provided *a really good installation is built*, then in most cases no one would be aware of the process at all.

Rainwater usually presents no difficulty. A hole 1·2–1·5 m (4–5 ft) deep is dug a short distance from a building or hard-paved area and is filled with gravel or stones—in some districts on chalk, gravel or sand even this is unnecessary. The problem areas are those on impermeable clay unsuitable for soakaways. Land drains must be used to channel the water into culverts, ditches, rivers or the sea.

A cesspit or cesspool is a large cylindrical tank in (or half in) the ground measuring from 1·5–3 m (5–10 ft) in diameter and 3–4·5 m (10–15 ft) deep. It should be watertight with an impervious cover and have a manhole for inspection and for pumping out every 3 or 4 months. On certain very porous soils filtration through a perforated base is still sometimes permitted but this is not recommended, not so much because of any hazard to health but because it is a common cause of algae and weed growth, particularly on inland water. In any case, the Water Resources Act 1963 forbids the discharge of raw effluent into the substrata where water gathering grounds may be affected.

A septic tank is a large brick-built chamber built into the ground through which the effluent filters, settles and is broken down by bacterial action. After the material has passed through a settlement compartment it is filtered through another container filled with clinker. As with cesspits, some local authorities permit secondary treatment to take place by subsurface irrigation but this is not advised for inland sites on enclosed water. It might be possible, however, where treated liquid can be directed safely to the sea. The increasing use of detergents causes trouble in cesspits and septic tanks and may, in consequence, require plant of a larger volume and the more frequent removal of waste.

Several proprietary systems are marketed for the treatment and disposal of sewage using conventional processes. Most of them are prefabricated, pre-cast concrete boxes which require the minimum of on-site construction. Discussions with the manufacturers should answer four main questions: how large, where, what type and how much? It is, nevertheless, essential at an early stage, to ensure that any system is approved by the Local Authority for it will usually be responsible for approving the design and location as well as the regular removal of sludge.

Chemical closets (see also Chapter 11, pages 289–290)
With the new laws governing the discharge of effluent, the processing of waste from craft will become a major management responsibility. Even in marinas where adequate toilet facilities are available there will still be a need to provide emptying points for chemical closets from the heads of visiting craft. This is usually achieved by the owner (or if a service, by the marina staff) emptying the container into the mains sewer, or its equivalent, via a trapped gully. These are housed in special small cabinets (approximately 1·8 × 1·8 × 2·1 m (6 × 6 × 7 ft)) near the moorings and contain a stainless steel slop sink at about 500 mm (20 in) above ground level and a flexible

spray-jet hose for washing down both the sink and the empty container. A heavy-duty shelf beside the sink should be provided upon which the containers can be rested. Chemically-treated effluent is not always suitable for discharging into septic tanks as this destroys the bacterial action of the system. In these cases the effluent can either connect direct to a soakaway or, better still, be discharged into a collecting tank which is then taken away and disposed of by the Local Authority. These closet stations are also useful to caravan owners who can also make use of them, although they will differ from the ground-level sewage disposal stations found at caravan sites.

Location and landscape treatment

Cesspits, septic tanks, emptying closets and, perhaps public conveniences, will all benefit from careful landscaping and screen planting. (See Chapter 7, page 221.)

Electricity

In general the main uses for electrical power in marinas fall into the following categories:

1 Marina construction
2 On-shore buildings
3 External lighting and security systems
4 Boat building, repair and service
5 Boat handling and storage equipment
6 Service to boats
7 Lock gates and special equipment

Electricity supply during construction

Whilst this is largely the responsibility of the contractor, a safe and efficient installation is in everyone's interest. Electrical supplies for construction sites should be given the same careful consideration as permanent installation. Power is needed for the following:

1 Plant and tools
2 Construction techniques: concrete curing, drying out, etc.
3 Lighting and heating
4 Space and water heating
5 Cooking

Records show that the dangers with electricity are greatest in wet or damp conditions, such as are common on marina sites. *Cabinets for electrical equipment should be strong, weatherproof and mounted well above ground level.* They should provide the following:

1 415/240 V medium voltage supply (this is much the most common supply in Britain although 110 V is more common abroad)

Authority's cable

Intake fuses

Current transformers

Meter

Other fuses for sub-main control gear

2 Reduced voltage supply, 110 V single and three phase

Step-down transformers

Outgoing circuit control gear

3 Portable units with socket outlets and adequate circuit protection

On-shore buildings

The on-shore supplies of electricity are no different from those expected in any other development. The Electricity Authority for the area must be consulted and will advise on installation. It may demand a transformer chamber and certain special safety measures, particularly with damp-proof cable, flexible connections and fusing arrangements.

The medium voltage distribution system is normally 415 V (line voltage) and 240 V (phase voltage). Hotels, factories and public buildings are usually supplied with a four-wire distribution (three-phase and neutral). Larger buildings may be supplied with 3,300, 6,600 or 11,000 V supply. With the normal electricity service the incoming cables of high voltage are laid in the service trench below the footway with about 600 mm (2 ft) cover and away from gas mains or GPO cables. These incoming cables are taken to the distribution boards within the individual buildings. Upon the board will be the following equipment:

1 A sealing chamber to prevent moisture returning from the cable
2 The Authority's cut-out
3 The meter
4 The consumer unit

The consumer is responsible for all the equipment on the distribution board except the meter.

From the consumer unit travel the fuseways supplying the various circuits.

Wiring and materials

Wiring runs should be as short as possible and the total system will need to be of above-average quality in most marina conditions.

PVC cable of the flexible type is a suitable material as it insulates well, is very resistant to solvents and is non-flammable.

1 Conduit or trunking. Essential along walkways or, in fact, in most situations where PVC cable is to be used. Where cables *have* to be laid beneath carriageways or buildings they may be drawn through 100 mm (4 in) diameter earthenware ducts
2 MICC. A mineral-insulated, copper-covered cable which may be used without further protection in locations such as underwater, beneath ramps from quay to walkways, or elsewhere where protection is needed from water, acids or abrasion
3 Socket outlets. These will probably be of the standard design to receive a 3-pin rectangular plug with the fuse rating determined by the calculated loading. They will incorporate a spring-loaded shutter as a safety device. (International plug standardisation is still awaited but 13 amp 3-pin fused plugs and socket outlets to BS 1363 are usual throughout UK.)

Wiring layout

This is best designed by an electrical engineer who will determine the pattern of loading and specify the correct cable capacity for each purpose. Demand will vary by season, weather and diurnal requirement. Until recently in Britain some marinas were still being developed with only a minimum on-shore supply, but needs and expectations now demand a supply that is comparable to any other recreational facility. Even in winter, requirements can be quite high for heating, lighting and repairs. Some owners leave a thermostatic tubular heater aboard during winter months to ensure their vessel remains dry.

An estimate of the total demand may not be critical in an urban environment where the supply is unlikely to be inadequate, but in rural, remote or virgin sites the supply authority must be persuaded to meet the demand for construction, initial and future needs. It is also important that the supply is permanent and underground from the beginning. It is tempting to accept a temporary, perhaps overheard, supply in the desire for an early start, but once connected in this way the temporary expedients have a habit of becoming permanent solutions. Situations will vary greatly but the following schedule lists some areas and facilities which are likely to depend upon an electrical supply:

Onshore	Offshore	Shoreline
Marina construction	Piers and walkway lighting	Boat handling equipment
Marina buildings	Floodlights, underwater and special effects	Bunkering
lighting and heating	Power for:	Customs and harbour master
ventilation, air conditioning	moorings	Security points
refrigeration and cooking, etc.	lighting, heating	
Lighting of grounds	TV/radio	
access roads	power tools	
car parks	telephone	
special effects	Police and fireboat	
security	Lock gates	
Boat building, boat repair	Approach channel and other offshore (mains)	
Marina workshop	lighting	
Transport workshop		
Battery generation		

8.C Electricity supply requirements

Except for sites which are remote or have abnormal ground conditions, the average marina should not present greater difficulties in electrical layout, wiring or equipment than any other site. Special care in installation is, of course, required in wet conditions and materials may have to be above average in quality. The supply to moorings and other off-shore areas may require special thought and unusual fittings but nothing beyond the scope of an experienced electrical contractor.

Wiring to the berths will be distributed from waterproof on-shore chambers which will incorporate the requisite fusing and earthing arrangements. From here the distribution cables of MICC or sheathed and insulated flexible PVC are taken, in conduit, along ducts provided in the walkways. Flexible joints or loops will be necessary between pontoon sections. The dangers of flooding can be overcome by installing within the codes of practice for such conditions. Advice on metering will be necessary from the supply authority. When one console is shared by several berths, individual check meters will usually be included for each slip. Another system, less precise, and less satisfactory, is to fuse each outlet to the wattage specified and paid for by each boat owner: this limits the maximum power but not the duration of intake.

Overhead cables are sometimes seen in older marinas but they are unsightly and dangers arise unless they can be run where boat masts have no possibility of contact.

In designing the layout it is important to ensure that all the off-shore section of the system is separated from the on-shore supply by means of an isolating transformer. This will normally be positioned where the cable leaves the shore, perhaps near a central ramp. From the transformer chamber the supply will divide along the walkways, each leg being provided with a circuit breaker.

Information boards

In some marinas, where serious boating competitions at national or international level are expected, an electronic information board will be necessary. Few people other than the judges can get out (or are allowed out) into competition water. Information on times and positions is communicated from the stewards' boats to a control room beside or 'within' a visual information display board. This is strategically positioned on-shore for maximum visibility so that spectators, using binoculars, can see the craft (preferably from an elevated position—the clubhouse roof for instance) and the display board without moving their positions. Such boards are usually capable of time-elapsed display with immediate winning-time registration. They are operated from a teleprinter typewriter keyboard and can be fed automatically from punched tape.

The proportion of marinas that require such elaborate equipment is small, but where they are necessary they add to the spectators' enjoyment and are of great assistance to sports writers, commentators and television crews. Whilst small consoles recording simple information can be installed within clubhouses, larger displays would be expected at major sailing centres. The National Water Sports Centre at Holme Pierrepont on the River Trent near Nottingham, England, has an information board 29 m × 4·2 m (95 ft × 14 ft) and weighing 14 tons. It has over 5,000 lamps which can consume 200 kw and is positioned on the opposite side of the rowing course nearly 300 m (1,000 ft) from the main spectator stands.

Lighting

7. It should be noted that B.S. C.P. 1004 is particularly useful for a marina development as it covers most circumstances that are likely to be encountered. *Lighting for roads with special requirements (Group F)* C.P. 1004: Part 8: 1967 is of especial interest as it deals with street lighting in the vicinity of docks and navigable waterways.

Lighting[7] deserves special attention. The atmosphere, appearance and, to a large extent, the function of a marina changes considerably after dark. The main purposes of artificial lighting should be:

1   To ensure safety for pedestrians, vehicles and those aboard
2   To guide craft safely in and out of the harbour, its environs and approach channels
3   To enhance security
4   For public amenity and atmosphere
5   To illuminate direction signs and notice boards
6   To display goods for sale or hire

The overall design must bear in mind all the seasonal conditions of the marina after dark—not only the warm summer evenings but the cold and windy conditions that often prevail. A marina can be a dangerous and gloomy place without adequate artificial light and piers and walkways present obvious hazards at night-time, particularly to the visitor unfamiliar with its layout.

It should achieve the above objectives unobtrusively, so that the equipment used looks as good by day as by night and integrates with the total design concept without interference with day or night views. Important vistas from the water as well as the land must not become cluttered with too many columns.

258

In using individual standard lamps and floodlighting on the piers, the emphasis should be on illuminating the *edges*. This can be effectively emphasised by the use of white or fluorescent paint along verges of piers, quaysides and handrails.

Particular attention should be paid to lighting those areas likely to be used at night such as:

Walkways	Access roads
Ramps and steps	Car parks
Main footpaths	Gatekeeper/information office
Public lavatories	Vulnerable security points
Boat park/hardstanding	Lock/harbour entrance
Boat handling equipment	Lifebelts, hose reels

Except perhaps for car parks and access roads, the light sources are best kept low and numerous because high, glaring light is less effective and may present navigational confusion. Lighting of the on-shore grounds, boat-handling equipment and car parks, etc. should be at least to the standards necessary in other public places, not only for safety and convenience, but for the protection of property as marinas seem to suffer from theft and vandalism more than most places. Perhaps more enterprise and experiment needs to be shown in the lighting of marinas. Interesting and pleasant effects are often seen in lighting water and fountains in parks and other places. Under-water illumination, floating water-level flood lighting, controlled colour changes and low-level illumination of quay walls are ideas which have been recently suggested for harbours. This may be appropriate only for urban sites and even then may only be used on certain occasions.

Lighting the water

Whether it is still or moving the water surface will reflect the lighting on land provided that the land and water levels are similar. Where the water level is much lower, either permanently or by tidal movement then a 'shadow' will be cast on the water and moorings alike. This darkened margin, gaining neither from direct nor reflected land lighting will require special attention. Stepping from firm land onto moving pontoons and from light into darkness can be unnerving and requires lighting designed to bridge the gap. The normal low-level walkway lamp may be insufficient and in addition it may be helpful to illuminate the quay walls and the ramps (particularly the top and bottom) leading from it. Such lights should be glare-free and aim to light horizontally so as to train the light onto the wall at all levels of the tide. This type of system is usually better than installing lighting *within* the quay wall which is vulnerable to inundation and corrosion or lighting *on* the wall which, for similar reasons, requires to be on columns which may look inappropriate and spoil both on-shore and off-shore views.

Lighting the grounds

It is unnecessary and uneconomic to light evenly the whole ground area. The ambient (i.e. background) lighting in most marinas is likely to be low and derive from sporadic sources such as moonlight and the clubhouse etc. In form and function lighting divides into utility or service needs and amenity lighting. In considering the total provision of illumination, calculations should be based only on the former, allowing the decorative or amenity supply to be extra to the normal standard, not subtracted from it. The lighting of access roads up to the ownership boundary is the responsibility of the local authority (usually the city, borough or county council) and discussions with them and the Electricity Board will ensure continuity of illumination and design of equipment across the boundary line.

An early decision must be made regarding the actual type of light sources to be used. The choice is wide and will affect the cost, appearance and maintenance of the system. The principal characteristics of the more commonly available lamps are as follows:

8.D Lamp types

Type of lamp	Brightness	Type of colour	Life	Lamp cost	Operating cost	Principal uses
Tungsten general lighting	Very high	Good	Short	Low	High	Floods, signs
Tungsten reflector	Beam	Good	Short	Low	High	Small floods
Tungsten halogen	Very high	Good	Short	Medium	High	Floods
Low-pressure fluorescent tubes	Very low	Very good	Short	Low	Low	Signs, some floods, tunnels
Low-pressure sodium	Low	Yellow	Long	Medium	Very low	Streets, tunnels, floods
High-pressure Mercury (uncorrected)	High	Blue	Long	Medium	Fairly low	Floods
High-pressure Mercury fluorescent	Medium	Good	Long	Medium	Low	Streets, floods
High-pressure sodium	High	Yellow	Fairly long	High	High	Streets, floods
Mercury–Iodide	High	Good	Fairly long	High	High	Streets, floods
High-tension Cold cathode	Low	Many	Very long	High	Medium	Signs

Telephones

The number and location of telephones will vary according to the status of the marina. If outside instruments are considered necessary it is usually sufficient to place one at the end of each walkway. Alternatively, where changing rooms, lavatories and lockers are grouped together, one or two instruments may be incorporated in kiosks or booths outside. Public telephones may be included within or near the clubhouse and instruments for members are usually available in the club itself. An internal telephone system can be a most useful aid to staff efficiency in a large marina and particularly helpful during an emergency.

Public address systems

Although not always popular public address systems do allow quick communication for convenience and emergency use. For yacht clubs and specialist sail training centres they may be especially useful but when installed in ordinary marinas the management will need to ensure restraint if annoyance to marina users and nearby residents is to be avoided.

Alternative energy sources and conservation

Since the first edition of this book the need to conserve energy and perhaps to consider on-site generation has grown considerably.

Seaboard and estuarine sites enjoy natural advantages in this respect and it is provident to utilise it. The benign influence of the sea raises the average annual temperature above the equivalent inland sites and contributes a levelling effect on seasonal temperatures—cooling the summers and warming the winters.

Solar energy

The weather along many coastlines tends to be different from that being experienced inland at the same moment: enjoying sunshine when the hinterland is overcast or suffering sea mists when sunshine prevails inland. Its annual record for temperature and radiation is however rather higher for most countries than its inland equivalent. Many marinas are well located to take advantage of solar energy, for coastal sites are usually free from overshadowing and enjoy a clear atmosphere for solar rays.

Marina buildings may take advantage of the boating season being coincident with longer days and warmer sunshine. Regional solar tables and radiation charts are available in most countries. The marina buildings most likely to benefit from solar heat are restaurants, dry boat stores, committee rooms, marina offices and information centres or clubhouses. Ancillary buildings such as hotels, holiday flats, public houses, offices, customs or harbourmasters' accommodation may also be considered, as well as outside facilities like swimming pools.

Feasible systems include solar water and space heating, solar cooling and electric power generation but for most marina purposes the choice will fall between various types of panel collector, either integral with the roof or free-standing.

Wind power

The location of marina sites may warrant consideration of wind power as a generator. Techniques and various proprietary systems of collection and conversion will need to be chosen.

Tidal or wave energy

This method is not, at present, a serious contender as a power supply for marinas it demands large-scale application and most devices are at present experimental. Any proposal for off-shore provision would be no doubt carefully sited and marked to minimise danger to commercial and recreation vessels.

Energy conservation

This increasingly important constituent is particularly essential in exposed areas. As well as the site consideration of orientation and protection from the elements the buildings' external envelope should be designed to minimise heat loss. Double glazing, effective insulation for roof and walls the control of solar gain and the conservation of hot water are all essential components rather than optional extras in modern construction.

8. Statutory services: Check list

General

See page 240 (Logical Procedure Chart)
Contact appropriate authority:
   local river, coastal, harbour
   local government
   local utility company
   local supplier/distributor (for LPG etc.)
   local contractor or sub-contractor (for installation etc.)
Establish likelihood of supply and get firm date in writing
Investigate capacity to serve present needs and future expansion
Survey ground condition and levels for buried services
Find out condition of existing services
Check made-up ground
Examine obstructions, tree roots, water table, etc.
Consider easements
Determine pedestrian paths and 'desire lines'
Relate service runs to footpath pattern
Check depth and relative positions of services
Consider grouping services
Determine order of laying the services
Examine trenching methods (preferably with contractor)
Consider import of granular fill
Prepare general layouts (individual and collective)

Water

Read General Check List above
Calculate total consumption needs now and to serve future expansion (see Check Lists, pages 149–150)
Establish adequacy of supply, quality, pressure
Remember fire-fighting needs
Establish local water board and fire service requirements (i.e. materials, layout, pipe sizes, metering, back syphonage)
Is main storage tank necessary—if so, what size, height, design?
Design layout for
   on shore
   off shore
Determine berth to tap ratio, pipe diameters and tap positions
Consider design of walkway ducts, flexible joints, heat expansion, type of lagging, type of tap (e.g. lost key, non-concussive)
Determine system of on-shore to off-shore piping in tidal conditions
Remember frost, locate drain cocks (lowest points in system)
Get local water board/fire service approval of layout and details

Fire fighting

Read General and Water Check Lists above, also Fire Prevention section of Chapter 13, pages 331–332
Ensure adequacy of mains supply
Locate couplings, extinguishers, fire points, fire boat, hose reels, sprinklers, alarms, etc.
Match combustible products with correct extinguishing material (e.g. water, powder, chemical, foam liquid, etc.)
Arrange servicing and maintenance programme
Instruct staff in operation and duties

Mains gas

Read General Check List above
Determine pipe runs, material, diameters and depth
Position valves and traps for condensate
Arrange inspection and regular maintenance with Board

Liquefied petroleum gas

Consult national supplier (local branch) and fire service
Consider need for cylinder (Butane) and bulk (Propane) supplies
Estimate total demand of each remembering construction equipment, fork lifts, workshop
    needs, generators, etc.
Agree tariffs, rebates, etc.
Consider storage of cylinders including 'empties'
Determine means of handling—forklift, trolley, etc.
Location and design of bulk or cylinder store
Consider layout, pipe sizes, jointing, protection, etc.
Consult Codes of Practice (provided by supplier)
Obtain instructions and safety advice for staff and users

Main drainage

Read General Check List above
Consult local drainage authority
Is existing, or will proposed type be combined or separate system?
Obtain plans of existing layout.
Determine pipe sizes, material, jointing, depth, falls, ventilation
Position manholes, interceptors, etc.
Consider trenching methods, consolidation and back-fill.
Prepare overall layout
Leave system open for inspection and test by local authority.

Soakaways, septic tanks, etc.

Determine volumes involved now and for future expansion
Locate positions—anywhere convenient for soakaways but leeward of prevailing wind for
    cesspools and septic tanks
Consider proprietary systems but ensure local authority approves
Comply with safeguards under Water Resources Act 1963
Warn users about detergents
Submit final scheme for local authority approval and arrange with them about sludge
    removal and maintenance
Arrange for landscape treatment and planting

Chemical closets

Determine likely total volume
Decide whether self-service or marina service
Locate emptying points to discharge to mains, septic tank or collection tank—if last, then
    arrange about removal
Arrange for landscaping and planting

Electricity

Read General Check List above
Contact local electricity board
Determine likely total demand now and for further expansion
Consider supply needs during construction (voltage, current, single and three phase)
Is a transformer chamber needed?
Consult Board about choice of cable(s)
Agree tariff(s) with Board
Allow for damp conditions
Decide on underground or overhead supply—if former consider trench and depth
Consider trunking and conduit
Arrange (with electrical engineer) the wiring layout
Ensure good quality materials and workmanship
Determine the design and type of consoles, socket outlets and meters to moorings
Remember flexibility between pontoons and ensure neat shore-to-moorings system in tidal
    basins
Submit final scheme to electricity board for approval
Prepare maintenance programme

Lighting

Prepare overall lighting plan and standards of illumination
Consider capital, running and maintenance costs
Decide on both functional and amenity lighting
Consult appropriate B.S. Codes of Practice (see page 263)
Emphasise danger points, steps, ramps etc. (see pages 257–258)
Consider effects upon navigation, local environment and long distance views
Decide suitability of high mast lighting, floodlighting, fluorescent paint and special effects
    within grounds and on or under water
Decide switching system (manual, automatic, photo-electric etc.)

Remember security aspects, along boundary and at berths
Consider needs at approach channels, harbour entrance, lock, etc.
Ensure tall equipment does not obstruct or spoil views
Pay particular attention to car parks and access roads
Consult with local authority where its lighting meets marine responsibility
Prepare maintenance programme

Telephone and public address systems

Determine cable runs
Locate telephones in public, well-lit areas
Internal telephones for staff (and public?) use for communication and emergency
Consider telephone security link with G.P.O. or police

8. Statutory services: Bibliography

General

'AJ Handbook of Building Services and Circulation', *The Architects' Journal* 1970.

British Standards Institution:

Ducts for building services C.P. 413: 1973.

Graphical symbols for pipes and valves B.S. 1553: Part I: 1949.

Light gauge copper tubes (light drawn) B.S. 659: 1967.

Malleable cast iron and cast copper alloy screwed pipe fittings for steam, water, gas and oil B.S. 143: 1968.

Mechanical ventilation and air conditioning in buildings C.P. 352: 1958.

Building Research Station/Department of the Environment, *Simplified Tables of External Loads on Buried Pipe Lines*, H.M.S.O., London 1970.

Burrows, W. N., *Coordination of Underground Services on Building Sites*, Public Works and Municipal Services Congress, London 1970. *Reprints available from the D.O.E.*

Design Council, The, *Street Furniture from the Design Index*, The Council, London 1973.

Coordination of Underground Services, Film (Seven minutes). *Available from The Film Library, Room 211, Neville House, Page Street, London S.W.1.*

Institutions of Civil Engineers and Municipal Engineers, Joint Committee of the, *Location of Underground Services*, The Joint Committee, London 1963.

Ministry of Public Building and Works/Directorate of Building Development, *Co-ordination of underground services on building sites: 1 The common trench*, H.M.S.O., London.

Oates, J. A. (ed.), *Pipes and Pipelines Manual* (Fourth Edition), Scientific Surveys Ltd., Beaconsfield 1972.

Alternative energy sources and conservation

Allen, R., *Proceedings of the Solar Heating and Cooling for Buildings Workshop*, Washington D.C. *Part I* 1973, *II* 1974.

American Institute Of Architects, *Dwelling Design and Site Planning for Solar Energy Utilization*, AIA Research Corporation, Washington 1975.

Department of the Environment, *Solar Energy: An Annotated Bibliography*, Public Services Agency Library, Croydon 1976.

Szokolay, S. V. *Solar Energy and Building*, The Architectural Press Ltd., London 1975; Halsted Press, New York 1975.

Westinghouse Electric Corporation, *Solar Heating and Cooling of Buildings. Phase 0. Final Report*, 4 vols, Baltimore 1974.

Electricity and lighting

British Standards Institution:

Aluminium street lighting columns B.S. 3989.

Concrete street lighting B.S. 1308.

Electric lamps B.S. 161.

Electric street lanterns B.S. 1788.

Electric wiring systems C.P. 321: 101.

Electrical installations C.P. 321.

Fused plugs and shuttered socket outlets B.S. 1363.

Polyvinyl chloride insulated cables B.S. 2004.

Provision of artificial light C.P. 3, Chap. VII.

Provision of electric lighting in dwellings C.P. 324: 101.

Provision of electricity and service cables for small houses C.P. 322: 101.

Plug and socket outlets B.S. 546.

Steel columns for street lighting B.S. 1840.

Steel conduits and fittings B.S. 31.

Switches for domestic and similar purposes B.S. 3676.

Street lighting C.P. 1004.

The testing approval of domestic electrical appliances B.S. 3456.

Department of the Environment, *Electrical Distribution Systems External to Buildings*, M. & E. No. 2, H.M.S.O., London 1968.

Department of the Environment, *Electrical Installations in Buildings*, (Maude No. 1) H.M.S.O., London 1971.

Electric sign Manufacturers Association, *Journey into light*, The Association, London 1964.

'Electricity on Building Sites', *Electrical Times*, IPC Electrical-Electronic Press Ltd., 1972.

The Electricity Council, *Electrics 1972/73: Handbook of Electrical Services in Buildings*, The Council, London 1972.

The Illuminating Engineering Society, *I.E.S. Technical Report 6: The Floodlighting of Buildings*, The Society, London 1964.

The Institution of Electrical Engineers, *Regulations for the Electrical Equipment of Buildings*, The Institution, London 1970.

Miller, H. A., *Guide to the Wiring Regulations*, Peregrinus, London 1970.

National Playing Fields Association, *Floodlighting of Outdoor Sports Facilities*, The Association, London 1970.

Fire

British Standards Institution:

Classification of fires B.S.–EN 2: 1972.

Dry powder portable fire extinguishers B.S. 3465: 1962.

Fire hose couplings and ancillary equipment B.S. 336: 1965.

Fire protection for electronic data processing installations C.P. 95: 1970.

Heat sensitive (point) detectors B.S. 3116: Part I: 1970.

Hydrant systems C.P. 402.101: 1952.

Portable carbon dioxide fire extinguishers B.S. 3326: 1960

Portable fire extinguishers of the foam type (chemical) B.S. 740: Part I: 1948.

Portable fire extinguishers of the foam type (gas pressure) B.S. 740: Part 2: 1952.

Portable fire extinguishers of the halogenated hydrocarbon type B.S. 1721: 1968.

Portable fire extinguishers of the water type (gas pressure) B.S. 1382: 1948.

Portable fire extinguishers of the water type (stored pressure) B.S. 3709: 1964.

Precautions against fire C.P. 3: Chapter IV: 1948.

Precautions against fire in flats and maisonettes (in block over two storeys) C.P. 3: Chapter IV: Part I: 1971.

Precautions against fire in shops and departmental stores C.P. 3: Chapter IV: Part 2: 1968.

Precautions against fire in office buildings C.P. 3: Chapter IV: Part 3: 1968.

Portable fire extinguishers for buildings and plant C.P. 402: Part 3: 1964.

Sprinkler systems C.P. 402.201: 1952.

Department of the Environment and Fire Officers' Committee Joint Fire Research Organisation, *Fire Research 1970: Report of the Fire Research Steering Committee*, H.M.S.O., London 1971.

Department of Trade and Industry and Joint Fire Research Organisation, *Movement of Smoke on Escape Routes in Buildings: Symposium* 4, 1969.

Fire Offices Committee, *Rules for automatic sprinkler installations* (Twenty-ninth Edition).

Fire Protection Association: (Fire safety data sheets):

Automatic fire alarm systems F.S. 6005.

Fixed fire extinguishing equipment F.S. 6004, F.S. 6006, F.S. 6007.

Portable fire extinguishers F.S. 6001, F.S. 6002.

Fire Protection Association, *Fire Prevention Design Guide*, Planning For Fire Safety In Buildings Series, The Association, London 1969.

Hobson, P. J. and Stewart, L. J., Fire Research Note 958: *Pressurisation of escape routes in buildings*, Joint Fire Research Organisation 1972.

H.M.S.O., Fire Note 5: *Fire venting in single storey buildings*.

H.M.S.O.: (Principal statutory documents)

Building Regulations 1965.

Factories Act 1961.

Fire Precautions Act 1971.

London Building Acts.

London Building (Construction) Bylaws.

Offices, Shops and Railway Premises Act 1963.

Public Health Act 1936.

Joint Fire Research Organisation, *Fire in hotels*, Fire Research Paper 23, 1969.

Joint Fire Research Organisation, *Relative fire frequency of different industries*, Fire Note 7, 1966.

Joint Fire Research Organisation, *U.K. Fire and loss statistics*, 1970.

Langdon-Thomas, G. J., *Fire Safety in Buildings: Principles and Practice*, A. & C. Black Ltd., London 1972.

Gas

British Standards Institution:

Domestic butane-gas-burning installations in boats, yachts and other vessels, C.P. 339: Part 3: 1956.

Domestic butane-gas-burning installations in caravans and other small non-permanent buildings, C.P. 339: Part 2: 1956.

Domestic butane-gas-burning installations in permanent dwellings, C.P. 339: Part 1: 1956.

Domestic propane-gas-burning installations in permanent dwellings, C.P. 338: 1957.

Installation of metering and meter control for town gas, C.P. 331: Part 2: 1965.

Installation pipes for town gas, C.P. 331: Part 3: 1965.

Installation of service pipes for town gas, C.P. 331: Part I: 1957.

Gas Council, The, *Data Sheets for Architects, Builders and Heating and Ventilating Engineers*, The Council, 1969–70.

Liquefied Petroleum Gas Industry Technical Association: (Codes of practice)

Installations of Fixed Bulk L.P.G. Storage at Consumer's Premises, No. 1, 1969.

Maintenance of Fixed Bulk L.P.G. Vessels at Consumers' Premises No. 8, 1969.

Recommendations for prevention or control of fire involving L.P.G. No. 3, 1968.

Safe handling and transport of L.P.G. in bulk by road No. 2.

Heating and ventilation

British Standards Institution:

Asbestos-cement flue pipes and fittings, light quality, B.S. 567: 1973.

Cast iron boilers for central heating and hot water supply, B.S. 779: 1961.

Cast iron flue pipes, B.S. 41: 1973.

Central heating by low pressure hot water, C.P. 341.300–307: 1956.

Heating and thermal insulation, C.P. 3: Chapter VIII: 1949.

Mechanical ventilation and air conditioning in buildings, C.P. 352: 1958.

Pre-cast concrete flue blocks for gas fires (of the domestic type) and ventilation, B.S. 1289: 1945.

Power driven circulations for heating plants, B.S. 1394: 1971.

Room heaters burning solid fuel, B.S. 3378: 1972.

Thermal insulation in relation to the control of the environment, C.P. 3: Chapter II: 1970.

Ventilation, C.P. 3: Chapter I (C): 1950.

Department of the Environment, *Heating, Hot and Cold Water, Steam and Gas Installations for Buildings*, (M. & E.) No. 3, H.M.S.O., London 1968.

Gas Council, The, *Data Sheets for Architects and Heating, Insulating and Ventilating Engineers*, London, April 1970 onwards.

Kell, J. R. and Martin, P. L., *Faber and Kell's Heating and Air Conditioning of Buildings* (Fifth Edition), The Architectural Press Ltd., London 1971.

Water and drainage

Asher, S. J., *Water Supply and Main Drainage*, Crosby Lockwood & Son Ltd., London 1961.

British Standards Institution:

Asbestos cement flue pipes and fittings, light quality, B.S. 567: 1973.

Automatic flushing cisterns for urinals, B.S. 1876: 1972.

Ballvalves (excluding floats) piston type, B.S. 1212: Part I: 1953.

Ballvalves (excluding floats) diaphragm type, B.S. 1212: Part II: 1970.

Building drainage, C.P. 301: 1971.

Calorifiers for central heating and hot water supply, B.S. 853: 1960.

Cast manhole covers, road gully gratings and frames, for drainage purposes, B.S. 497: 1967.

Capillary and compression tube fittings of copper and copper alloy, B.S. 864: 1953.

Cast iron baths for domestic purposes, B.S. 1189: 1972.

Cast iron pipes, bends, branches and access fittings, B.S. 437: Part I: 1970.

Cast iron spigot and socket pipes (vertically cast), B.S. 78: Part I: 1961.

Cast iron spigot and socket fittings, B.S. 78: Part II: 1965.

Ceramic wash basins and pedestals, B.S. 1188: 1965.

Ceramic washdown W.C. pans (dimensions and workmanship), B.S. 1213: 1945.

Cesspools, C.P. 302.200: 1949.

Combination hot water storage units (copper) for domestic purposes, B.S. 3198: 1960.

Concrete cylindrical pipes and fittings, B.S. 556.

Copper and copper alloy traps, B.S. 1184: 1961.

Copper indirect cylinders for domestic purposes, B.S. 1566.

Copper tubes (heavy gauge), B.S. 61: Part I: 1947.

Copper tubes (light gauge), B.S. 659: 1967.

Copper tubes to be buried underground, B.S. 1386: 1957.

Domestic electric water-heating installations, C.P. 324.202: 1951.

Draw-off taps and stop valves for water services, B.S. 1010.

Fire clay sinks, B.S. 1206: 1945.

Floats for ballvalves (copper), B.S. 1968: 1953.

Floats for ballvalves (plastics), B.S. 2456: 1954.

Frost precautions for water services, C.P. 99: 1972.

Galvanised mild steel indirect cylinders, B.S. 1565.

Metal lavatory basins, B.S. 1329: 1956.

Metal sinks, B.S. 1244.

Polythene pipe (Type 32) for cold water services, B.S. 1972: 1967.

Polythene pipe (Type 50) for cold water services, B.S. 3284: 1967.

Sanitary appliances, C.P. 305. 1952.

Sanitary pipework above ground, C.P. 304: 1968.

Small sewage treatment works, C.P. 302: 1972.

Thermal storage electric water heaters, B.S. 843: 1964.

Wastes for sanitary appliances, B.S. 3380.

Water supply, C.P. 310: 1965.

W.C. flushing cisterns, B.S. 1125: 1969.

W.C. seats (plastics), B.S. 1254: 1971.

Department of the Environment Film Library, *Coastal Sewage Dispersion Studies*, Film (eleven minutes).

Department of the Environment, *Advisory Leaflet 24: Laying Rigid Drain and Small Sewer pipes*, H.M.S.O., London 1971.

Department of the Environment, *The Report of the Technical Committee on Back Syphonage in Water Installations*, H.M.S.O., London 1973. *See Chapter II 'Protection of water Used in Docks and Harbours'*.

Department of the Environment, *Third Report of the Working Party on the Design and Construction of Underground Pipe Sewers*, H.M.S.O., London 1971.

Little, Harry R., *Design Criteria for Solid Waste Management in Recreational Areas*, U.S. Environmental Protection Agency, The United States Government.

Ministry of Agriculture, Fisheries and Food, *Leaflet 44: Mole drainage for heavy land*, H.M.S.O., London 1960.

'Pumping equipment for Water Supply', *Surveyor*, October 1973, Vol. 142. Conference paper.

'Sewage, Irrigation and Land Drainage', *Surveyor*, October 1973, Vol. 142. Conference paper.

Twort, A. C., *A Textbook of Water Supply*, Edward Arnold Ltd., London 1963.

9 Boat handling and storage

Boat handling

| | Neap | Spring |
Port	tides	tides
London	4·8 m (15¾′)	6·5 m (21½′)
Liverpool	4·6 m (15¼′)	8·3 m (27′)
Bristol		
(Avonmouth)	6·6 m (21¾′)	12·2 m (40′)
Hull	5·0 m (16¼′)	6·3 m (20¾′)
Glasgow	3·3 m (10¾′)	3·9 m (12½′)
Southampton	—	4·0 m (13′)
Belfast	2·1 m (6½′)	2·9 m (9½′)
Antwerp	3·8 m (12′)	4·9 m (16′)
Hamburg	1·7 m (5½′)	1·9 m (6¼′)
Rotterdam	1·5 m (5′)	2·0 m (6′)
New York	—	1·5 m (4¾′)
Wellington		
(N.Z.)	0·8 m (2¾′)	1·2 m (4′)
Mediterranean	Nil	Nil

9.A Range of tides at some ports of the world

Selecting the equipment

Boat-handling equipment and boat storage are so closely allied that it is necessary to consider them as aspects of the same question. Together they are an extension of the whole marina concept which is in itself primarily a handling and storage problem.

In any marina or harbour for pleasure craft there is a need to launch and retrieve at least some of the vessels as a minimum requirement. The amount of equipment, its cost, position, type and relationship with other facilities will, at the lowest level, be an important management decision and with large developments will probably warrant a feasibility study involving most of the project team. As an illustration of the close relationship between storage facilities and boat-handling equipment both are available as proprietary systems usually needing minimum adaptation to site conditions and may be purchased by the marina operator in the same way that a factory owner might purchase a warehouse and stock-handling system. At whatever level the operation is mounted the decision on the magnitude and sophistication of such equipment will stem from a series of value judgements according to:

1 The type of marina
2 The number and size of berths and craft anticipated at each stage

At many dinghy or sailing clubs on inland sites boats are either manhandled or launched and retrieved on trailers and in these cases a good ramp may be all that is necessary. At the other end of the league table the demand for handling equipment for a coastal marina of over 1,000 berths with a boat-building and repair yard attached would be very considerable.

It is not necessarily a good idea to cater for large vessels on first opening the marina, particularly if this denies service to the average boat owner. Large craft have a greater range allowing them a large choice of marina services and it is not easy to capture the maintenance and repair custom from their normal yard. An early decision is needed however if equipment requires to be set into the structure of the development. As this applies most often to the heavier machinery a conflict may arise between installing cheaper equipment initially in the hope that custom will come, and waiting until custom has built up before adding the facility at greater expense later. An example of compromise in this situation would be to build the *dock* into the waterside and install the hoist a few seasons later or reinforce the quay wall in order that the derrick or forklift can be accommodated in a year or two.

Existing harbours adapted to marina purposes present their own problems and it is difficult to generalise. Usually, however, it is not easy to install equipment into an ageing fabric and old harbours do not generally berth sufficient craft to justify a full range of plant. Handling and storage represent the fastest growing facilities after the demand for the moorings themselves and any medium-to-large marina which is to be competitive will need to incorporate some facility of this kind.

Equipment is often purchased and installed without sufficient thought being given to its real contribution in terms of individual work-load and profitability. Dennis Sessions, the authority on marina economics, recounts how he encountered the dissatisfied managers of one American marina who complained that their expensive (British-made) hoist was useless. Upon inspection this proved absolutely true for it was in a state of complete neglect, only rarely used and never maintained. There was no boatyard, no repair service and very little demand from owners for launching or retrieval. Such attitudes increase overheads and waste capital, space and labour.

As with many marina components it needs to be decided by the management whether each separate area of operation or each piece of equipment must be directly profit earning or may be regarded as a loss leader. This dilemma is most common in deciding upon equipment to cater for large craft, for whilst its cost may be heavy and its use limited below an economic level, the large-boat owner spends the most money. It is in this respect that affiliation with a boat yard brings the greatest benefit.

In matching the equipment to the marina it is useful to bear in mind the three main categories giving rise to the need for handling:
1   Launching and retrieval
2   Service, repair, maintenance, refitting and surveying
3   Storage
Other reasons include emergencies, marina maintenance and self-service but these cover a much smaller field of operation. Within these generalisations a host of different types of tackle may be found, some being standard proprietary equipment and others purpose-made to suit an individual need or overcome some site difficulty. The design of such individual pieces often shows great ingenuity, however, for the purpose of description and definition the following list of equipment covers most of the more common, standard pieces.

Boat handling equipment

1   Winches
Used individually winches simply pull boats on trailers up ramps or boats on cradles up an inclined rail. They are also used as part of more complicated systems of launching and hauling. Modern electric motors are so safe, compact and powerful that a neat and simple winch system is still a good, inexpensive proposition for the modern marina and, for small craft clubs, they may be self-operated.

2   Gantries
These are fixed frameworks supporting one or two overhead trolley-rails, the craft being cradled and hoisted in slings suspended from them. The simplest types straddle a docking area, lifting the craft from water to trailer. In tidal basins boats are sometimes winched up a ramp at low tide until below the gantry frame. Gantry systems are often linked with storage or

9.1 This old winch still pulls boats up the shingle at Dunwich, Suffolk boatyard areas, especially for larger craft. There is a tendency for them to be superseded in many areas by travelling lifts for the largest craft and forklifts for the smaller ones which can handle masted and keeled boats more easily.

3   Railways

These are usually used for fairly heavy boats and consist of a substantial structure supporting railway lines upon which runs a cradle supporting the vessel as it is winched up the slope and out of the water. Roller tracks are an alternative. The whole construction is either fixed to a heavy duty concrete ramp or else to capped pile supports in concrete timber or steel. The cradles are adjustable to each keel shape and are sometimes self-powered but more usually winched up by motor and cable either direct or through a series of pulleys and sheaves.

Railways, like gantries are lessening in popularity because modern hoists and lifts are faster and require less staff and maintenance. However for large, masted boats they should still be given serious consideration particularly in gently shelving situations. Unlike most other handling equipment they are almost invisible from a distance, and, where necessary, their rather heavy-engineering appearance from close-to could be readily improved with light cover-plates or panels.

4   Platforms and lifts

This is a vertically-moving staging upon which rests a cradle to support the keel and hull, the elevated craft being then transferred to a trailer or dolly. Lifts are suitable for deep-water situations, steep banks and bulkhead walls. In locked marinas or inland sites their vertical travel will be short and in other areas the height of lift is related to the tidal range.

The framework is usually timber or steel piles to which are attached the lift guides. The platform and cradle is lifted from beneath the floating vessel by a winch and cable system attached to the platform at four or more points. The latest designs are neat and unobtrusive with nothing projecting above the raised platform which only rises about 2 m (6 ft) above ground level to cater for the deepest keels.

5   Hoists

This type of equipment originated on Lake Michigan in the late 1940s and quickly became popular. The early models were tractor-towed but self-propulsion has improved manoeuvrability and reduced operating space. They are basically mobile gantries which, having straddled a wet dock and lifted the craft by slings, can then transport it over uneven ground to a repair yard, storage bay or trailer. The more expensive models are a remarkable advance on most gantry systems having a telescopic 4·2–8·5 m (14–28 ft) wheelbase and the great advantage of an opening boom-gate in one crossbeam to cater for masted vessels or those with high-flying bridges. The size-range spans from 8–50 tons weight of craft and 4·9–21·3 m (14–70 ft) in length. They are very fast and gentle workers, one-man operated and include many fail-safe features. Power is usually electric/hydraulic. Being mobile they can stand under cover in winter to avoid deterioration.

6   Cranes

Fixed or mobile cranes are used to move craft of up to 20 tons and step the

9.2 A fixed gantry boat-hoist at San Francisco. The bank surface is unfinished

9.3 A. H. Moody & Son's hard-working Renner Hoist at Swanwick Marina, near Southampton, England

9.4 This 'dead' 'Travelift' boat hoist handles boats up to 5,443 kg (12,000 lb) in weight and 13·71 m (45 ft) in height. It can be towed by any fork-lift truck or light vehicle

9.5 A 23,000 kg (50,000 lb) Renner Comporter

9.6 A 5-ton crane at Charleston, South Carolina

9.7 Do-it-yourself dinghy crane at a San Francisco marina

masts of larger boats. Caterpillar tracks or pneumatic tyres are used on the movable types. Stationary cranes are usually high masted with block and suspension cables. Small, simple self-operated jib-cranes are popular for dinghy launching.

7 Derricks and shearlegs

These are more common and more useful in boat yards than marinas. Sometimes secondhand equipment from naval yards or docks is used, but even with adequate slings the one-point suspension can make awkward handling and two or more men are usually required to control any swinging or slipping. They may be unusable in windy conditions.

8 Davits

These can be employed singly for handling small craft or the engines of larger boats or, more commonly in pairs for lifting or launching medium-sized vessels. The swinging types are the best although, being independently winched by hand or motor, they need to be carefully operated.

9 Fork lifts

Particularly where linked with a multi-storey dry storage system this equipment can be very versatile. Standard plant can easily be adapted by the manufacturer to serve either as fixed quayside lifts or mobile high-mast trucks stacking to 10 m (33 ft) or more. The larger vehicles can lift 6,800 kg ($6\frac{1}{2}$ tons) to a height of 4 m (13 ft)—greater heights being unusual as heavier craft are restricted to the lower racks. Adjustable fork spacing and length together with swivel-pads enable fork lifts to adapt to most shallow drafted or keeled boats; masted craft present no problem except of course for multi-storage and outboard engines can remain in position. Forklifts can of

9.8 This fixed hoist together with a fork-lift truck can lift any unmasted boat up to 9·14 m (30 ft) from the water and stack it on to a 7·62 m (25 ft)-high storage cradle

9.9 The same fork-lift boat-hoist in operation at Sanford, Florida

9.10 Given a comfortable gradient and a good natural surface, really small craft may not need a ramp at all

9.11 This roller ramp is designed for use in tidal waters. Outboard boats are handled and launched easily by guiding the boat on the rollers which are installed on the ramp section. Trailers or dollies are used to transport the boat to the top of the ramp or dock. The ramp is held in position by strap hinges on the dockside and supported on the waterside by two small floats which also serve as boat landing facilities. All floating ramps should be installed in sheltered waters

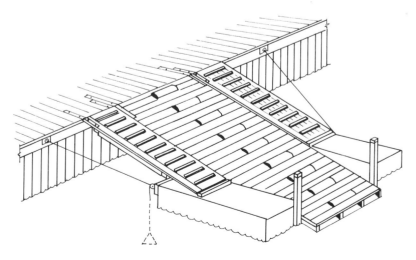

course be used for many marina jobs apart from getting craft in and out. Small boats can be moved complete with cradle and materials and stores and pontoons can be carted about with ease.

10 Tractors and trailers

The combination of tractor and trailer is the basis of the 'haul-out' marina. Either the owner's or the company's trailer is used. For large craft it is usually the latter where trailers are purpose-made to cater for most types of boats. Keel craft, however, require special deep-framed trailers which are not so common. Fairey Marine's Yard on Southampton Water is the best known example in the United Kingdom. The system is versatile and inexpensive requiring little or no dredging and no piling or bulkheads. All that is required (and all that is visible) is a ramp of about 1:10 with a length depending upon tidal range and a width of approximately 9·14 m (30 ft) per vehicle.

11 Launching ramps

Whilst not strictly 'equipment' the great family of ramps, slopes and slipways are the commonest form of getting afloat and vary from a natural sandy beach to the semi-commercial types supporting heavy-duty marine railways. They may be 'fixed' or floating and made of concrete, timber, steel planking or stabilized earth. Varieties include roller ramps, floating pontoon types, articulated ramps or those constructed of concrete or precast non-slip blocks.

Simple, self-launching ramps should have the gentlest of inclines and the inclusion of self-operating winches will be at the developer's discretion. The blend at the top of the ramp must not catch loaded trailers or vehicles and the access from road or car park must allow for backing and turning vehicles and boats. The degree of slope in fixed ramps is generally 5° (1:18)–15° (1:6). A greater range is possible if rollers or winches are incorporated. On steep inclines a bollard or pulley is sometimes positioned at the ramp top. Fixed ramps become over-long and less practical with a big tidal range, whilst hinged ramps should not incline to a dangerous degree at low water. All ramps, particularly floating ramps should be installed in sheltered waters.

12 Other items

In addition to the equipment already described there is a variety of individual equipment such as turntables and yacht hangers etc. Cradles, dollies and trailers may need adapting for certain craft and ingenuity may be needed to receive or deliver from some types of low loaders and special trailers but this may be preferable to turning away custom.

Whilst not strictly boat-handling equipment, the normal transport and machinery of the marina workshop should not be forgotten. The harbour's own lorries, trailers, vans and runabouts will need capital to purchase, house and maintain. Equipment within the boat yard is likely to be a major item and may be considered separately unless it is treated as an integral part of the marina project.

Installation

All lifting and hauling sites must allow for ample depth of water to cater for the maximum likely draft and the shore construction must be designed for

9.12, 13 Launching ramps at Key West, Florida and Shilsole Bay, Seattle

9.14 Roller-launches at Mission Bay Park, San Diego, California

9.15 A caretaker's 'Cushman' electric runabout at Lake Munroe, Florida. In a large marina the total length of piers and quays may be several miles and such vehicles are a necessity. Goods and staff models are available

9.16, 17 Some kind of work boat is indispensable in most marinas. The paddled punt (9.16) has limited uses but the Rotork 'Sea Truck' (9.17) has an unusual degree of flexibility and can handle a variety of heavy or bulk cargoes

the considerable point loads imposed by craft and equipment. Purpose-made docks for gantries and hoists are best designed to the manufacturers requirements. The least expensive kind are piled finger piers but these may not conform to regulations governing projections from the shore line. A neater and more integrated method is the recessed well which generally requires bulkheading on three sides, usually by sheet piling. Sometimes a corner may be used where one side of the hoist would travel on shore and the other on a pier.

Ramps will need hard-wearing non-slip surfacing with regular clearing of weed and slime below the water line and special arrangements for icy conditions. Coverings to protect ramps or even electric wiring within the concrete are possibilities but are not usually worthwhile because in fresh-water if the ramp is frozen so too is the sailing water, and in coastal areas the problem is uncommon for salt-water will only freeze at about 29°F (− 1°C), depending on salinity and requiring placid water—conditions unlikely to be met with in most sailing areas. It may be necessary however to warn users about the slippery ramps whether caused by ice, algae or seaweed, because it is easy to slip into the water under these conditions and difficult to climb out—particularly if the person is alone.

Safety with all equipment must be a major consideration with regular servicing assisting in this aspect, in addition to increasing the machinery's efficiency and longevity.

Most boat-handling equipment is robust, well made and quite expensive. It pays to look after for it is expected to work hard and reliably in an exposed environment.

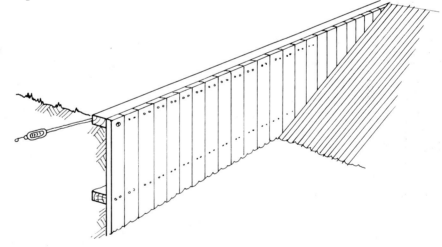

9.18 In the development of some launching ramps it is often desirable to cut back into a bank, away from the shore line, to form the gradual slope of the ramp. To hold the earth embankment in place on each side of the ramp, small bulkheads are constructed. In some cases the bulkhead should extend into the water some distance, in order to protect the underwater section of the ramp from erosion by currents and wave action

9.19 The collection bay at Sanford, California, where owners leave their boats to be dry-stored or pick them up at 30 minutes notice

Dual use

Modern handling equipment is fascinating when in action. People really enjoy seeing it at work and should be allowed to do so within the limits of safety and security. A point worth reiterating is the benefit of sharing equipment between the marina and any adjoining boat-building or repair yard. Except perhaps for a few hours each year in the high season, equipment is rarely used to capacity and dual use is unlikely to create a problem because the maintenance peak is Easter to mid-June and the sailing peak from then until mid-September.

Of all marina functions the moorings and the boatyard have most to gain from each other and such a combination is well worth exploring during the initial planning of the project.

Boat storage

Advantages to owners and management

One of the greatest benefits to be derived from adequate boat storage provision is that by enabling boats to remain at the site, the marina continues to attract visitors off season. This is particularly necessary in areas where one cannot rely upon a 'passing trade' of visiting vessels.

In many marinas in America and nearly *all* marinas in the UK no adequate off-season storage or protection is available and largely because of this the whole expensive development 'dies' for half of every year, boats being trailed or sailed home or to other basins where storage and maintenance are offered.

The economic success of almost any small craft harbour hinges upon the length of its season. In the battle to extend the harbour's annual operational life the provision of good secure storage and quick reliable handling is arguably a greater draw than that other prerequisite—a lively clubhouse. According to the Meteorological Office in the six winter months October to March inclusive there are on average 82 days when more than 6 hours of safe daylight sailing can be enjoyed in the Solent area of England's south coast. The figure for inland sites would probably be well over 100. The off-season attraction of being able to launch one's craft quickly, sail for a few hours and return it to safe storage would give advantage to most marinas and, with modern vessels, handling and storage should be easily organised. The joys of frostbite sailing are not however universal for whilst owners in mild climates may enjoy off-season visits, it is usually for social or maintenance reasons.

Long-term storage

The simplest form of storage is for craft to remain at their floating moorings either in the open or at permanent covered berths. This is not very satisfactory although better than an off-shore anchorage outside the harbour. Boats wintering at wet berths may need to be moved to other slips whilst piers and walkways receive off-season maintenance. In sheltered conditions boats floating at marina moorings should be safe in winter, but in more open tidal basins they may run the risk of damage from ranging and buffeting during severe storms.

Out-of-water storage varies from hauling on to the beach, hardstanding or dinghy park to setting on legs or cradles for larger craft. Large craft may be either weatherproofed and left in the water or brought ashore onto

9.20 Severe storms are always a danger in Florida. Heavy construction at Fort Lauderdale

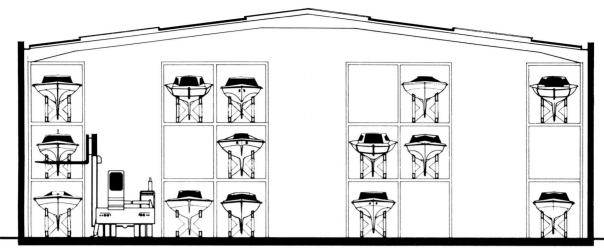

9.21 This covered store houses boats on three levels. The hoist, handling keeled boats, is of heavier duty than the fork-lift type

9.22 This multi-stack boat store in Florida can hold 240 vessels. It also contains a boat sales area, offices and maintenance bays. The scale of such buildings may bring planning problems

9.23 An interior view of the boat store

9.24 Local enterprise: a covered mooring at San Diego

9.25 Multi-level storage of small craft in California

9.26 Simple painted timber storage racks for small boats at Del Rey Yacht Club, California

hardstanding. Few will be under cover as this involves costly and spacious structure. Modern craft suffer far less harm from being out of water than their older counterparts, whose strakes opened up and timbers dried out.

Keels and masts are a difficulty with all covered storage systems. If the mast can be stepped the problem is eased because most medium-sized boats will then store under a 4 m (10 ft) roof and this can be lowered further if the keels are housed in a docking pit.

Choosing the system

The sophistication of the storage system should be related to the quality of the marina as a whole and management will need to determine its policy according to the marina's status. Many marinas include hardstanding where owners may store and maintain boats themselves. This may be essential in a simple municipal development, but poor policy in a luxury harbour where owners need encouragement to use the covered boat store and service yard.

Where hardstanding is provided it may be used as car parking in the season when more boats will be in the water. It is an advantage to be close to the clubhouse, chandlery or some form of shelter. It will need to be decided if a nominal charge is to be made. The covered areas of floating berths are usually charged at 50 per cent more than uncovered berths and usually cater for the larger vessels whose equipment would otherwise deteriorate more rapidly.

Boat stacker storage

With more advanced handling and storage arrangements a distinction should be made between those which house craft on a long-term basis and the boat-stacker type, designed to allow in-and-out movements on daily demand throughout the year. The latter is ideal for grp outboards up to about 7·5 m (25 ft) but there seems no reason why unmasted keel boats could not be similarly stored on a multi-storey principle in a staggered pattern. The need to unmast the craft would make frequent launchings impractical, but land-use economies may validate such a system, particularly in costly urban areas. There is little doubt that boat-stacker systems will grow in popularity particularly where power craft predominate. Their rapid increase on the mild southern seaboards of America stems more from convenience and security than from shelter but the added benefit of weather protection should increase the demand for similar arrangements in harder climates. After a telephone call the owner's boat is launched and gassed-up waiting in a collection bay within minutes. Upon returning, the outboard and even personal gear can safely remain aboard. A single forklift can handle 20–25 movements each hour depending upon carrying distance and average height of lift.

An American solution

At Sanford in Florida, the realisation that over three-quarters of the 240 boats catered for were fibreglass, less than 9·1 m (30 ft) in length and that about 10 per cent were shallow-draft power-boats 9·1 m–12·1 m (30–40 ft) long the management has covered 85 per cent[1] of its range by installing the following equipment:

1 A generous 18·2 m (60 ft)-wide slipway where most vessels may be launched

2 The 'fork' section of a heavy-duty forklift truck has been purchsed, adapted and placed in permanent position in an angle of the quay. This can lift any vessel up to 9·1 m (30 ft) overall length out of the water whatever its level. When raised the boat is in position awaiting:

3 A mobile medium-duty forklift truck with a special 7·3 m (24 ft) lift. This carries the boat into the storage shed where boats are stacked up to 4 'shelves' high

4 A 5,400 kg (12,000 lb) Travelift hoist deals with larger boats for storing, launching and minor maintenance

Thus, not counting the slipway which is necessary anywhere, three simple pieces of equipment and a multi-purpose tractor deal efficiently with almost all needs. It must be admitted, however, that very few keeled and masted boats are kept in the area, although they could be handled if not dry-stored with the Sanford equipment.

The storage problems at Sanford, as in most southern States, are simplified by the climate which is never really severe enough to warrant the laying up of boats in winter. This very popular and financially successful venture provided at $1 per foot per month in May 1968 for the owner, at half an hour's notice, to have his boat fuelled and put in the water to await his arrival, and to be taken from the water and stacked at his departure 9.19.

Once decided upon, the siting and erection should be straightforward and an accredited storage system will include installation advice from the manufacturer. More difficult is the market and financial research to establish

1. The other 15 per cent comprised 8 per cent over 12 m (40 ft) boats at covered wet berths and 7 per cent under 4·2 m (14 ft) using slipway and trailer.

Out-of-season dry storage of boats is difficult on cramped sites incapable of expansion or where waterfront is prohibitively expensive. In America ancillary inland sites, some miles from the parent marina, are increasingly being bought for laying up boats between seasons. The advantages are purely financial: one marina operator in Florida claims a land-cost saving of $85,000, being the difference between $15,000 actually spent and $100,000 for a similar waterfront site. (See Hardman, Tom 'Are Dry Land Marinas Profitable?' *Boating Industry* August 1964).

The disadvantages however are also considerable. Ancillary-site security would involve either full-time staff or sophisticated alarm systems and impregnable fencing. Boat haulage to and from the inland site, particularly if large space-consuming boats are involved, may also be expensive and require additional insurance cover, and the permission of boat-owners may not always be forthcoming.

Although planning for bulky boat-storage buildings and high-security fencing may be tolerated in a waterside environment, planning permission for such large structures may not be easily obtained some miles inland. Furthermore, the character of the marina may be adversely affected by the removal of 'dead' boats causing an out-of-season 'dead' marina. This may be unavoidable in severe winter conditions, but otherwise the principle of encouraging year-round activity in the marina by social events, self-maintenance and other inducements would in general appear to be a sounder principle.

See also: 'Dealer Guide to Storage' *Boating Industry* November 1973

Hardman, Tom 'Dry Land Marinas' *Boating Industry* August 1963

the system's possibilities in the context of the total development. A further essential with buildings of the warehouse class (which such stores are likely to be) is an early visit to the local authority to ensure that the building's bulk is acceptable in planning terms.

Other retrieval and covered storage systems include overhead single and double rails and light marine ramps and railways, usually powered with a light electric motor. These are often designed to be linked with multi-stack or ground level storage sheds, running to them or through them on a looped system.

Small boat provision

It is generally uneconomic for management or owners to use these storage arrangements for sailing dinghys and other boats less than 4 m (14 ft) *but provision for them elsewhere is very important both to enliven the development and to encourage young owners whose craft and resources are likely to grow larger.*

Whereas the cost of providing berths for small boats is relatively more expensive than for large ones this does not necessarily apply to storage, where large craft can sterilise three or four times their own area—acceptable perhaps where space is plentiful but not otherwise. However there is less need for large boats to leave the water and when they do it is usually for service rather than storage.

Architectural considerations

The architectural possibilities of large storage silos are limited, for economy usually dictates a simple cubic form of light metal framing. The type and colour of cladding may offer alternatives where, if necessary, the building's impact may be lessened. Sometimes the boat-carrying framework is independent from the building envelope but in heavier structures the columns support both the boat cradles and the building. Flooring is usually granolithic or similar jointless screeding and must be accurately laid to assist in forklift handling—not always easy to reconcile with adequate drainage.

The principles of storage and handling are safety, speed, reliability and economy. Modern systems usually ensure the first three and the last depends upon a sufficiency of work.

9.27 The provision of covered mast-storage racks would overcome this untidy waste of space and prevent damage to boat-owners' equipment

A factor attractive to owners is the low insurance premium charged for craft in marina-based covered storage. The storage of items other than boats requires thought. Lockers on walkways or within utility buildings will hold small gear, tools and clothes—owners should be warned about aerosol cans, petrol and some paint strippers. Outboards are usually stored separately but may be incorporated within the trailer store.

Masts are sometimes chained to horizontal frames adapted from covered bicycle racks.

For small boats plain, neat arrangements are best. Dinghy parks are often a mess and trailer parks are worse. Simple racks such as the back-to-back E (9.26) save space, cut launching distance, give more shelter, improve security and reduce damage.

---

9. Boat handling and storage: Check list

Handling equipment

Consider handling and storage together, including all methods of launching and retrieval

Design adequate vehicular access and manoeuvring space

Relate equipment to marina needs regarding timing, cost, type, relationship

Seek manufacturers' advice

Examine suitable proprietary systems and compare with adaption of standard equipment or purpose-made designs

Evaluate structural needs at design stage

Examine nearby marinas and ascertain needs of potential patrons

Determine (at design stage) whether to have piled (built out) hoist, wells or recessed (built in) type

With launching ramps in sheltered water decide whether fixed or hinged and determine the incline, width and material

Consider rollers and non-slip surfaces

Decide on winches and whether hand-operated or electric

Investigate pros and cons of gantries and hoists versus platforms and lifts

Tie in boatyard needs with those of marina moorings (dual use)

Prepare maintenance programmes and safety measures

Remember non-boat equipment, i.e. workshop, transport, etc.

Storage

Establish value of boat storage in terms of land-use economics

Determine policy regarding long-term storage

Consider self-maintenance and provision of space, materials and equipment

Architectural considerations

Assess likely bulk of storage buildings and assess impact on local and long-distance views

Consult local planning authority

Consider planting, screening, impact and colour of materials

Remember non-boat storage—gear, lockers, trailers, cradles etc.

---

9. Boat handling and storage: Bibliography

Connell, Maurice H. & Associates Inc., *The River Port Marina for the City of Sanford, Florida: Study and Report for a Master Plan*, Maurice H. Connell & Associates Inc., Miami 1964.

Mack, Andrew R., *Boat Handling Equipment in the Modern Marina*, National Association of Engine and Boat Manufacturers, New York 1959.

Outboard Boating Club of America, *Boating Facilities'* Vols. 1–6, The Club, Chicago 1959–64.

Outboard Boating Club of America, *Small Boat Launching Ramps, Dock and Piers: Construction Hints, Drawings and Specifications*, The Club, Chicago 1958.

10 Bunkering facilities

1. For the control of petrol and oil pollution see Chapter 11, pages 292–294.

The need for bunkering facilities depends directly upon the number of boats with engines which are accommodated within, or which visit, the marina. Vehicles too may influence the decision if a double-aspect station is being considered.

Most oil companies disapprove of floating refuellers or direct filling by road tanker on operational and safety grounds. Petroleum products are highly inflammable, smelly and dirty and, because they float on water, their use at marinas is made more dangerous from possible spillage.[1] A layer of petroleum on a water surface constitutes a deadly hazard from ignition and instant fire-spread is likely amongst closely-moored boats. *Safety precautions are essential and apply to construction, location and operation.* They are highly technical and should be decided and *approved* in the early planning stage. Facilities at many existing marinas are poor—often the result of late and inadequate consideration.

The risk of fire spread should be minimised by providing a boom to form a bund around the bunkering area. One method is to use 101×76 mm (4×3 in) canvas-linked timbers arranged in an arc, with an opening section giving access and exit to the boat being refuelled. The positioning of the area within the marina will need to give attention to minimising the smell, the overall appearance, the manoeuvring of boats and refuelling tankers, tidal range and the wave and current condition throughout the year. If a combined facility for cars and boats is considered then the forecourt arrangements will be similar to the normal land-based station except for the special care needed in delineating and safeguarding the edge of the quay.

A split-level arrangement can give advantages for it allows sales for both vehicles and boats from a single station with the possibility of other items like ice, bait and vending machines being incorporated quite neatly within one integrated design. Oil companies will naturally be more interested in a two-sided station and the introduction of other sales features may turn a necessary but unprofitable service into an attractive proposition.

When designing bunkering installations provision should be made for an operator's kiosk containing all the necessary electrical switch gear associated with the pumps and situated in such a position that from it the whole refuelling operation can be supervised. In addition provision should be made for storage of bulk or canned lubricants, and bottled gas which are products also controlled under the petroleum regulations. The provision of an adequate fresh water main at the bunkering position is an advantage, the dispensing of which could be by a recoiling hose unit if costs will allow.

10.1 A good example of the 'split-level' fuelling station at Sanford, Florida. The canopy provides shade and illumination

10.2 A low-level view of the same installation

10.3 Old boats refitted as bunkering stations are not really satisfactory. A purpose-built fuelling station is better and safer

Pumping arrangements

The location of spirit tanks is governed by recommendations contained in the Petroleum (Consolidation) Act 1928 and administered by the Petroleum Officer or Fire Officer responsible to the local licensing authority. The tanks fills must be within 9·14 m (30 ft) of the road tanker access position and a minimum of 4·26 m (14 ft) from site boundaries or any building within the site unless the building concerned is constructed of extra fire-resisting materials to the satisfaction of the local licensing authority. Diesel and gas oil tanks are not covered by these conditions although the distance of 9·14 m (30 ft) maximum from the tanker off-loading position still applies.

The positions of spirit pumps is governed by the same regulations but in addition the suction pipe from tank to pump should rise a minimum of 38 mm in 3·04 m ($1\frac{1}{2}$ in in 10 ft) and lines should not exceed 30·5 m (100 ft).

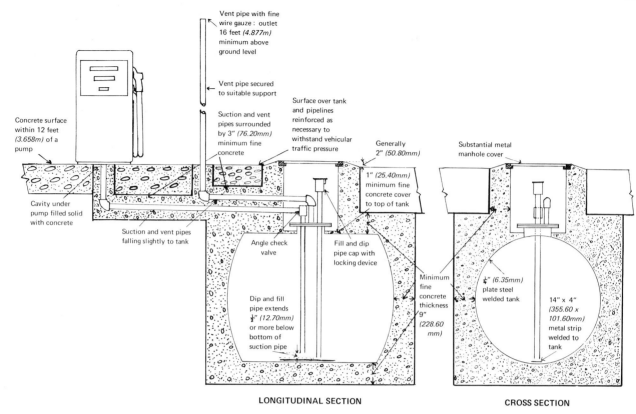

Vent pipe with fine wire gauze : outlet 16 feet *(4.877m)* minimum above ground level

Vent pipe secured to suitable support

Surface over tank and pipelines reinforced as necessary to withstand vehicular traffic pressure

Suction and vent pipes surrounded by 3" *(76.20mm)* minimum fine concrete

Concrete surface within 12 feet *(3.658m)* of a pump

Cavity under pump filled solid with concrete

Suction and vent pipes falling slightly to tank

Angle check valve

Fill and dip pipe cap with locking device

Dip and fill pipe extends ½" *(12.70mm)* or more below bottom of suction pipe

Generally 2" *(50.80mm)*

1" *(25.40mm)* minimum fine concrete cover to top of tank

Minimum fine concrete thickness 9" *(228.60 mm)*

Substantial metal manhole cover

¼" *(6.35mm)* plate steel welded tank

14" x 4" *(355.60 x 101.60mm)* metal strip welded to tank

LONGITUDINAL SECTION

CROSS SECTION

10.4 Typical lay-out of a petrol tank installation

Notes: Each tank to be surrounded except for the manhole opening by fine concrete not less than 228·60 mm (9 in) thick

Manhole chamber walls to be of fine concrete not less than 228·60 mm (9 in) thick

Tanks and pipes below ground to be protected against corrosion

Specification of fine concrete:

1 part sulphate-resisting Portland cement

2 parts dry clean sharp graded sand to pass 4·76 mm ($\frac{3}{16}$ in) mesh

4 parts clean gravel or crushed stone to pass 19·05 mm ($\frac{3}{4}$ in) mesh

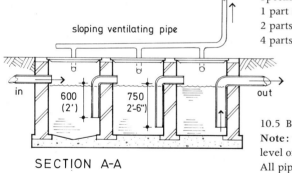

sloping ventilating pipe

in

600 (2')

750 2'-6"

out

SECTION A-A

10.5 Brick intercepting chambers for retaining petroleum

Note: the ventilating pipes from the chambers must not be joined together below the level of the lowest gully connected to the interceptor.

All pipes within the chambers through which liquid passes should be of iron, and all brickwork should be rendered with cement mortar, unless engineering bricks are used

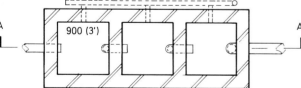

A

900 (3')

A

PLAN

10.6 Pontoon, office and ramp: bunkering service at San Diego

Pumps can work over 30·5 m (100 ft) but this is not recommended if the maintenance of pump motors is to be minimised. Booster pumps may be required over this distance. In many cases it may not be possible to locate the storage tank below the level of the pump and a falling suction line will result. To avoid build-up in such situations a pressure reducing valve must be fitted in the line prior to connection to the pump. Petroleum regulations prohibit the use of flexible lines between tank and pump, preventing pumps being fitted on to a floating pontoon attached to a fixed jetty unless the pumps have separate tanks on the pontoon which are replenished from the main storage tanks at set positions of the tide.

Within marinas, additional problems occur due to dispensing problems in refuelling boats, which are increased further if the bunkering location is subject to tidal conditions. Boats of many sizes and layout have to be catered for, with fuel-tank fill pipes of varying diameters situated anywhere on the deck and often obstructed by deck equipment and rigging. A tidal range as great as 12·19 m (40 ft) may have to be contended with. All these problems cannot be met or overcome without the design of special dispensing equipment. Standard filling-station pumps with longer hoses or swing arm attachments are used on most marinas, but the lack of hose length and obstruction of the swing arm by rigging makes operation difficult and often unsatisfactory. The use of pumps with long recoiling hoses may overcome many of the problems. A light-aircraft refuelling pump is manufactured by the Tokheim Corporation UK Division, Newark Road, Glenrothes, Fife, Scotland. This can be adopted for marine work with 15·24 m (50 ft) hose and electric rewind mechanism.

Fuel tanks

Storage tanks for petroleum spirit, diesel, gas-oil, etc., should be between 1,000–3,000 gallon capacity depending on demand to be met and any rebates that may be offered by the supplying company on bulk deliveries of fuel. Petroleum spirit is stored in cylindrical tanks of approved specification installed normally below ground and encased in minimum of 152 mm (6 in) sand or 1:2:4 concrete with adequate provision made to prevent

Tanks	Capacity	Size	Approximate Cost
Cylindrical fuel	4,546 litres (1,000 gallons)	3·5 × 1·4 m (11 ft 6 in × 4 ft 6 in)	£140
Storage tanks for	9,092 litres (2,000 gallons)	3·3 × 2·0m (10 ft 9 in × 6 ft 6 in)	£200
Installation above or below ground	13,638 litres (3,000 gallons)	4·1 × 2·1 m (13 ft 6 in × 7 ft 0 in)	£240
Single compartment	18,184 litres (4,000 gallons)	5·4 × 2·1 m (17 ft 9 in × 7 ft 0 in)	£270

Pumps	
Single spirit pump	£340–£350
Duo spirit pump	£670–£680
High-speed Derv pump	£370–£400
Spirit blender pump	£820–£930
Adapted aircraft refuelling pump with 15·24 m (50 ft) recoiling hose for spirit or Derv	£580–£610

2. Prices current in May 1972.   10.A Approximate costs of bunkers[2]

corrosion. If installed above ground the storage tank must be contained within an enclosure large enough to contain the full capacity of the tank should a leak occur. Tanks for diesel and gas-oil storage can be either cylindrical or rectangular and installed above or below ground. If above ground, bunded enclosures must be provided as for spirit tanks.

Fire precautions

Adequate fire extinguishers must be provided and kept properly maintained in positions adjacent to the pumps and tanks in accordance with the Petroleum License conditions. Means of access for fire-fighting appliances must be allowed for, both to the bunkering area and to suitable access positions throughout the marina development. Advice on suitable equipment and means of access should be obtained from the local fire officer.

10. Bunkering facilities: Check list

Procedure

Contact oil companies at the early planning stage
Assess number of power craft and vehicles accommodated and visiting
Negotiate terms for fuel supply, equipment, maintenance, bulk concessions
Determine type and capacity of tank(s)
    petrol
    diesel
    gas-oil
Can there be a combined station for craft and vehicles?
Consider double-aspect and split-level forms
Decide between land-based and floating stations
Consider above-ground bund and below-ground concrete casing
Locate station to minimise smell
        to ensure acceptable appearance
        to ease the manoeuvring of craft and tankers
        to obtain most sheltered conditions
Consider the effects of tidal conditions
Get advice from oil company about pumps, valves, spirit boosters, recoiling hoses and all special equipment
Check level and length of suction pipes
    minimum 38 mm in 3 m ($1\frac{1}{2}$ in in 10 ft)
    maximum length 30·5 m (100 ft)
Consider forecourt layout and quay edge design
Decide on lighting arrangements
Consider the integration of operator's kiosk
           vending machines and other sales
           switch gear
           lubricants, bottled gas, water
Prepare maintenance schedule

Safety

Contact fire and licensing authorities
Obtain requirements on location, construction and operation
Combat dangers from direct filling
        floating refuellers
        spillage, ignition and fire-spread
Collect information on fire-boom requirements
Check regulation distances of station from boundary
        station from building
        tanker from holding tank
Take special care about quay-edge delineation and hand-railing
Ensure adequate means of access for fire appliances

10. Bunkering facilities: Bibliography

Association for Petroleum Acts Administration, The, *Code of Practise*, The Association, Dudley, Worcestershire 1969.

Fire Protection Association, The, *for advice and pamphlets.*

Home Office Model Code of Principles of Construction and Licensing Conditions:

Section 1 *Storage of Petroleum Spirit in Cans, Drums etc.*

Section 2 *Petroleum Spirit Filling Stations.*

Section 3 *Electrical Equipment and its Installation.*

Section 4 *Petroleum Spirit Pumps*

Section 5 *Attended Self Service*

Section 6 *Unattended Self Service*

Section 7 *General Notes*

Section 8 *Conditions of License.*

British Standards Institution:

Cast and forged steel flanged screwed and socket-welding wedge gate valves (compact design) sizes 2 in and smaller, for the petroleum industry B.S. 3808: 1964.

Electrically bonded hose and hose assemblies for kerbside dispensing pumps B.S. 3395: 1972.

Forged steel pipe fittings, screwed and socket-welding for the petroleum industry B.S. 3799: 1964.

High pressure hose couplings for petrol, oil and lubricants B.S. 2464: Part 2: 1969.

Petroleum measurement tables (British edition) Handbook No. 15.

Petroleum measurement tables (Metric edition) Handbook No. 16.

Quick-acting hose couplings for petrol, oil and lubricants B.S. 2464: Part 3: 1968.

Steel pipe flanges and flanged fittings (nominal sizes $\frac{1}{2}$ in to 24 in) for the petroleum industry B.S. 1560: 1958.

Vertical steel welded storage tanks with butt-welded shells for the petroleum industry B.S. 2654: 1973.

Monopolies Commission, The, *The Supply of Petrol to Retailers in the U.K.*, H.M.S.O., London 1965.

Petroleum (Consolidation) Acts 1928 and revisions 1968.

Petroleum (Regulation) Acts 1928 and 1936.

Petroleum (Transfer of Licenses) Act 1936.

Petroleum (Mixtures) Order 1929.

Sedgewick, J. R. E. and Westbrook, R. W., *The Valuation and Development of Petrol Filling Stations* (Revised edition) The Estates Gazette Ltd., London 1969.

Town and Country Planning Use Classes Order 1963.

The Waleran Report: Petrol Stations, Report of the Technical Committee, H.M.S.O., London 1949.

See also the local requirements of: River Boards and Conservancies; Port and Harbour Authority regulations; Petroleum Licensing Authorities' Regulations; Model and Local Bye-Laws.

11 Pollution control

1. Royal Commission on Environmental Pollution, *First Report*, Cmnd. 4585, H.M.S.O., London 1971; *Second Report*, Cmnd. 4894, H.M.S.O., London 1972; *Third Report*, Cmnd. 5054, H.M.S.O., London 1972.

2. Consents in operation in August 1971 allowed 8,620,000 tons of sewage and wastes to be dumped into waters around England and Wales alone.

3. Department of the Environment and The Welsh Office, *Report of a River Pollution Survey of England and Wales 1970*, Vol. 1, H.M.S.O., London 1971; Vol. 2, H.M.S.O., London 1972.

Marinas are affected by pollution in several ways depending upon their nature, size and location. It is unfortunate but, when considering the water areas throughout the world, marinas are most likely to be developed where pollution is relatively high.

Between the coastline and the open seas there are the so-called tidal waters: as the Ashby Report on environmental pollution pointed out[1] these waters are shallow, close to towns and to industry and used not only for recreation but as a convenient and cheap sink for domestic and industrial wastes. Perhaps the least affected areas are the more steeply shelving coastal sites where dissipation lessens the problem but even here, as can be seen in 11.10, 11.11 recreation is having to share the water with enormous quantities of sewage and trade wastes. Big estuaries suffer even more in this respect for most of them have become great centres of industry and they are not subject to all the legislation and surveillance which (in the United Kingdom) exists to control pollution in rivers. Britain is fortunate in being surrounded by seas which are subject to strong currents and a considerable tidal range. Nevertheless, very large quantities of effluent and wastes are discharged into the sea.[2] The Department of the Environment's 1970 Survey[3] showed that the seven largest river authorities in England and Wales between them shared over 184 km (114 miles) of tidal river which was described as grossly polluted and 322 km (200 miles) that was considered to be urgently in need of improvement. Pollution affects marinas in two ways. Directly, by causing immediate and obvious unpleasantness and indirectly by the recognisable harm that it does to the natural scene. Plants, birds and fish are affected and the whole environment that patrons and visitors come to enjoy is depressed and debased. Sailing is very much an environmental recreation and any assault upon environmental quality immediately lowers the appeal of the sport.

Pollution is either generated by the marina itself or imposed upon it from outside. The former is a management responsibility and if the harbour is being adversely affected it may be the management's duty on behalf of its patrons to make representations to the offenders, if known, or to complain to the local authority or the Department of the Environment. Pollution and its effects will be an important factor in site selection. With an established marina it may be necessary for the management to monitor the situation by obtaining readings from the water authority both from within the harbour and from the sailing waters nearby.

Pollution and its prevention is now of enormous public interest. No developer can disregard its importance and no authority can permit a scheme likely to add to its effects in any way. Inland sites are particularly vulnerable

11.1 Sewage suction-pump disposal unit, Chicago

11.2 Refuse is often a problem and warning notices seem little help

11.3 Complete with livestock: litter and garbage in the Southern States

to abuse because of their size and location. However, even the largest water areas have their problems. All five of the Great Lakes of North America for example suffer in this way and all of them come within the top fifteen of the world's largest lakes. Lake Erie had six of its thirty-two public recreation areas closed down by 1965, the water being considered unfit for contact by humans—and this despite its huge area of 26,000 km² (10,000 miles²). Perhaps this is not so surprising when one considers the almost unbelievable story of the River Cuyahoga, which flows into it. This stretch of water is officially classified as a fire hazard because in 1969 its surface oil combined with river-bed sewage gases, ignited and burned down two bridges.

Contamination is a particularly serious problem in yacht harbours because pollution, in its broadest sense is what people have come to *avoid* whilst at the same time by their numbers and mood they are quite likely to *create* it. Problems created locally may be solved locally, but pollution often originates from where its impact is felt, and often outside the jurisdiction of the suffering authority. Pollution is diverse and cumulative, it can be unhealthy, unaesthetic or both. A paper cup on the ground causes no physical harm but looks bad: carbon monoxide is invisible but deadly. This chapter deals with several aspects of pollution separately. It will be the task of the management to provide adequate means of preventing or minimising any nuisance.

Sanitation (for drainage see also Chapter 8, pages 252–254)

The possibility of connecting into a main sewage system in remote areas may be an important factor in getting planning permission and the cost of providing an alternative sufficient to satisfy the local authority may become a critical economic question. Where main drainage is available the sewage system within the on-shore grounds of the marina does not usually present problems any greater than those found in other types of development, except perhaps in made-up land where settlement is still likely and the land is low enough to warrant sewage pumps.

The provision of baths, showers and WCs for entertainment, sporting or club facilities will be governed by the nature of the activity and the numbers of people anticipated at peak times. It is not sufficient only to provide these things within the clubhouse unless special arrangements are made for their permanent use by non-members, including use at night-time. In addition to private amenities, an average provision of 1 WC and 1 shower for each sex per 20 berths, would be an adequate provision with this being adjusted to average boat size and local conditions. The problems of sewage disposal abroad, however, are considerable (see Chapter 8, pages 252–254). Methods of control vary from asking crews not to flush their lavatories whilst in the marina and trusting them not to do so, to actually locking and sealing the heads of all boats entering the basin. This may be an extreme measure in a tidal sea-shore marina, but in a locked, freshwater basin it is probably essential.

The question of sanitation in harbours is vexed and contentious. Some authorities take great pains, others less so but there seems little doubt from the views of medical officers and health inspectors, that the presence of untreated sewage does not present a really serious health hazard. The objection is one of aesthetics rather than hygiene but this view may be tempered according to conditions and climate. Although the amount of marine pollution from pleasure boats is minute compared with sewage outfall and chemicals there is every reason not to add to the grave situation now facing most of the world's oceans.

Nevertheless, chemical closets emptied into the river or sea do far greater harm than crude sewage. In fact all methods of treatment by chlorination, antiseptics, germicides and other chemical additives by individuals or authorities are questionable as there is a likelihood of the treatment being a greater pollutant than the sewage. An alternative to sealing a craft's lavatory is to empty it by a suction-pump unit. This draws the effluent from a craft's holding-tank, afterwards washing the tank through and discharging the contents into the main sewage system. Holding-tanks aboard smaller vessels are still not common in Britain and a point for emptying and cleaning chemical closets may therefore be necessary.

Rubbish disposal

Rubbish disposal is a most important consideration and one upon which, fairly or not, the whole image of the marina may be judged. Few things are more likely to detract from custom than the presence of rubbish or litter scattered about the harbour. Even when adequate provision for litter disposal is made, frequent emptying of the receptacles and *tidying up around*

11.4 A neat inexpensive solution. The bags withstand wind, rain, soggy rubbish, vermin and wasps

11.5 The water hyacinth: pretty, but a nuisance

11.6 Water hyacinth choking the launching ramp at Sanford, Florida. The marina was quite free the previous evening, the weed being brought in with the morning tide

them is essential. A further penalty of neglecting this is the encouragement of rats and nothing is more off-putting than to see this kind of livestock in the harbour. With the increase of disposable plastic and metal containers the amount of floating rubbish is growing. The removal of rubbish from the water is usually by manual means: an unpleasant but not too difficult task, as there is usually one corner of the harbour into which it collects.

Disposable paper bags on racks leave no receptacles to be cleansed and are therefore preferable to bins. But whatever system is adopted it will need regular daily attention—probably twice daily in summer when flies and wasps are about and the marina is at its most active. Strategic placing of litter containers is important and they are probably best sited at the land end of each walkway. It is easy to underestimate the cubic capacity required, for much on-board catering involves pre-packed food. Approximately $0 \cdot 015$ m³ ($0 \cdot 5$ ft³) per berth daily is recommended.

The positioning of disposal stations will depend upon the layout and the average number of berths per pier. The maximum size of receptacle will vary with the system adopted but if $0 \cdot 08$ m³ (3 ft³) is about the most that can be carried by one man, then 3 or 4 such bins would be needed at each 20-berth pier-end assuming a daily collection. Probably 3 would normally suffice with 4 provided at peak periods.

Vehicular collection by truck, trailer or low loading platform eases and speeds up collection. There may be difficulty along floating walkways, handcarts or 'dollies' being sometimes used and at one American marina quite large containers are winched from the floating walkway to the top of the quay wall by a breakdown truck which self-loads and deposits them at a central collection station where they are removed by the local authority's pick-a-back container system. Central collection points within the marina site—usually near the service yard—are quite common because whilst the local authority will often agree (or is obliged) to collect refuse from a marina it may not be willing to pick up from many small containers, particularly along the floating walkways.

A considerable advantage with disposable sacks, bags or linings is that the weight, bulk and cleansing of the receptacle is removed and nothing has to be returned to the disposal stations thus halving the total travel distance—new sacks being provided when the full ones are removed.

Algae

Although this is not a serious problem in sea-shore marinas or those on tidal or fast-flowing rivers, algae can cause concern in lakes, locked marinas or in locations where the flow of water is unreliable. Serious problems have arisen in the past at Lake Zurich and quite recently at Lake Washington in Seattle. Algae will grow in most water which is not kept moving by natural or artificial means but it is more common in fresh than salt water. The problem, however, is not usually serious until nutrients such as animal manure, fertilisers or sewage are introduced into the water. When it rains much of the fertilisers run off farms into lakes and rivers and a caravan or camping site, a faulty drain from a nearby housing estate or direct pollution by cattle, may all provide the nutrients essential for algae growth.

The results can be very serious. At Seattle the pier was covered with a stinking red scum and it cost several million dollars to divert the sewage of half a million people. The scheme was finished in 1967, and as the nutrients were cut off, the harmful algae rapidly decreased. There is increasing concern with regard to this problem in the British Lake District in general and Lake Windermere in particular, where Dr. John Stockner, an American limnologist who has studied the Zurich and Seattle cases, has sounded a warning that one hot summer could turn Windermere into a scummy abhorrence, smelling of hydrogen sulphide. Apart from being a health hazard, algae causes danger to fish and animals, and spoils the appearance of the water, giving it a bloom that blows on the wind and carpets the piers and shore. The most effective controls are to ensure that nutrients do not seep into the water and that sufficient flow or circulation is maintained. Spraying is expensive and often kills fish and plant life.

Petrol and oil[4]

The pollution of seas and rivers was the subject of the Jeger Report, an intensive study published in 1972.[5] As a result of its recommendations to control pollution in commercial harbours and docks, more stringent anti-pollution measures may be applied in future to marina developments.

The strict fire regulations regarding bunkering are referred to on page 282. In controlling petrol spillage for reasons of fire-spread the pollution question is also largely dealt with. When installed above ground, storage tanks for petroleum, diesel and gas oil must each be surrounded by a bund wall, enclosing a space large enough to contain the full capacity of the tank should a leak occur. Petrol traps or gulleys are usually required in vehicle service and wash down areas.

Apart from smelling and affecting plant and animal life, even a very small spillage of oil will cling to craft at the water-line, eventually forming an oily black band that is most unpopular with boat owners.

In addition to its responsibilities for minimising oil pollution within the marina and preventing its spread into outside waters, the marina management now has a new obligation to prevent serious oil contamination in open

5. Royal Commission on Environmental Pollution, *Third Report: Pollution in some British Estuaries and Coastal Waters*, Cmnd. 5054, H.M.S.O., London 1972.

4. In 1970 the Marine Exhaust Research Council (U.S.A.) carried out a thorough research programme into the effect of exhaust gases upon an unnamed American lake. From the results of this study some interesting observations were made by a simple application of the facts as they relate to recreational boating.

Engine operation on Lake X had consumed 340,552 gallons of fuel and oil *per year* for the previous 4 years. Based on an average consumption of 5 gallons per hour, this translated to 68,000 engine hours annually, or a fleet of 680 fishermen devoting 100 hours each to their hobby. In spite of these unusual conditions, examination of Lake X water and sediment by gas chromatography revealed no evidence of contamination by hydrocarbons found in exhaust water. Additional studies of the water and the sediment showed that phytoplankton and bottom organisms, which are necessary in a healthy ecological chain, were not affected by exhaust water hydrocarbons. Hydrocarbons in exhaust water are toxic, but the concentration necessary to kill fish or damage the ecological balance would require extremes far beyond the normal use of marine engines on rivers and lakes. To achieve with outboard motors in Lake X the same degree of toxicity which had proved dangerous to fish in the bio-assay experiment, would require the operation of 18,000 engines 24 hours a day for a period of one year.

Reducing the above activity to an eight-hour fishing day and a six-month fishing season raises the number of outboard motors needed to 108,000. This would require 77 boats per acre to produce a toxicity problem in Lake X. Quite obviously, this is a ridiculous and unreal parallel. An area of danger does exist at fuelling stations where quantities of gasoline and oil are spilled in careless operations. This source of pollution indicates the value of a training programme for marina personnel. But generally speaking, the conclusion reached was that the normal use of outboard motors constitutes minor pollution compared to the major contributors in industry, agriculture, and municipalities.

11.8, 9 The SLURP unit by Ambler Engineering, Walsall, Staffordshire, has been developed by Esso. It is normally used in connection with a boom to divert and collect the oil at the banks. It can also be used alone where wind concentrates oil at the edges of a lake, pool or harbour

11.7 Oil booms are a valuable first line of defence, particularly for tidal basins. This type, by RFD Ltd., London, has inflated tubes supporting an under-water screen and regular mooring lines. The screen's curve encourages a lineal inshore flow where the oil may be collected by a skimmer or soaked up with straw and lifted out

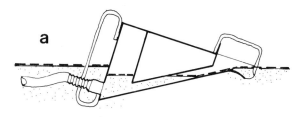

a

(a) Fast pumping gives a deep weir immersion suitable for the rapid recovery of thick oil slicks

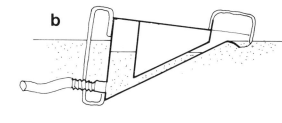

b

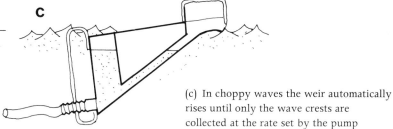

c

(c) In choppy waves the weir automatically rises until only the wave crests are collected at the rate set by the pump

(b) As the slick thins out the pumping rate should be slowed down. The liquid level inside the SLURP rises and the skimmer tilts to reduce the weir immersion

waters from entering the marina. In some areas the local authority will formulate a strategic plan. Locked marinas or those on enclosed waters or with a narrow entrance channel are in a much stronger defensive position and emergency measures involving a floating boom across the entrance or the closure of locks may prove sufficient protection even if this means stopping all movement to and from the harbour during a crisis. Elsewhere, on open water, if the problem is soluble at all, it may involve an extensive barrage or coastal defence, probably beyond the means of the marina management, although it should know the position and perhaps press for adequate counter-measures.

However good the safeguards may be, a plan for remedial action should be ready, together with the necessary equipment. Some owners will wish to have their boats taken from the water when warning is given, and the management must be prepared for this. Chemical and detergents must be used only with great care and understanding and the manual or mechanical method is still the safest. Some recent mechanical appliances for tackling petrol spillage and contamination are helpful. One that is effective and inexpensive is the Self-Levelling Unit for Removing Pollution (SLURP) 11.8, 9. This is an oil skimmer for inland and sheltered waters, streams, rivers, lakes, harbours and estuaries. It is the size of a large suitcase and weighing 28 kg (60 lb), it floats and can be used with any on-shore tank and self-priming pump. It is self-adjusting to suit the pumping rate and the wave conditions and costs about £350 complete with bouyant hose and attachments.

Noise

Noise is a very real and intensive form of pollution. Water reflects noise and with hard surfaces and buildings embracing the harbour area, a sound-box effect can easily be created. Elemental noises rarely seem a problem. The sound of wind and waves is a sailor's lullaby but other sounds are not. Transistors, televisions, car doors, engines and loud voices may all need control by persuasion or regulation from the marina management. The

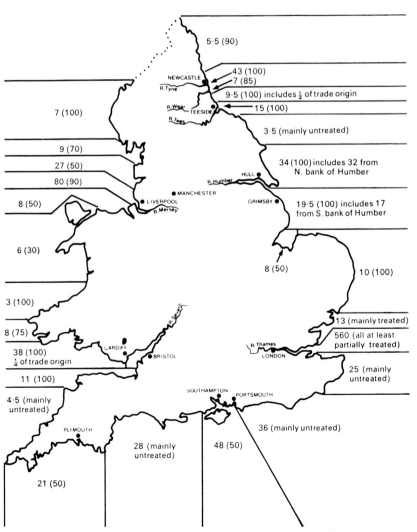

11.10 Discharges of sewage to estuarine and coastal waters in million gallons per day (percentage untreated in brackets)

continuous creaking of pontoons against piles can also be disturbing. The slapping of halliards and running rigging against metal masts is a perennial annoyance—particularly as it is curable with one minute of time and two pennyworth of string. One manager tours his moorings at regular intervals issuing £1 on-the-spot fines and he claims to have the quietest marina on the south coast.

Noise control should begin at the planning stage during which thought should be given to locating noise sources away from bedroom areas and moorings. The Wilson Report[6] emphasised the contribution towards the reduction of contractural noise which can be made by developers, architects and civil engineers, as it is their designs which largely govern the processes that the contractor must use. Designers should consider, in preparing their designs, whether there are alternatives which will involve less noise in construction. They should be specially careful to avoid the use of very noisy processes such as sheet steel piling, if the circumstances of the site

6. Committee on the Problem of Noise, *Final Report*, H.M.S.O., London 1963. Reprinted 1971.

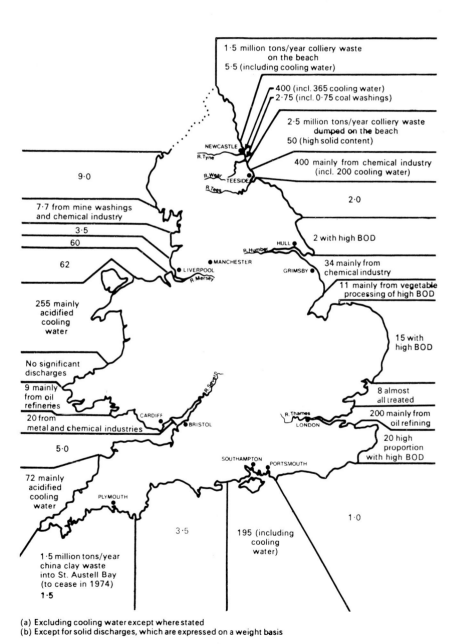

(a) Excluding cooling water except where stated
(b) Except for solid discharges, which are expressed on a weight basis

11.11 Discharges of trade wastes to estuarine and coastal waters, in million gallons per day

and the costs make this feasible. Dredging too, can be very noisy over considerable periods of time.

Some contractors are well aware of the importance of maintaining good relations with the people living and working nearby. It is quite common nowadays for contractors to provide viewing platforms with descriptions of the building to be erected and statements of the progress of the work. Similarly a contractor can help to make noise more tolerable to nearby people by explaining to them, perhaps by personal letter, why he has had to use noisy processes, the problems that he has had to overcome, and the steps he has taken to avoid disturbing the public more than is necessary. The problem is not confined to highly populated areas. Marinas located on flat rural estuaries or in quiet coastal areas may find that noise can travel far enough to disturb residents a considerable distance from its source.

The law relating to the control of noise in the United Kingdom falls broadly into three categories:

1 Local bye-laws
2 The Noise Abatement Act 1960
3 Common law

Local bye-laws cover specific issues upon which it has been felt necessary to make special legal provision. Marina managers should note the following:

1 Music (except religious services) is not permitted within 91 m (100 yd) of houses or hospitals *if a request to desist is made*
2 No annoying or disturbing sound is permitted between 11 p.m. and 6 a.m.
3 Radios (which, of course, includes TV sound) and record players are not permitted to form a source of annoyance
4 Restrictions are made on noisy selling, advertising or animals

7. Committee on the Problem of Noise, *op cit.*

The Wilson Report[7] points out that the noise from motor-boats on rivers, lakes and the sea has begun to cause complaint and in some areas to become a nuisance. With the growing popularity of motor-boating and water-ski-ing the nuisance is likely to become worse and more widespread unless preventive measures are taken. Some local authorities have made bye-laws under Section 249 of the Local Government Act 1933, or Section 76 of the Public Health Act 1961, requiring motor-boats to be fitted with "silencers suitable and sufficient for reducing, as far as may be reasonable, the noise caused by the escape of exhaust gases from the engine", or simply with "effectual silencers". Some local and navigation authorities look to the enforcement of speed limits as a means of controlling noise.

The Ship and Boat Builders' National Federation is considering the formulation of standards of noise for marine engines and that motor-boat racing organisations in this country subscribe to international rules which require all boats to be silenced to the satisfaction of the race organisers.

The Wilson committee thought that if, in spite of these measures, the noise from boats threatened to become serious and widespread then the most effective control would be to introduce noise limits on the lines of those recommended for motor vehicles and then review them every two years.

Legal aspects (see above for legal aspects of noise)
The legal aspects of pollution are complex and often involve international law. The principal Acts affecting United Kingdom waters are as follows:

1   The Oil in Navigable Waters Act 1955–71. These Acts make illegal the discharge of oil of any kind within United Kingdom waters

2   The prevention of Oil Pollution Act 1971 increases the maximum penalty for illegal discharges of oil to £50,000

3   The Intervention on the High Seas in cases of Oil Pollution Casualties Act 1969 makes it easier for governments to intervene to protect their coasts when an accident takes place

4   The Convention on Civil Liability for Oil Pollution Damage 1969 makes owners responsible for damage caused by oil spills

5   The Merchant Shipping (Oil Pollution) Act 1971 provides for the United Kingdom to ratify the above convention

6   The Convention on the Control of Marine Pollution by Dumping from Ships and Aircraft 1972. This relates specifically to dumping but contains pledges to take steps to prevent sea pollution

7   The Clean Rivers (Estuaries and Tidal Waters) Act 1960 gives river authorities powers to control outlets and discharges into tidal waters

8   The Rivers (Prevention of Pollution) Acts 1951 and 1961 gives the Secretary of State power to make "tidal waters Orders"

9   The Public Health Acts 1936 and 1961 both allow and control the discharge of effluent and wastes into public sewers

10   The Water Resources Act 1963 controls discharges into underground strata by wells, boreholes and pipes

11   The Deposit of Poisonous Waste Act 1972 makes it an offence to deposit poisonous wastes which threaten pollution of underground or surface water supply

12   The Salmon and Freshwater Fisheries Act 1923 and 1965 protects and preserves all species of freshwater fish

13   The Sea Fisheries Regulation Act 1966 allows bye-laws prohibiting deposits within the 3 mile limit

14   The Coast Protection Act 1949 regulates dredging and depositing spoil below the high water mark

11. Pollution control: Check list

General

With outside pollution detect origin and cause
Consider the effects—health
                            aesthetic
                            cumulative
                            legal
Decide policy on closets aboard
                        sealing of heads
                        holding tanks
                        emptying cabinets
See section on chemical closets in Chapter 8, pages 253–254
Consider mobile suction units

Sanitation

See drainage section in Chapter 8, pages 252–254
Contact Public Health Authority
Review existing and future likely conditions
    within harbour
    in nearby sailing waters
Ensure marina drainage system cannot contaminate harbour or surrounding waters
Provide disposal system for berthed craft
Establish means of monitoring

Rubbish disposal	Ensure adequate collection and disposal system in respect of capacity, location, lighting, maintenance, screen planting, etc.

Rubbish disposal

Ensure adequate collection and disposal system in respect of capacity, location, lighting, maintenance, screen planting, etc.
Instruct and deploy staff regarding duties
Integrate system with local authority collection methods
Consider proprietary container systems
Calculate likely volume of refuse from each source (craft, clubhouse, staffrooms, etc.)
Arrange for adequate enclosed central storage bin
Ensure adequate cleansing including hose-down facility
Provide adequate access for collection vehicles

Algae

Inland sites: check on all forms of nutrients seeping into enclosed fresh water (fertilisers, faulty drains, etc.)

Petrol and oil

Consult oil company and licensing authorities
Devise means of preventing spillage from fuel stations
Provide petrol traps where necessary
Acquaint patrons with oil spillage effects and dangers
Establish whether local authority emergency provision is adequate
Set up emergency action plan for harbour waters
Consider pump-skimmer and evacuation of craft from water area
Ensure adequate and effective insurance cover

Noise

Instruct builder to minimise contractual noise
Contact neighbours and explain situation
Ensure that patrons are noise conscious by persuasion or regulation
Take measures regarding transistors, TV, loud-hailers, public-address systems, starting guns, slapping halliards, creaking pontoons, boats' engines, generators, etc.

11. Pollution control: Bibliography

Aldous, Tony, *Battle for the Environment*, Fontana Books, London 1972.
Barr, John, *The Assaults on our Senses*, Methuen & Co. Ltd., London 1970.
Beranek, L. L. *Noise Reduction*, McGraw Hill, New York 1960.
Boyes, R. G. H., 'How to Remove Oil from Land', *Contract Journal*, November 1972, Vol. 250.
British Standards Institution:
   Sound insulation and noise reduction C.P. 3: Chapter III: Part I: 1960; Part 2: 1972.
   The storage and on-site treatment of refuse from buildings C.P. 306: Part I: 1972.
Bugler, Jeremy, *Polluting Britain: A Report*, Penguin Books Ltd., London 1972.
Central Advisory Water Committee, *The Future Management of Water in England and Wales*, Department of the Environment, H.M.S.O., London 1971.
Cole, H. A., 'Pollution of the seas', *Chemistry in Britain*, Vol. 7, No. 6, 1971.
Committee on the Problem of Noise, *Final Report*, H.M.S.O., London 1963. Reprinted 1971.
Connolly, Kathleen (compiler), *Marine and Coastal Pollution Bibliography No. 148c. Available in the Department of the Environment Library.*
Department of the Environment and the Welsh Office, *Report of a River Pollution Survey of England and Wales 1970*, Vol. 1, H.M.S.O., London 1971.
Department of the Environment and the Welsh Office, *Report of a River Pollution Survey of England and Wales 1970*, Vol. 2, H.M.S.O., London 1972.
Ducsik, Dennis W. (ed.), *Power, Pollution and Public Policy*, M.I.T. Press, Cambridge, Mass. and London 1972. See Chapter 3, 'The Crisis in Shoreline Recreation'.
Escritt, L. B., *Water Supply and Building Sanitation*, Macdonald and Evans Ltd., London 1972.
Food and Agriculture Organisation of the United Nations, *Seminar on Methods of Detection, Measurement and Monitoring of Pollutants in the Marine Environment*, FAO Technical Conference, Rome 1970.
Hamblin, Lynette, *Pollution: The World Crisis*, Tom Stacey Ltd., London 1970.
Harris, C. M. (ed.), *Handbook of Noise Control*, McGraw Hill, London 1957.
Isaac, P. C. G. (ed.), 'The Treatment of Trade Waste waters and the Prevention of River Pollution', *University of Durham Bulletin* No. 10, 1957.
Klein, Louis, *River Pollution*, Vol. 1, *Chemical Analysis*, Vol. 2, *Causes and Effects*, Vol. 3, Butterworth & Co., London 1969.

McLoughlin, J., 'Control of the Pollution of inland waters', *Journal of Planning and Environmental Law* June 1973.

McLoughlin, J., *The Law Relating to Pollution*, Studies in Environmental Pollution, Manchester University Press, Manchester 1972.

Medical Research Council, *Sewage Contamination of Bathing Beaches in England and Wales*, Memorandum No. 37, H.M.S.O., London 1959.

Ministry of Housing and Local Government, *Taken for Granted: Report of Working Party on Sewage Disposal*, H.M.S.O., London 1970.

Ministry of Housing and Local Government, *Refuse Storage and Collection: Report of the Working Party on Refuse Collection*, H.M.S.O., London 1967.

Moulder, D. S., *Current Research on Marine Pollution*, Marine Pollution Documentation and Information Centre, Marine Biological Association of the U.K., Plymouth 1971.

Nelson-Smith, A., *Effects of oil on Marine Plants and Animals*, Water Pollution by Oil (Proceedings of a Seminar held at Aviemore, 1970) Institute of Petroleum, London 1971.

Owens, M., *Chemical and Pesticide Pollution. Water Pollution as a World Problem*, Europa Publications Ltd., London 1970.

Parkin, P. H. and Humphreys, H. R., *Acoustics, Noise and Buildings*, Faber & Faber Ltd., London 1958.

Pritchard, D. W., 'The Movement and Mixing of Contaminants in Tidal Estuaries', *Proceedings of the First International Conference on Waste Disposal in the Marine Environment*, Berkeley, California 1959, Pergamon Press Ltd., Oxford 1960.

Raymont, J. E. G., 'Some aspects of pollution in Southampton Water 1971', *Proceedings of the Royal Society*, Series B, 1972.

Royal Commission on Environmental Pollution, *First Report* Cmnd. 4585, H.M.S.O., London 1971.

Royal Commission on Environmental Pollution, *Second Report: Three Issues in Industrial Pollution*, Cmnd. 4894, H.M.S.O., London 1972.

Royal Commission on Environmental Pollution, *Third Report: Pollution in some British Estuaries and Coastal Waters*, Cmnd. 5054, H.M.S.O., London 1972.

Staudinger, J. J. P., *Disposal of Plastic Waste and Litter*, S.C.1. Monograph No. 35, Society of Chemical Industries, London 1970.

Wright, Curtis, 'The Right Oil Retention Boom', *American City*, October 1972. Vol. 87. *Available in the Department of the Environment Library*.

Young, D. D., 'Prevention of Water Pollution by Solid Domestic and Industrial Wastes', *Public Health Engineer*, September 1973, No. 5.

12 Economics

The economics of marinas is, arguably, the most difficult and contentious aspect of all. It is certainly the most important, for no matter how practicable in all other respects, without financial credibility the project will not begin, or worse, if it does, will soon fail. A really hard-headed approach is essential and no individual or committee should contemplate developing a marina out of sentiment. The leisure business is notoriously tough, marinas are relatively unknown territory and American experience is not directly comparable.

To embark on a marina enterprise out of a liking for sailing is like becoming a stock farmer out of fondness for animals. It is also true however that if one's *sole* purpose is to make the maximum profit on one's money then a straightforward marina development is best avoided. An ideal attitude would consist of 80 per cent business acumen and hard work 10 per cent interest in boats and 10 per cent instinct for gambling.

The first absolutely essential rule is to ensure that sufficient funds are available—and will remain so. More new companies (not only marinas) fail because of under-capitalisation than for any other reason—inadequate accounting is a close second. It may appear mercenary but money is the very earth in which the project will grow—deprived of sufficient nourishment it will fade and die, particularly when it is still young.

The most significant factor affecting the economics of marinas is the great changes that have taken place in the concept of management and finance. It was probably Dennis Sessions, now of Marina Management Ltd., who submitted the UK's first planning proposal as recently as 1947, as an extension to a boatyard—at that time the only sensible application envisaged for such a facility.

Since then, not only have the categories of marina grown to include all those covered in Chapter 3 but the style of investment has altered entirely. The simple shipwright's-dock image has been superseded by a host of more complex fiscal arrangements. So unusual are some of these that it may sometimes be suspected that the tailoring of investment yields and tax outlets to parent companies are more carefully devised than is the ultimate benefit to the boat owners or the local environment. This transition has quite recently brought marina development to the attention of large, rich entertainment companies (and large rich entertainers too) not, ironically as high-speed money spinners but as ideal, slow-maturing, capital gaining absorbers of profits from more lucrative but more highly-taxed fields.

Sources of revenue

Like all business ventures, the financial aspects of marina development may be simplified into the two basic issues of how money is obtained and upon

12.1, 2 Amateur methods have their charms if not their profits

Facility	Profit ratio percentage	Gross income expressed as parts of one pound
Slip rentals	40%	20p
Winter storage	27%	10p
Repairs and maintenance	27%	12p
Sales		
New boats and engines	17%	10p
Used boats and engines	9%	8p
Chandlery	23%	13p
Petroleum products	15%	9p
Boat hire	15%	4p
Restaurant and bar		4p
Groceries, ice, vending machines	15%	2p
Fishing tackle, bait		4p
Boatel units		4p
		£1.00

12.A Sources of income of the average American marina. This table is based on reports received from 190 expanded marinas sited in many different parts of America. The profit ratio column shows the average percentage of net profit made on each operation after deducting costs of labour, materials and other charges specifically incurred by that facility. The gross income column averages out the amount in new pence each item contributed to every pound taken by the marina

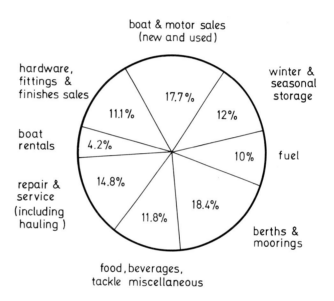

12.3 This pie diagram summarises the results of an American survey of the income sources of marinas, and is useful in illustrating the relationship which exists between the provision of facilities and the economic feasibility of the development. It is interesting to compare this survey with the gross income column (also a percentage) shown in 12.A. They are very similar, only differing by a per cent or two.

In reviewing individual and collective profits there are three main methods of assessment:

1. The percentage profit on each activity (or sale): for example, a boat bought at £500 is sold for £550 (10 per cent)

2. The collective annual profit for each sector: ten such sales will net £500 profit (100 per cent)

3. The contribution that 2 makes to the total marina income (17·7 per cent in the pie diagram)

what it is spent. Behind this over-simplification lies what is very often a highly complex structure of borrowing, investment and marketing.

The importance of sufficient capital has already been stressed. This advice raises the question of likely sources of revenue. It must be admitted that in America, despite thousands of financially successful projects, marina development is still fairly low in the league tables of most finance houses. In Britain, with so little experience of this type of development, the situation is even more difficult.

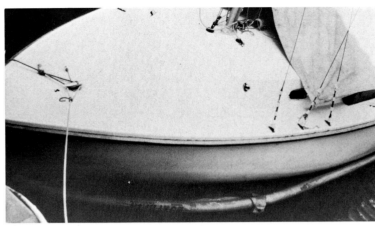

12.4 Boat building and repair can be a valuable service to client and proprietor alike. At this yacht harbour on the River Hamble the marina company took *on average* in 1972 over £800 from each owner in refits and repairs alone

12.5 This hull bath is one of many small services offered by a lively management at

Huntington Marina, Long Island, New York

12.6 Ramshackle covered moorings at New Orleans

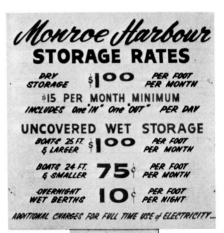

12.7 Dry storage rates at Sanford, Florida

It is at the fund-raising stage that the pilot study by the project team will prove its worth, for its function is to establish the development's viability in all aspects—not least in financial terms. In proving the scheme to themselves the developers will have the most valuable prospectus for convincing a potential backer. Likely sources of funding will include the clearing and merchant banks and development and insurance companies. Joint enterprises of public and private consortia may involve pension funds, local authorities and government agencies.

Marina projects seem, particularly in America, to lend themselves to a multitude of options and alternatives dependent upon the technical advice given by specialists. Money may be forthcoming from any or, in some instances, all of the following sources:

1 The Federal Government
2 The State
3 Development boards
4 The County
5 The City
6 Private investment

Within the last mentioned category, the possibilities of financial contributions are very varied and may include local sailing clubs, sporting interests of industrial firms or public bodies, firms in the entertainment field, hotels or pure speculative investment by any source from estate developers to petroleum companies.

Marinas invariably raise the value of neighbouring land—another reason for purchasing ahead, for to pay for land which has trebled its value within a year *because of one's own development* is a cruel experience.

The 'bulk-buying' principle was used in purchasing land in 1964 for the largest marina development in the Los Angeles area, where the average price per acre was $2,500, the majority of this value being attributable to the waterfront strip. In 1968, when about one-third of the berths were operational, two 10-acre sites were sold for $240,000 each, one for a housing project and the other to a motel/vacation concern. Such financial methods of land purchase not only benefit the developers directly, but also help to ensure the success of their marina by creating its own catchment.

The loan period sometimes brings difficulties with marina finance because it may take several seasons of operation before healthy returns are recorded and some loans, particularly from clearing banks, tend to be too short-term to meet amortization demands during the early years when the expense/income gap is at its widest.

Grants

Whilst direct grants for 'pure' marinas are unlikely, grants of varying percentages of the total cost may be given in Britain and certain other countries for ancillary aspects such as slum clearance, new or rehabilitated housing, historic preservation, industrial building and equipment, conservation, flood prevention, river and harbour development, youth and adult training schemes, promotion of tourism and wildlife protection. Where such proposals are envisaged as a constituent of the total scheme it would be worthwhile including them within a grant aid application.

With so many permutations possible in financial partnerships and the size and time-span of loans it is impossible to generalise upon the outcome of any approach for funds. The initial loan is generally the most difficult to get, for it is success that brings financial confidence. State aid abroad is quite common. Belgium for example offers direct grants and the current involvement by the French Government is enormous.

The Federal Government in America supplies half of the finance for the capital costs of marina developments which form part of approved recreational schemes; the American National Parks build marinas and boat facilities to be run by private concessionaires in return for a marginal (2 per cent) share of the annual profit. In Malta, the Government has undertaken complete responsibility for the cost of marina works sharing in the land development with a private enterprise consortium. The State of New York's State Boating plan has been made possible by the issue of a Government Bond: the amount of the bond being based upon the calculated income to the Government from the past, present and projected sales of marine petrol, which is subject to a Federal and a State tax.

In the UK direct grants of any kind are rare and there are no state-built marinas. Public money *is* spent however—by local authorities, harbour and waterways boards and official bodies promoting physical recreation or countryside activities. Usually, however, a development is initiated by private persons who raise money from the usual sources—banks, building societies, private money or by forming a company and offering shares. The sources from which money may be raised is closely linked with the develop-

Revenue	$			$		$	
Sale of bonds	2,000,000	94%					
From **** re							
extra work	5,626.06	0·3%					
City budget							
Deficit $90,288							
Paving $24,896							
Sidewalk $4,400	119,584	5·7%				2,125,210.06	100%
Expense							
Disc on sale of							
bonds	20,000						
Interest on bonds	21,222.28						
Bond sale expense	43,776						
Old debt							
elimination	132,949.44			217,947.72	10%		
Land acquisition	4,100						
Engineering fees	80,641.80						
Survey	8,604.78						
Borings	3,792			97,138.58	4%		
Dredging and							
bulkhead	980,292	64%					
Breakwater	97,032	6%					
Misc. facilities and							
structures	256,276	14%					
Floating docks,							
piers, piling	306,038	16%		1,639,640	81%		
Streets, parking,							
sidewalk	68,042						
Flood and street							
lights	5,522						
Water and sewer							
mains	24,834						
Sewage pumping							
station	13,068			111,472	4·6%		
Miscellaneous:—							
Gypsum spreading							
—overtime	1,506						
Gypsum	5,292						
Gas and repairs	640						
Landscaping	1,550						
Storm sewer—							
relocation	1,892						
Advertising	304						
Sundry misc.	460			11,550	0·4%	2,077,748	100%
Cash in hand						47,460	
Less: to be							
completed:							
Dock contract	12,858						
Sidewalks	4,400					17,258	
To close to							
contingency							
account						30,200	

12.B Fiscal Report on a lake-front marina project. This account summarises the income and expenditure of an actual lake-front marina development carried out in America, the figures being updated to May 1977. It is fairly typical of most projects of 200–400 moorings. This harbour specialised in shallow-draught grp power boats. Its sheltered inland location is indicated by the relatively low percentage cost for the breakwater

Outline Accounts for a Typical Inland Marina Development (area—4 hectares (10 acres); number of craft—250)

Production

		Cost	Sales (100% Efficiency)
1st year	—	—	—
2nd year	2 men	6,874	17,280
3rd year	4 men	13,746	34,560
4th year	6 men	20,620	51,840

Moorings 250 at £260 per annum = 65,000

1st year	—
2nd year (50%)	32,500
3rd year (75%)	48,750
4th year (100%)	65,000

	Year 1	Year 2	Year 3	Year 4
Labour sales	—	17,280	34,540	51,840
Moorings	—	35,500	48,750	65,000
	—	52,780	83,290	116,840

Overheads

	Year 1	Year 2	Year 3	Year 4
Manager	4,500	4,500	4,500	4,500
Office Girl	—	1,900	1,900	1,900
Wages	—	6,847	13,746	20,620
Advertising	2,000	2,000	2,000	2,000
Stationery	—	600	700	1,000
Postage	—	200	300	400
Telephone	—	800	1,000	1,000
Audit and Legal	—	1,000	700	900
Workshop Expenses	—	800	1,000	1,200
Heat and Light	—	1,500	1,500	1,500
Rates	—	3,000	3,000	3,000
Insurance	—	900	900	900
Sundries	1,000	2,000	2,000	2,000
Motor Car—Tractor	—	4,000	—	—
Depreciation	—	—	1,332	1,332
	7,500	30,074	34,578	42,252
Loss	7,500	—	—	—
Financing 10%	750	—	—	—
Profit	—	19,706	48,732	74,588
C/Fwd	£8,250(loss)	£19,706	£48,732	£74,588

12.C This hypothetical project for 250 berths on an inland site illustrates a safe method of funding over a fairly long period. The year yield of nearly 15 per cent is reasonable and the capital asset created during that time will show a substantial growth on the original construction cost.

Notes:

No taxation has been taken into account.

Financing has bccn taken as 10 per cent but more favourable terms could probably be negotiated.

All professional fees have been ignored.

ment objectives and the marina type; whoever the initiators may be they should be clear what their objectives are.

This may seem obvious but it is not uncommon to find even established companies with only the vaguest concepts of the firm's function, its strengths and weaknesses, future plans and improvements and how it compares with its rivals.

The *motive* for development and the *type* of marina are closely linked and these in turn largely govern the financial arrangements. 3.2 (p. 39) showed four main categories of developer and ten bases for development. With so many variables it is not surprising that the objectives and hoped-for

	Total	Year 1	Year 2	Year 3
4 hectares (10 acres) at £12,500 per hectare	50,000	22,224	16,666	11,110
Financing 10%	5,000	2,224	1,666	1,111
Inland Waterway—Construct Entrance—30 m (100 ft) at £266 per metre (Dredging—Sheeting—Landscaping	8,000	8,000		
Financing 10%	800	800		
Basin—Dredging to approximately 3 m (10 ft)—1·8 hectares (4½ acres). Removing 60,000 m³ (78,000 yd³) of	60,000	60,000		
material at £1.00 per metre³ Financing 10%	6,000	6,000		
Stabilising, Piling & Landscaping to 3 m (10 ft) width 573 m (1,750 ft) at £100 per metre**	57,300	57,300		
Financing 10%	5,730	5,730		
250 moorings at £800 per mooring: Including Fresh Water Points (1 point per 6 berths) Electricity—Basin Lighting—Provision of Dock— Car Park for 375 cars—Landscaping Whole Area	200,000	88,890	66,666	44,444
Financing 10%	20,000	8,889	6,666	4,444
Repairs, Service and Sales Building—550 m² (6,000 ft²) at £109 per metre²	60,000	40,000	20,000	
Financing 10%	6,000	4,000	2,000	
Boat Lift (2nd year)	16,000		16,000	
Financing 10%	1,600		1,600	
Plant and Machinery (Second-hand)	10,000		10,000	
Financing 10%	1,000		1,000	
	£507,430	£304,057	£142,264	£61,109

NOTES:

No taxation has been taken into account.
Financing has been taken as 10%, although it is hoped
that much more favourable terms could be negotiated.
All professional charges and architect's fees, relating
to the instigation and development of this project, have
been ignored.

**Assuming rectangular shape.

	Nett Trading Loss YEAR 1 C/Fwd		Nett Income YEAR 2		Nett Income YEAR 3	
		8,250		19,706		48,732
	312,307	(loss)	122,558		12,378	
			Loss B/F	312,307	Loss B/F	466,094
			Financing 10%	31,230	Financing 10%	46,609
			C/Fwd	£466,095		£525,081
Percentage Yield on Capital Employed		2·7133% (loss)		4·0564%		8·4926%

12.D This balance sheet for a hypothetical inland marina typifies the sort of financial problems likely to be encountered during the early years of many marina developments that are not reliant upon more profitable elements within the complex. The site is 4 ha (10 acres) in extent and 250 berths are envisaged. A manager would be required from the inception and probably an office girl in the second year. Between them they would be responsible for letting the moorings and for all office administration. The rest of the labour force listed is assumed as selling at 100 per cent, an ideal situation, unlikely because of fluctuations in demand working against this degree of utilisation. It is fair to predict however that all maintenance can be undertaken by resident staff during the 20 per cent or so of lost time period. The maintenance of grounds has been excluded, being dealt with by part-time labour

Year 5 and Year 6 taken exactly as Year 4

Balance C/Fwd	525,082
Nett Profit Year 4	74,588
	450,494
Financing 10%	45,050
	£495,544

Percentage Yield on Capital Employed	13·08%

Nett Profit Year 5	74,588
	420,956
Financing 10%	42,096
	£463,052

Percentage Yield on Capital Employed	13·87%

Nett Profit Year 6	74,588
	388,464
Financing 10%	38,846
	£427,310

Percentage Yield on Capital Employed	14·86%

profitability range widely from the primarily recreative principles of local authorities to the professional marina management companies who strive for a reliable increment on capital. Whilst no one wants to make a *real* loss, profits may be very obscure. The 'loss leader' marina acting as a town's amenity to promote commerce or complementing a hotel complex are examples.

In America small marina projects are sometimes used as stalking-horses for gradual residential development where a direct application would be refused. When construction is complete berths may be sold outright or on long leases for short-term profit. Individual components may be leased or sold on a concessionaire basis or the whole site sold as a going concern to one or more management companies. It is this wide scope which makes general advice so difficult.

Item No.	Item description	Quantity	Unit	Budget price				Total price
				Unit	Ext.	Unit	Ext.	
1, 2	Landfill and beach, including estimate of dredging *	13·9	Acres				$640,000	*
3	Bulkhead						296,000	*
4	Fixed berths and piers	94	Berths				182,000	**
5 & 6	Floating berths and piers	220	Berths				312,000	
6	Landings, floating, with gangway, at park, piers and channel	8	Each				16,000	
7	Utility buildings, toilets and showers	3	Each				150,000	**
8	Sanitary sewer and manholes	1	System				40,000	
9	Sewage pumping station, pneumatic ejector and force main	1	System				40,000	
10	Water main, fittings and hydrants	1	System				34,000	**
11 & 14	Heavy duty paving, streets and plaza parking lot	160	Cars				84,000	
11	Curb and gutter incl. planting areas for trees						34,000	
12	Electrical service, underground primary circuit						60,000	**
13	Sidewalks						20,000	
14 & 16	Street and parking plaza lighting						40,000	**
15	Drainage conduit (allowance)						42,000	
16	Large culvert at bulkhead, for harbour flushing (allowance)						52,000	
17	Launching ramp	200	Ft. W.				42,000	
18	Hoist and hoistway (allowance)						30,000	
19	Stabilized parking area, cars or trailers						18,000	
20	Control tower structure, part of marina centre or detached (allowance)						60,000	
21	Landscaping (allowance)						40,000	*
							$2,232,000	*
	Contingency reserve 15% *						334,000	*
							$2,566,000	*

* These figures may vary considerably with Hydrographic Survey information and final design.

** Recommendations in text for alternates for economy.

12.E This typical estimate sheet itemises expenditure on the first (314 berth) stage and 5 subsequent (100 berth) stages of an eventual 814 berth development. Stages 2–6 each envisaged 1·4 ha (3½ acres) 2-pier extensions. To overcome the high cost of repeated dredging this was eventually completed in 2 rather than 6 stages

Increments 2, 3, 4, 5, and 6

Land, 3·5 Acres	$140,000 *
Bulkhead 400 Lin. Ft.	52,000 *
Piers, 2, Total Avg. 100 Berths	160,000
Beach, 400 Lin. Ft. Incl. Park	—
Parking Space, 200 Units	17,800
Dry Berth Bldg. Site ¾ Acre (Displaces 75 Parking Units)	—
Utility Bldg. (1) Allowance	50,000
Sewer and Water	10,000
Lighting and Electrical Service	8,000
Street Paving, Drainage and Sidewalks	11,400
Landscaping, Allowance	4,000
	453,200
Contingency Reserve 15%	67,980
	$521,180

* Refer to Footnote on Estimate Sheet.

The total Project Cost Estimate is Compiled by Multiplying the above incremental cost by 5·0 and adding the first increment estimate.

THUS: Estimated 13·9 Acres,

First Increment	$2,566,000
5 × 521,180	2,605,900
Grand Total	$5,171,900

Survey costs, architect-engineer fees, legal and financing costs are not included in this total, but probably may be largely taken care of by the contingency item in the above estimates.

Percentage profits

Because undertakings vary so much only the most general examples can be given, but assuming that Stage I of a project is for 400 craft in a tidal basin then the following could be a typical breakdown of construction costs:

	Cost £	Percentage
Land	160,000	16
Wet dredging	400,000	40
Reclamation drainage, roads, car park	90,000	9
Bulkhead walls and ramps	150,000	15
Floating piers and walkways	90,000	9
Statutory services	90,000	9
Information office	10,000	1
Lavatories	10,000	1
Construction cost	£1,000,000	100
Fees at 10%	100,000	
Total	£1,100,000	

12.F Percentage costs

An average boat size of 7·5 m (25 ft) gives 3,000 m (10,000 ft) run of craft. Assuming a 1977 opening the going rate is likely by then to be £15 per foot per year (£50/m). Annual berth income at full capacity will be £150,000 and at 90 per cent capacity £135,000. Annual expenditure on maintenance, staff costs and depreciation may be taken at about 7 per cent of construction costs—£70,000 and this deducted from £135,000 gives an annual income of £65,000—a return of 6·5 per cent. Bearing in mind that interest on capital has yet to be paid this percentage by itself is not good enough and substantiates the widely held belief in the inadequacy of relying of berth rentals alone.

If, however, the same example is used as a basis upon which more profitable (or necessary) elements are superimposed when the total picture begins to look more healthy, thus:

	Capital cost £	Percentage	Annual profit
Construction costs and fees	1,100,000	6·5	71,500
Boat yard	200,000	30	60,000
Boat handling equipment	50,000 ⎫	60	54,000
Storage	40,000 ⎭		
Chandlery store	40,000	15	6,000
Fuel dock	10,000	5	500
Boat sales and brokerage	24,000	100	24,000
Clubhouse (Stage 1)	30,000	50	15,000
	1,494,000	15·5	231,000

12.G Capital cost and annual profit

Ancillary services have therefore built up percentage profits from 6·5 per cent to 15·5 per cent. Deducting 7 per cent again for maintenance, staff and depreciation against the capital cost of ancillaries (7 per cent of £394,000) the total annual profit becomes nearly £203,500—about 14 per cent, which is fairly respectable. If it is assumed that spare capacity was designed into the first stage to accept a 250-berth expansion in Stage II then an ultimate net profit could be expected (before interest repayment) of over 20 per cent.

It is often a good idea to build up a reasonable number of profit-making ancillaries not only because of their need or draw but to enable mooring charges to be kept to a minimum. Like many service industries, particularly in the leisure field, it is probably a better principle to capture a full clientele and attract them to spend money on options than to risk half-empty berths with a policy that maximises mooring costs.

Profitability of individual elements relies enormously upon mutual benefit. Within this strong symbiosis must be included certain *essential* functions which, by themselves show a poor return. The fuel dock is an example. Possibly the most speculative single facility (in America) of a marina complex is the clubhouse and most developers are happy to lease (not sell) land for this purpose at quite a low figure to ensure its success: they rarely sell, because knowing the financial vulnerability of such places, they wish to avoid a 'dead' clubhouse which would spoil the image of their marina.

Profiting from the land

If there is a golden rule to assist financial success in this field, it is *to purchase enough land (or to obtain the option of purchase) from the outset.* This will safeguard the possibility of expansion along the coast for the marina itself and open the possibilities of leasing or developing land for allied uses inland. More often than not, the sea walls, quays and bulkheads necessary for the marina development allow sites inland, which have previously been uneconomic, to develop because of the poor nature of the soil, and to be made suitable for building much more cheaply than before.

This previously undesirable area which the wise marina developer will probably have purchased at a low price, will, with the development of the waterside, grow in value. It is often economic, whilst the basic engineering is being considered, to extend the contract to include the building up and drainage of adjacent sites inland. By the time the marina is operating, the land which the developer originally purchased as a marsh will have been transformed into desirable building land overlooking the nearby amenity of the marina itself. Any extension will depend upon planning approval being granted but, provided the land is negotiable, then, at the very worst, it can be sold to offset losses.

Phasing construction and investment

In the phasing of construction and investment two alternative courses are open. Firstly, if the developer has all the capital readily available, then it must be right to buy the best expertise in the business, employ the necessary contractors and complete the development as quickly as possible in order that the maximum return on investment may be forthcoming with minimum delay. Some typical costings for this manner of approach are shown in the outline account on page 305.

Secondly, if capital is a problem and the developer wishes to make it available over a longer period of time, then it could be worthwhile to employ direct labour for the construction and spread the work over several years. Meanwhile, labour can also be made available to give service to the customer during this construction period. In the end, this will in all probability not cost more in terms of capital—assuming that labour costs increase no more than the general rate of inflation. Because of increasing costs, however, the money ultimately expended is almost sure to buy substantially less than it would if it had all been spent at the beginning. One advantage is that any trading profit made in this period will help to reduce the balance of any borrowed money. This second method may well suit a large company, with substantial profits which wishes to diversify and create a sizeable capital asset with a reasonable cash flow over a period of several years.

Finding where the profits are

The need for most marinas to offer a range of services has already been stressed. This economic requirement presents management with its main problem, for it will not, in effect, be running one business but a loose assembly of small operations obscurely linked by the common activity of boating. This fragmentation brings with it the need to know the profitability of each and their interdependence. A cost-accounting system should cover each activity individually. On the one hand this spells out the annual income and on the other it analyses outgoings—usually under the four main areas of expense outlined below:

1   Depreciation

Depreciation rates will vary with individual equipment and between each component of the marina itself. One method of calculation is to categorise these into groups and allocate a percentage for 'loading' each group according to its annual loss of value.

2   Overheads

Rates and other operating taxes, electricity and water supply, insurances, telephone and printing expenses are all examples of general operating or administrative costs and must be included when profit ratios are being calculated.

3   Maintenance

This includes costs of materials, spare part replacement and tools, but most systems exclude labour, which is accounted for separately.

4   Staff costs

This not only includes wages but also bonuses, employment and other taxes, welfare, insurance and clothing. An estimate will need to be made of the labour costs within individual activities. This is difficult because of the variety of jobs done by most marina staff.

Calculating profit ratios

In the marina previously exampled, and using the four main divisions of expenditure, a simple calculation to arrive at the annual profit ratio *on berthing only* might be as follows:

Staff costs	£32,000	%40·0	
Depreciation	14,000	17·5	£70,000 or 7% of construction costs.
Maintenance	24,000	30·0	(See previous calculation)
Overheads	10,000	12·5	
Annual outgoings	**£80,000**	100·0	
Annual income (pre-tax)	**£110,000**		
Gross profit	**£30,000**		

12.H Annual profit ratio.

Profit ratio as a percentage may be calculated by dividing gross profit by total outgoings and multiplying by 100:

$$\text{Thus:} \frac{30,000}{80,000} \times 100 = 37 \cdot 5\%$$

This figure is close to the 40 per cent given for average slip-rental profit ratios given in 12.A on page 301.

Whether the management of this hypothetical marina is satisfied will depend upon many other factors. The point is that, happy or not, at least they *know* and can make such adjustments as necessary.

The need for economic-ecological analysis

A developer's interest is largely economic and the questions he is most likely to ask his advisers are, 'How much will it cost? How much profit will it make? How quickly?' Although, like most recreational studies, the facts may be complex and interrelated they will be straight-forward compared with those relating to the much wider issues of the social, ecological and environmental impact of the development. When, to those wider issues is added the evaluation of the development alternatives and *their* implications the options presented are of factorial complexity. Nevertheless it is the sort of comparative analysis that must be expected to be undertaken nowadays, not necessarily by the developer but by the local authority or the government.

The potential developer or expanding owner must expect, at a public inquiry or elsewhere to present his case for a marina or its expansion against an increasingly knowledgeable array of local and national societies who will be submitting impressive evidence for the conservation of the site. The case for retention—or even reversion—may be broadly based and range from the preservation of a threatened plant or insect to the benefits of investing the money or resources in some, to them, more worthwhile cause.

Where the site is located in less sensitive surroundings the objections may be even more difficult to counter, for waterside areas are notorious for attracting rival demands for the siting of power stations, commercial ports or oil refineries. These major industrial uses have access to the evidence and counsel not normally available to ordinary developers. The sort of team a developer requires to advise on the physical development of a site is unlikely to possess the appropriate armoury to mount the thorough economic/ecological analysis required to counter the heavyweight evidence produced by large private concerns or government departments. In situations which demand it such expertise is best sought from specialists. There are now several consultants in Europe and America specialising in this field but their retention would not release the development team from all responsibility because data collection, initial explorations, planning input and detailed design descriptions would be expected from them as part of the brief to the specialist consultants. The following paragraph outlines the variety of topics most relevant to marina development and suggests ways of relating and comparing them.

Most water-land relationships combine to present challenging ecological situations. They frequently involve complex linkages between socio-economic and ecological systems, especially when related to recreational development. The natural environment itself presents some of the most delicate relationships to be found in nature, where land and sea-based species meet and inter-tidal organisms rely on the preservation of the precarious balance of natural factors. Even what appears as an unattractive expanse of useless swamp crying out for redevelopment may perform a vital role as a habitat for rare species, a reservoir for excess rainfall or a location unique in its geological formation. Most coastal stretches are interesting ecologically and estuarine ecosystems are acknowledged to be among the most fertile in the world.

Objectives and techniques

The object of such an analysis is to make those concerned with the environment aware of these intricate inter-relationships between the economy and the ecosystem and between economic development and environmental management. It is important to present some idea of the conceptual framework and methodology in a broad non-technical fashion. It is also important to demonstrate how existing methodology can be put to work on problems of marina development.

Four basic techniques found to be generally useful and having a high degree of validity for recreation design and development are as follows:

1   Comparative cost analysis

2 Input–output techniques
3 The gravity model (Spatial interaction analysis)
4 Activity complex analysis

These four techniques are fused in the economic-ecological analysis.

1 Comparative cost analysis

This provides a useful planning tool in evaluating the opportunities and potentials of a region as compared with others, as a location for marina—or any other—development. With the comparative cost technique it is possible to include more than strictly monetary costs. As an example, different locations will vary in damage suffered or benefit derived from loss or gain of species, pollution, alteration of the coastal regimen for good or ill, and so on. These factors can be assessed by 'valuing' the positive gain or negative loss in money terms. A cost/benefit analysis in fact.

2 Input–output techniques

This method provides a comprehensive quantitative description of the regional economy. Its strength lies in the presentation of the production and distribution characteristics of the region's industry. It has become, since its inception by Leontief at Harvard in the 1930s a basic tool for regional economic analysis.

The region is divided into its relevant economic categories (sectors). Data is collected for input-output formulation—origins of inputs, destinations of outputs. From this information an input-output flow table is compiled. The technique represents a double-entry book-keeping system revealing both purchases and sales within the economy. There are numerous ways in which the technique can be employed and it is most useful in providing a complete planning framework involving consistent description and projection.

3 The gravity model (Spatial interaction analysis)

Estimates of demand for services are required for planning many types of infrastructure—for instance the demand for services to be provided by a recreational complex. The method must take into account such variables as population clusters, the distance between them and their interaction, thus describing the effects at one place of events at another.

The term gravity model is taken from Newton's concept of gravity and extended into social science. It is used in traffic problems to prepare trip generation and distribution models and may be adopted to estimate recreational demands for various coastal areas. Since the calculations, whilst simple are tedious, computers are generally employed.

4 Activity complex analysis

Whist comparative cost analysis (see 1 above) is suitable for determining the cost of a single activity at a single site, very often *groups* of economic activities, like shops in a town, gain economic advantage when located in proximity to one another. A marina is just such a case, for its major relationship involves the common use of factors such as parking areas, breakwaters, ramps, moorings, boat handling equipment and so on. By sharing the annual charges on the capital investment each activity is able to operate at a lower unit cost. The sharing of labour and other direct service inputs leads to further savings.

Ecological studies
This is usually defined as the study of the relationships of organisms or groups of organisms to their environment.

To evaluate the *economic* consequences of alternate development policies it is necessary to follow a systematic means of projecting their *environmental* consequences. This requires consideration under such headings as:

1   Natural resource classification
2   The ecosystem and food chain
3   The division of the marine region into zones (land, intertidal, open water)

The two basic components of an ecosystem are:

1   The non-living (abiotic) substances organised into the following categories:
a)   Climate
b)   Geology
c)   Physiography
d)   Hydrology
e)   Soils
2   The living (biotic) substances organised into:
a)   Plants
b)   Animals

A thorough classification and analysis of the site under these headings can only be undertaken by a team of ecologists with widely ranging disciplines.

Combined economic–ecological analysis
To evolve adequate methods for evaluating various alternatives related to the development of coastal areas and the environment in general, it is necessary to bring together the four basic techniques outlined above and to add to them a systematic ecological analysis.

One method of linking the economic and ecological analysis is by using a modification of the linear system developed by mathematicians. Any methodology for analysis will be imperfect because the links governing social and natural phenomena are extremely complex and the variables intricately interrelated.

Nevertheless a logical system with a scientific basis is better than basing decisions on intuition and guesswork. It should be possible to approach the critical problems which arise from developing waterside areas much more effectively.

It is important to accept that there will be an ecological as well as a financial price to pay. 'Better' or 'worse' are emotive terms where marina development is concerned but 'change' is undeniable and it is to everyone's benefit to predict as accurately as possible what such change will involve.

Recalling the many poor and ill-advised decisions concerning resource development made in the past, it seems undeniable that simple but thorough models of relationships when properly applied to reliable data cannot help improving the quality of decision making.

12.8 Dredging the site for the Lymington
Yacht Haven, Hampshire, England, for
which the first phase of construction took
28 weeks.
12.9 The 250 berth first stage of the marina.
Hire of plant on such a scale can represent
a substantial part of the cost of a marina,
but some dredged material may be saleable:
see page 170.

12. Economics: Check list

Research data	Establish motive and objectives
	Identify market in terms of:
	key economic factors:
	catchment in respect of area, size, population, social class, travel times and modes
	Marina type:
	tidal basin, locked harbour, haul out marina, etc.
	Distance from other marinas
	Consider financial effect of development upon neighbouring land
	Tourists—attraction and potential to visitors and foreign boats
	Local social amenities ⎫ affecting marina
	Local recreational activity ⎬ (a) beneficially
	Local economic activity ⎭ (b) adversely
	Consider market research methods
	Obtain appropriate Regional Study (H.M.S.O.)
	Consider formation of company
	Interpretation and evaluation of data
Presentation	Prepare well formulated statement of total concept best suited to the opportunity in terms of location (if appropriate), facilities, service, image, prices (proposed rates) with conclusion, options and recommendation(s)
Development and operation costs	Initial overall cost-estimate by engineer/quantity surveyor
	Calculate individual and proportional costs of all major factors:
	Capital/physical:
	Land
	Dredging
	Quay walls
	Equipment
	Fees
	Revenue etc. invisible:
	Staff
	Services
	Maintenance
	Depreciation
	Insurance
	Rates, rent, VAT
	Loan repayment
	Advertising etc.
	Form of building contract
	competitive tender
	negotiated contract etc.
	Period of contract and phasing
	Likely increases during contract period
Sources of funding	Prepare statement of financial feasibility and obtain advice from accountants regarding possible backers, including:
	Banks (merchant and clearing)
	Development companies
	Insurance companies
	Public funds
	Private sources
	Pension funds
	Consortia
	Share issue etc.
	Consider joint enterprise of public/private consortium
	Ensure sufficient funding for:
	research and studies
	development
	fees
	period of early loss
	expansion

Check framework and timing of borrowing in relation to phasing of development
Prepare estimate of funding needs beyond development period
Prepare prospectus of financial potential for prospective revenue sources

Grants

Consider grants from:
  Central Government or Government agencies
  Local Authorities—regional, county or borough level
  Sports Council(s) for individual or collective sports or recreations
  Charities—check conditions, consider formation of charitable trust
Consider grants for:
  Rehabilitation of derelict land (MOH/LG circ. 30/60)
  Tree Planting—see MOH/LG circ. 39/63 also 61/63 and S.6 of Local Authorities (Land)
    Act 1963
  Marinas, including boat yards, check S.20(1)b and S.21(2) of Industrial Development Act,
    1966 and S.9 of Local development on derelict or unsightly land
  Conservation—Contact Nature Conservancy
  Flood prevention—Contact River/Harbour Authority
  Historic preservation—See Local Authorities (Historic Buildings) Act, 1962
  River, Harbour Development
  Development Areas—particularly where marina will provide considerable employment
  Tourism
  Wild life
  Training Schemes   Youth and adult (sail training)

Sources of income

Prepare breakdown of potential income in terms of
  percentage contribution to total income from each individual source
  individual profit ratios per annum
Include potential income from
  Berths (average Loa × average occupancy rate)
  Concessionaires
  Clubs
  Tenants
  Shops
  Boatyard (building, repairs, refits)
  Boat handling equipment
  Storage
  Boat sales
  Brokerage etc.

Operating costs

Prepare statement of estimated operating costs, expenditure and general overheads including:
  Depreciation
  Maintenance costs
  Amortisation (consider setting up fund)
  Staff salaries
  Taxation
  Insurance premiums
  Rates
  VAT
  Services (heating, lighting, gas, GPO, etc.)
  Loan charges etc.

12. Economics: Bibliography

Clawson, M. and Knetsch, J. L., *Economics of Outdoor Recreation*, The John Hopkins University Press, London 1966.

Eglin, Roger, 'Sailor Beware', *The Observer*, 17 September 1972.

Isard, Walter, *Ecologic: Economic Analysis for Regional Development*, The Free Press, New York, 1972; Collier-MacMillan, London 1972.

Lewis, William Cris, 'Public Investment Impacts and Regional Economic Growth', *Water Resources Research*, Vol. 9, August 1973.

National Association of Engine and Boat Manufacturers Inc., *Marina Revenues and Costs, Profit Ratios, Slip Rental Rates, Labour Rates*, The Association, New York 1967.

National Association of Engine and Boat Manufacturers Inc., *The Modern Marina. A Guide book for the Community and Private Investor interested in Marina Development*, The Association, New York 1963.

Parfitt, R., 'Management Information Part I: Regular Statements of Profit'; 'Management Information Part II: Budgetary Control', *Information Circular* No. 19, Parts 1 and 2. *Available in the Department of the Environment Library.*

Sessions, D. H., 'Economics of the Marina'. Paper presented at the *Symposium of Marinas and Small Craft Harbours*, Department of Civil Engineering, University of Southampton, April 1972.

Sessions, D. H., 'Marinas', *Yachting World* June 1966.

Sessions, D. H., 'Yacht Harbours', *The Yachtsman* May 1966.

Shuer, Max, 'Getting to the Grass Roots of Dereliction', *Surveyor* December 1972, Vol. 140. *Deals with Department of the Environment grant assistance.*

13 Marina management

There are many sources from which a management organisation might spring and many agencies to whom it may be responsible:

1 A local authority committee answerable to its electorate
2 A management board answerable to a parent company
3 A public company answerable to shareholders
4 A private company answerable to the owner
5 A charitable trust answerable to a board of governors

Such boards are assembled in many ways—one method which preserves continuity from the inception is to appoint the nucleus of the board from the original research and construction team. The lawyer and accountant are candidates to succeed to management positions and the engineer and architect too may continue as advisers on topics such as expansion and maintenance. Certain specialists, concessionaires and councillors may all contribute valuable skills. Joint enterprise by private and public interests may generate a good balance of local interests and technical experience.

Whatever its composition the concern of the board will be to initiate and direct policy, ensuring that it is implemented speedily and well. A prerequisite will be the formation of clear objectives to be achieved by a planning and budgeting system within an agreed time period.

The principles applicable to good marina management are substantially the same as those for most concerns of a similar size and nature. Charles A. Chaney[1] lists the main factors as follows:

1 Efficiency of operation and management
2 Order and cleanliness of the marina
3 Character of the personnel
4 Quality and amount of services offered
5 Rates charged for sales and services

Decisions taken at the planning stage will have a great influence on the marina's day-to-day management. The right type of development suitable in scale and design, with well chosen equipment and services will, of course, be much easier and economical to operate, protect and maintain. Considering a marina company from its inception it is probable that some early problems will be as difficult as the management is ever likely to face and, unless the company is already established or the individuals have previously worked together, this early period can be strenuous. Marina development evolves from many different circumstances and the variations in size, type and management are wide, but in the progression of decision-making during the design and building stages they are likely to share a basic similarity. The 12 steps outlined in the Management Diary on page 321 chronicle some fairly typical items encountered during the preparation, development and post-

1. Chaney, Charles A., *Marinas: Recommendations for Design, Construction and Maintenance* (Second edition) National Association of Engine and Boat Manufacturers Inc., New York 1961.

construction periods. Two things are clear from such a list. Firstly, there must be a strong rapport between the management and the professionals in the development team. Secondly, the breadth of understanding required from management to control and direct such a programme must be considerable. The professionals' advice stems from their specialist knowledge of one field of activity, *but this must be evaluated by management in terms of all the aspects of the project*. To give some idea of the extent and breadth of the workload involved, the breakdown on page 336 lists, under three main headings, the sort of items likely at some time or another to find their way onto the agenda at management meetings. The list may seem formidable but, of course, in small businesses many items may not apply or can be bracketed together. Nevertheless, most managements would benefit by more frequent reference to such a table to ensure that all the relevant items are being adequately covered within their firms. It is not a case of over-

13.A Diary of marina management

Approximate order	Subject matter	Principal responsibility
Step 1	Formation of marina company	Management board
	Appointment of development team	Management
	Agree scales of fees, timing, etc.	Management and development team
	Individual and collective briefing	Management
Step 2	Initial appraisal	Development team
	Site investigations	Engineer, etc.
	Feasibility studies	Development team
	Outline planning application	Architect
	Detailed user requirements	Management
	Financial structure: fund raising, grants etc.	Management
Step 3	Outline proposals	Development team
	Preliminary layout and alternatives	Architect, engineer
	Initial estimate of costs	Quantity surveyor
Step 4	Final layout	Architect Engineer
	Phasing proposals	Architect
	Interim cost estimates	Quantity surveyor
	Approval in principle	Management
Step 5	Design presentation	Architect
	Public inquiry preparations	Lawyer & management
	Itemised cost check	Quantity surveyor
Step 6	Working drawings	Architect engineer
	Submit detailed planning application	Architect, lawyer
Step 7	Prepare bills of quantities and specification	Quantity surveyor
	Approval of contractor(s) for tendering	Management
Step 8	Approve contractor's programme	Development team and management
	Initial site works	Engineer
	Final pricing policy	Management
Step 9	Out to tender	Architect
	Selection and nomination of successful contractor(s)	Management and quantity surveyor

END OF PRE-CONSTRUCTION PERIOD

Approximate order	Subject matter	Principal responsibility
	START OF CONSTRUCTION PERIOD	
Step 10	Start work	Contractor
	Execution and supervision	Contractor, architect, engineer
	Progress reports	Development team
	Approval of variations	Architect
	Interim payments	Quantity surveyor
	Staffing interviews	Management
	Publicity and advertising	Management
Step 10a	Certificate of practical completion	Architect
	Site inspection	Development team
	Occupation of site }	Management and staff
	Marina operational }	
	Defects liability period begins	—
Step 10b	Defects and maintenance work (usually 6 months)	Contractor
Step 11	Defects liability period ends	—
	Final handover	Contractor, management
	Final account	Architect, quantity surveyor
	END OF CONSTRUCTION PERIOD START OF POST-CONSTRUCTION PERIOD	
Step 12	Post contract policy	Management
	First half-yearly report	Staff and management
	Decision on future of development team	Management
	Periodic progress reports	

complicating the issues but of simplifying them. It is a sad fact but research has proved that large numbers of managements neglect important aspects of their business and some cannot clearly identify any specific aims or objectives.

An anomaly in modern management, which particularly affects the development of new marinas, is that in an age of specialists there exists *a growing* dearth of generalists which is above all what a marina manager needs to be. The person contemplating a new development often finds that his horizons must extend beyond his own speciality to include most, if not all, the aspects of management set out above. It may be too much to expect most aspirant or actual new marina managers to be aware of all the exploding advances in the management sciences but on the other hand to assume that a background of the services, boatbuilding or salesmanship will provide sufficient expertise to ensure advancement on all fronts in a relatively new, difficult and complex field of activity is the sort of unfounded optimism which frequently leads to decline.

It is only possible here to deal with those aspects of management which bear directly upon marinas and their constituent parts. The management of components such as hotels or boat-building may be controlled by the marina company which would invite specialist members on to the board. Reference has been made in Chapters 3 and 12 to the wide range of uses and facilities

13.1 Some of the services offered at Mariners' Basin, California. Whilst the sign itself is over-large, it is at least informative and interesting

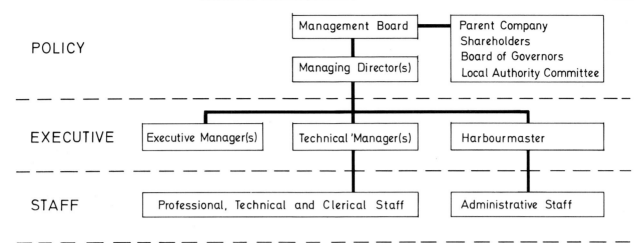

which may be ancillary to or part of the marina complex. The degree to which these auxiliary elements are allowed to grow and the manner and availability of their service is best kept under the control of the marina company. It is not uncommon for marina companies to sell part of their holding outright or to grant unconditional leases or concessions only to find that, having relinquished all control, the incoming company follows a policy in fundamental conflict to that of the marina itself. This can happen with any component, but seems most common with clubhouses. When it becomes an independent concern a clubhouse is sometimes at variance with the marina company, but more frequently, it goes into a decline from which, because of its autonomy, the marina management is powerless to rescue it.

The structure by which the marina is controlled and staffed will vary. The broad divisions of policy, executive and staff are outlined above. A review of all financial aspects is covered in Chapter 12 and it is only necessary here to reaffirm that the economic policy by which the firm is developed and run will be the single most important factor in its management.

Generalisations are not easy because of the very wide range of size and nature of marinas. The management of a small one-man business in a quiet rural area has little in common with a 3,000 berth complex offering everything from a choice of night clubs to a radar repair service, but what they do have in common is *that their success will be directly proportional to the quality of their management.* Nothing should deny success as being theirs nor, conversely, can any excuses by the management shift the responsibility from them in times of failure.

Book-keeping and records

According to the banks and credit companies 80–90 per cent of all business failures are due to inadequate record keeping. In America 52 per cent of all failures occur in retailing and 25 per cent in the luxury, hobby or recreational field. Any new business in America has only a 50/50 chance of survival, one out of three will last less than four years and only one in five celebrates its tenth birthday. British figures follow a similar pattern.

These figures are common knowledge in the financial world and are not a reflection of some recent trading slump for, with slight variations the situation has been going on for more than a century.

In recent years the University of Pittsburg has studied business failure, success and profitability. Its research shows that 90 per cent of good-profit firms keep proper records and 90 per cent of poor-profit firms do not. Similarly, 40 per cent of unsuccessful firms without adequate records were in trouble with the Inland Revenue, not for deliberate fraud however, for many did not claim legitimate allowances. The survey showed that 33 per cent of businesses had excessive fixed costs, a further 33 per cent were financially over-expanded and 50 per cent did not really know their actual market and type of customer. By far the biggest profit-reducing aspect and the cause of the majority of failures was the lack of proper records.

The leisure industry, perhaps because of its carefree context, often seems to disregard the basic principles of business efficiency whereas its susceptibility to the seasons, weather and other variables in fact demands a control tighter than in more conventional fields. Whilst good record-keeping cannot ensure success, failure to do so *nearly always* leads to inefficiency if not collapse. The following is a basic list of records which any company would be expected to keep:

1	Record of Cash Receipts	A daily account of money taken
2	Record of Cash Expenditure	A payments register for wages, petty cash etc.
3	Wages Book	Names, national insurance numbers, gross and net pay etc.
4	Account Book	Money owing from customers Goods owing from suppliers
5	Resources and Accounts Payable	How much is owed and to whom
6	Documentation File	Invoices, stubs, etc. in support of other records
7	General Ledger	The summary and integration of other records giving an accurate result of operations over a given period 　A running record of assets, liabilities and capital at any given time

These essential records are the eyes and ears of the business—showing the position of the firm at all times. Without them nothing is provable to the bank, the Government or creditors and with tax matters it must be remembered that the burden of proof is upon the management.

Providing the service

A factor of prime importance in marina management is to recognise that the sole purpose of a marina is to give a *service*. Service is the keystone of the business and all other efforts are embodied within it. The task of management is to provide this efficiently and economically for the boat owner, profitably for the company and sympathetically for the environment.

In a very real sense the boat owners and visitors themselves have a responsibility towards the harbour's success. Managers of 'open' or municipal marinas may be unable to exercise any control over the selection of those applying to rent a berth or join a club nor, perhaps, should they be encouraged to do so. In more private marinas and clubs an indirect selection does take place, if only by pricing and recommendation. How far management could or should interfere, particularly where a club is concerned in

13.2 Fishing is enormously popular and a 'natural' within any marina. Gear for sale or hire may be an obvious extension to chandlery

EARL MARKS
DIVER

PROPS PULLED • ITEMS RECOVERED
BOTTOMS CLEANED
BOTTOMS SURVEYED
TREASURES SALVAGED • LOW RATES

13.3 Management should encourage enterprise and offer all manner of services

13.4 April 7, warm and sunny—but not a boat in sight. Profits are strongly related to the length of season

restricting the membership for personal reasons is a very delicate matter but, in the great majority of cases it would seem most unwise to do so. Management restriction by the application of reasonable and agreed regulations is, of course, quite different and is dealt with elsewhere in this chapter (see page 333).

The list of facilities that can be made available is long (see Check List on page 149) and may be readily increased by including more specialised services. What is more difficult is deciding which to include and at what stage. The management will need to decide three questions:

1 What is possible?
2 What is popular?
3 What is profitable?

Such an examination should form part of the pre-development research programme because many facilities will be more economically installed (or provision made for them) during the construction period.

The provision of physical needs and comforts are specific and material but equally important are the less readily definable qualities brought about by the general attitude of the management and staff. It is generally held to be good practice to match the style of service to the style of business. This being so, *most marinas should offer a relaxed but disciplined air* such as one might find aboard a naval vessel during an open day. Marina patrons are usually rather individualistic people who spend their leisure in an unusual way. They would expect from the marina staff an attitude similar to that of a good crew. 'Cheerful efficiency' might be a good maxim with thoughtfulness as a welcome extra. A close relationship between management, staff, owners and visitors is really essential. This affinity is important in several ways:

1 It is much the best way of giving the development its individual character and personality

2 It allows management and staff to receive the credits and criticisms from patrons in the best possible way by allowing the boat owners constant access to the management and, conversely by keeping management in day-to-day contact with owners' opinions and ideas

3 It is the best possible form of advertising. The sailing community is remarkably close-knit and recommendations (or otherwise) spread quickly

4 Whilst staff should not intrude, owners should be made to feel that staff are available during reasonable hours and obtainable in an emergency

13.5, 6 2 stages in the Greater London Council's new town at Thamesmead. The first shows the 12 ha (30 acre) Stage I lake being dredged by drag line in June 1967. The second shows the same view in May 1970.
Thamesmead is 11·27 km (7 miles) down-stream from Westminster. This 12 ha (30 acre) man-made balancing lake provides for maximum safety by being only 1·4 m (4 ft 6 in) deep and gently shelving wherever the public have access. The lake is an engineering necessity, stabilising the level of the canal and ground-water drainage systems. The recreative provision shows opportunism at its best

Charging for services

The way in which payment for services is recovered from patrons will depend upon management policy. Some facilities such as small-craft launching ramps, self-operated winches and derricks may be free for everyone, berth holders, visitors and the general public alike. This may be because:

1   There is no practical means of enforcing charges
2   The management are pleased to offer free facilities which will encourage custom in other ways, publicise the marina and attract future berth holders and club members. Other facilities will operate on a different basis.

Boat stacker storage may (after a standing charge) allow one 'in' and one 'out' each day, after which a charge is made for each movement irrespective of craft size. Large craft may be launched or retrieved free of charge when undergoing a repair or refit. Otherwise owners or others would be expected to pay according to length or displacement. The arrangements will vary depending upon circumstances but a firm, suitable, simply and preferably unchanging system must be devised by the management which will need to ensure that the charges for services are tailored to fit the total economic structure of the development.

Staff

Page 307 depicts the type of staffing framework applicable to many marina companies. In small enterprises the heirarchy may be shortened by dispensing with the executive band and replacing their roles either with executive agents from staff section or specialist managers with executive duties.

Recruitment will normally be by advertising or by recommendation. As an occupation, marina work has much to offer by way of relative independence, an outdoor life, considerable variety, usually local employment and a pleasant waterside environment with, no doubt, ample opportunity for sailing. The level of remuneration varies but is usually rather lower (in the UK) than in the building industry for comparable work. Working clothes (including life jackets), or a clothing allowance are usually provided. Whilst a uniform, as such, may be unsuitable due to the variety of tasks, some identification or house style could be useful both for the convenience of the staff and patrons. In emergencies distinctly coloured overalls or some distinguishing motif would prove particularly helpful. Accommodation may be provided for a caretaker-handyman or club manager. Except for the club employees, staff numbers are usually fairly constant throughout the year, for most maintenance tasks are undertaken during the off-season months and boat repairs and re-fit services are often busier in winter than summer.

The number of staff employed is determined by the roles they play, the standard of service offered, the number of berths and the range of facilities. No marina company is likely to be big business, in the way that many manufacturing companies are. A marina with a permanent staff of 20 would be very large in Britain. Operating a marina involves hundreds of tasks of great diversity. With a small number of employees it is inevitable that they will each be expected to tackle a multitude of jobs. For many, both staff and management alike, this will be the main appeal of operating a marina. Work is unlikely to be repetitive—or not for long. Management will look for staff who are enterprising, hardworking and who can be left with confidence to work alone or together with the minimum of supervision. Some may be tradesmen, primarily carpenters and engineers, others will be general staff capable of a wide range of duties. Marina operation is not unlike a cross between crewing a passenger ship and running a building business and indeed staff are frequently drawn from these sectors for they are capable of filling successfully most of the roles in harbour operation, except perhaps the maintenance of the soft landscape.

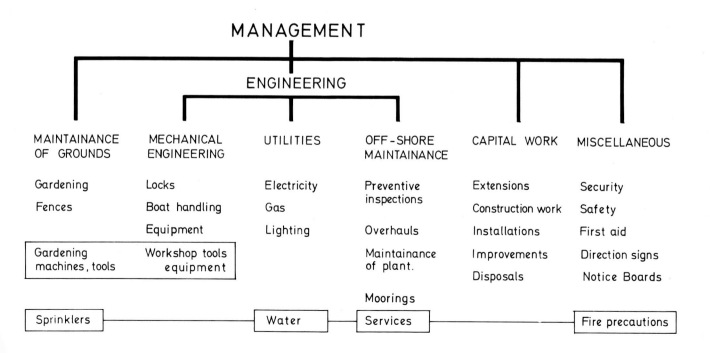

A breakdown of staff and their responsibilities in the marina would generally be as follows:

Management	Harbour master	Technical	Specialist	Social
Policy etc.	Lock operation	Maintenance	Sales	Clubhouse staff
See the	Craft movements	Engineering	Brokerage	
Management	Office	electrical	Insurance	
Diary on	administration	mechanical	Chandlery	
pages 305–306	Reception	Transport		
	Bookings			
	Rental collection			
	Publications			
	Information			
	Security			
	fire			
	accidents			
	general safety			

13.B Staff responsibilities

Union membership should be respected by management although a marina is an unlikely place to be able to enforce restrictive practices or rigid demarcation. The conditions of service, working hours and holidays etc. will be determined by management as in any other business. *Working in exposed conditions in winter weather will require the management to provide comfortable, warm staff rooms and offices for groundstaff, watermen and security personnel.*

Occupancy of berths

It is often said, and quite rightly, that a marina offering nothing but berths and moorings will fail. It is also true, however, that this basic facility still makes the largest single contribution to gross income and has the largest profit percentage of all the normally offered services. (see 12A and 12.3, page 301). Rental prices, the differential for varying berth sizes and seasonal adjustments, are all important to ensure that the pricing policy is keen but reasonable, in tune with charges in the surrounding catchment and commensurate with the breadth and quality of service offered.

As berth rentals play a key role in marina economics, evolving a policy to ensure maximum berth occupancy is important. This requires skill and experience. Harbourmasters will be able to record boat movements from day to day and therefore build a picture of occupancy rates on a monthly basis and compare their performance season to season or year by year, Perhaps operators could play the 'occupancy game' on the following allocation of marks:

1	A filled berth with an active crew	3 points
2	A filled berth with an absent crew	2 points
3	A long-empty but paid-for berth	1 point
4	A long-empty unpaid-for berth	0 points

The one point difference between the first and second is because active owners bring far more business than those who rarely appear. Reserving the right proportion of berths for visitors is another important aspect. Too few and the opportunity of introducing new custom will be lost. Too many can be a costly mistake if they are not taken up. It is a good principle to remember that:

1 Crews on the move spend more than crews at their home harbour
2 Families, friends and visitors are customers to be encouraged
3 Today's children are tomorrow's boat owners
4 Handicapped people are anxious to use and pay for facilities helpful to them
5 With demand at its present level there is no such thing as a rival marina

Every time a boat sails it leaves an empty berth. An understanding between the berth holder and the operator can benefit both, for if due notice is given, a rebate for each full week of absence during the active season (May–September inclusive) could be agreed allowing the operator to re-let the berth to visiting craft. The harbourmaster must put the terms (time and money) in writing to both owner and visitor. (See item 7, page 333).

Waiting lists

The waiting list is another tribulation. Too long a wait creates frustration and bad will but to refuse to accept additions may offend too. One marina operator has worked out that his *average* berth holders stays eight years and has confined his waiting time to a quarter of this—two years. As the marina has about 240 berths, the two-year waiting list has about 60 people on it.

Whatever system is adopted those waiting must be informed of their position at least every season and their interest in staying on the list—or otherwise—recorded. It may be tempting from time to time to allow friends or free-spending owners to jump the queue. This must be resisted. Even worse is accepting a premium to shorten or eliminate waiting time. People get to know of these things and, understandably, take offence.

A policy on 'inheritance' of berths must also be formulated. This sort of information needs to be clearly set down in the marina regulations, for although trivial rules should be avoided, a clearly stated policy prevents argument and possible litigation. (See page 333 for possible inclusions).

Security

Vandalism

Security is a major management responsibility. *This is largely what boat owners are paying for and loss or damage to craft and gear should be an acute embarrassment to the management which owes it to its clients to pursue a policy of effective and continuous vigilance.* Because of its watery and often remote position a marina and its craft are vulnerable to both natural and malicious damage. Protection from the elements is usually inherent in the design but safeguards against theft and vandalism are often poor. Boatyards are particularly susceptible to entry and with tools and materials lying about, the costs of damage and disruption can be very high. Good security should start at a deeper level than gates and barbed wire. Liaison with the police, the adequate lighting of grounds, the education of owners and reliable communications, both within the harbour and from it to the outside world, all form a sensible security groundwork. Security arrangements once put into operation will need a regular overhaul just as much as the equipment and buildings.

The presence of lock-keepers, harbour masters and customs officials help as a deterrent and so do night watchmen and security patrols. The dangers are as great in winter when things may be quiet as they are in summer.

Cutting boats adrift, 'entering' craft and joy-riding are frequent forms of nuisance and arson, boat scuttling and theft are also common. It is very difficult to control entry to the marina. Many genuine visitors may be unknown to the staff and with boating people one cannot go by looks. Security gates either to the grounds or to the berths are not really much use even when keys are limited to owners and staff only.

Security fencing needs careful detailing. Black plastic-coated nylon can be reinforced with back-up planting of small trees and shrubs, preferably evergreen.

Bearing in mind the increase in crime and vandalism and the particular vulnerability of marinas it is becoming increasingly necessary to adopt more modern systems of security control. The four principal methods are as follows:

1	External audible alarm	This covers many types of alarm and may be used alone or in addition to other means of protection
2	Transmission of signal to company's central station	An alarm signal feeds back to a screen within a central console
3	Sending pre-recorded messages to GPO telephone operators for transmission to police	A break in the circuit triggers a message to the GPO
4	Direct signals to police	As above but direct to the police station or patrol cars

Systems 1 and 2 require security staff whereas 3 and 4 can operate entirely automatically. With 2, 3 and 4 the area to be safeguarded is usually divided into separate circuits—buildings, grounds, moorings (or separate piers), boundary fence and so on, each of which can be separately identified to narrow the field of search.

The new type of sports and leisure centre is not dissimilar to a marina complex for it too has central buildings around which are located golf clubs and playing fields similar to a marina's grounds and moorings. Unlike many marina managements the directors of sports centres have recognised their security difficulties and have taken much more effective measures to counter the problem with a tight and sophisticated alarm system feeding back to a central screen. The system is often tied into the general electric wiring network of the whole complex. This allows individual failure of lights and equipment to be registered on the control console. Thus a maintenance check is linked into the security system. Efficiency is improved and staff time is saved because one person, centrally located, operates a total security and fault-finding coverage for the whole development.

Any security system should be integrated with the management's insurance policy and should, of course, reduce the premiums. Above all, the company should not skimp on insurance and during times of rapidly increasing costs, should frequently review their policy coverage. Management should also remind owners of *their* insurance responsibilities and the likelihood of decreased premiums once their craft are marina-moored. Not all accidents or other damage will be a management liability and in some cases it may be difficult to apportion responsibility. There is a growing course of action by marina managements to require potential patrons to sign statements absolving the company from security responsibilities before a berth is made over to them. This practice seems quite wrong except in very special circumstances and even when signatures are obtained it is most doubtful if the court would find in the company's favour. It is also meaningless for most management to write things in small print about not being responsible. A marina management *is* responsible or else it is taking boat-owners' money on false pretences. Management must also have a firm policy on prosecutions if culprits are caught. A good talking-to in the manager's office may be all that is needed in some cases but for others, the police must be contacted.

Fire prevention (for Fire *Fighting* see Chapter 8, pages 247–250)
Fire protection must be given serious thought from the outset. Buildings will need to comply with the regulations applicable to their category in terms of the following:
1 The means of escape in case of fire
2 The fire resistance of material
3 The provision of fire alarms
4 The type and location of fire-fighting equipment
5 The provision of sprinkler systems
6 The width, clearance and bearing pressure of fire paths
Water provision should be by 100 mm (4 in) mains with properly positioned and approved couplings. It is not enough to assume that the Fire Brigade's suction hose will be available. Side-by-side mooring makes fire-spread easier. However, boats are moored in marinas largely to protect them from damage and it is the management's duty to provide *and maintain* a first-class defence system. The local authority or fire service will usually determine the minimum requirements. Either a mains system of hydrants and hose reels or a mobile arrangement of extinguishers is usually adopted. In large marinas, fire patrol vessels and special staff are employed.

The following guide lines on fire protection in marinas are based on American experience of the problem:
1 Ensure that some member or members of the marina staff is responsible for fire prevention. It is better to have a rough and ready method which is understood and can be brought quickly into operation, than a complicated system for which no one is really answerable
2 Fire or explosion aboard boats moored in the marina is the biggest single danger. If attempts at immediate control fail, the vessel is best abandoned, and should, if possible, be cut adrift to burn out in a safe corner of the basin
3 Fuelling stations are danger points and permanent equipment for fighting petroleum or diesel fires should be readily accessible in these areas
4 Quick contact with the land and water-borne fire services is essential. In locked marinas special arrangements need to be agreed to prevent delay to fire-fighting boats
5 All equipment should be well maintained and tested regularly
6 Consult the local authority and the Fire Brigade at an early stage: their requirements may affect the design and layout of the marina
Marinas are very vulnerable to damage by fire. In 1973 in the United Kingdom there were nearly 1,000 fires recorded in ships, river craft and marine structures. Materials such as bottled gas, petroleum, polystyrene, sailcloth and timber are all commonly found in yacht basins and whilst water is the traditional means of extinguishing fires it can also act as an effective means of fire spread. It cannot be emphasised too strongly the

amount of fire damage that is caused by children. As a source of ignition of fires *in buildings* children are responsible for well over 20,000 fires a year, the next biggest source being cooking appliances (8,400). As to *outside* fires the dangers from children are even worse as they cause over 60,000 fires annually, the next worse source being rubbish burning (15,000). With their fascination for boats, marinas are doubly vulnerable to fires caused by children and whilst no one would want to see an unreasonable attitude, managements should take particular precautions against the dangers.

The above figures apply to the United Kingdom, but the pattern is much the same in most countries. At Essex Marina, north of New York, 83 boats were destroyed in 1966 and at Miami, 25 were destroyed in 1969.

Safety

Apart from security, fire-fighting and fire precautions it is important for management to consider safety in more general terms. Coverage of the problem may be divided into:

1 Preventive measures designed to minimise mishaps
2 Methods and equipment to deal with an emergency

Preventive measures should include:

1 Adequate lighting
  handrailing
  lifebelts
2 Non-slip surfaces
3 Internal telephones/public address systems
4 Speed limits for craft and vehicles
5 Advice to boat owners on marina equipment
  control of children
  wearing of life-jackets
6 Weather information service

Coping with accidents may necessitate:

1 Staff duty officer
2 First aid post or equipment
3 Fire-fighting equipment
4 Quick contact with doctor/hospital/ambulance service
                        Police
                        Lifeboat service
                        Lifeguards
                        Coastguards

Apart from dealing with emergencies within the marina boundary, management may be expected to cope with mishaps occurring outside, not only to its patrons but the public in general. Harbours have a traditional role of helping in an emergency by providing a means of contact, if not by providing the service themselves.

Marina regulations

Marina regulations should not be drafted as an arbitrary list of restrictions but should be carefully formulated by the management as an extension of its marina policy. Regulations should, of course, be kept to an essential minimum and be regularly reviewed. Few patrons would object to sensible rules and guide lines, but over-restrictive, unnecessary or outdated restraints are an annoyance. Forming a charter for a new marina is difficult because the best systems stem from operating experience. It is best, therefore, *to confine the rules of a new marina to an irreducible minimum sufficient to satisfy the needs of security, safety and insurance and then to subject them to frequent review until the development has settled into an established pattern.* Club rules may be different from those of the marina depending upon its degree of independence.

Marinas vary so much in their size, structure and objectives that each will need to evolve its own legislation, however, some of the more important issues that a management may wish to consider are as follows:

1 Compliance with marina regulations
2 Restrictions on various types of craft, e.g.
   houseboats
   commercial craft
   length overall (LOA)
   multi-hulled craft
   storage and mooring of dinghys
3 Compliance with navigation laws
4 Lock regulations
5 Operator's right to inspect
   seaworthiness and general condition of craft
   fire equipment aboard craft
6 Operator's acceptance of responsibility for
   damage to craft and property
   injury to persons within the marina
7 Operator's right to lease berths during absence
   prohibition (or otherwise) of sub-letting berths
8 General safety regulations
   notification of entry/departure from harbour
   lighting on entering/leaving or within the harbour
   speed restrictions within the marina
   safety precautions during refuelling
   control of bottled gas
   obstruction of walkways, footpaths etc.
   wearing of life-jackets
   supervision of children
9 Pollution control
   disposal of effluent, sealing of heads etc.
   disposal of rubbish
   spillage of oil
   fouling of grounds by animals
   noise: transistors, slapping halliards, bilge pumps, engines, generators,
   animals, children
10 Use of boat-handling equipment
11 Mooring elsewhere than at a berth
12 Living aboard or in vehicles (hire restrictions)
13 Control of repairs to craft within the marina
14 Restrictions on trading within the harbour
    ban or restrictions on advertising
15 Use of club premises

16   Disorderly behaviour

17   Payment of fees: how much, when and to whom

There is a need, however, to balance restrictions with helpful advice, for positive counsel makes much better reading than lists of rules. It is pointless for example, to forbid rubbish being thrown in the water unless the operator has provided adequate refuse points. The information useful to owners (and particularly visitors) will vary but might include the following:

1   What the marina has to offer

2   Information on weather and tides

3   Where to find local shops, doctor, dentist, postal services, telephones etc.

4   Transport information: bus and rail timetables, local cab services

5   Navigation chart of local waters (hung in club and information office)

6   Nearest service garage

7   Local entertainments, restaurants and pubs

Both the rules and the information should be presented in a clear and simple booklet and kept separate from other literature such as application forms. This is part of the marina image and its contents and design are very important. It is much better to consult a graphic designer than the local printer. At St. Katharine's Dock, London, one simple card suffices for membership, information sheet, guarantee, rule book and security aid, for without it a craft cannot leave the harbour—it is, so to speak the key to the lock. As well as general rules and local information it may be necessary for management to acquaint owners about security, fire prevention, handling bottled gas, and so on. Once again helpful and positive guidance is much more acceptable than a negative list of 'don'ts'.

Pros and cons of marina management

Chapter 12 has given an overall prospect of the present financial situation in the marina field but, as was emphasised there, it is not likely to return quick profits. For those contemplating development or expansion it may, therefore, be of interest to consider the implications not only of the present-day situation but the long term expectations. Commencing with the general situation in the mid-70s and looking towards the future the considerations seem to be as follows. The total number of staff is likely to be small, therefore, everyone *should* know everyone else. The location is generally

13.7 'Getting away from it all' at a Detroit marina. It seems unbelievable that management can encourage this kind of thing

13.8 Whilst regulations should be kept to a minimum, some control is necessary if standards are not to drop below a reasonable level

pleasant, the work congenial and the atmosphere pleasant by most standards. The industry is young and growing, the demand is rising and exceeds supply. Even competition is to be welcomed rather than feared, for to some extent, the very provision of moorings creates the need for more because the sport as a whole gains by an improvement in facilities, which in turn is likely to increase boat ownership.

As against these advantages the whole field in Britain is financially unproven, fiscal support is hesitant and the attitude of the public and officialdom is often hostile and restrictive. The comforting probability is, however, that while the advantages will probably continue for many years the disadvantages will, in time, probably lessen. Success demands a wide range of skills from management and social as well as business proficiency is essential.

America, so often the barometer of trends and promotions throughout the world, has, during the last twenty years, set a great pace in marina development. The growing interest in environmental issues will no doubt demand particular care from potential developers, but this is to be welcomed by those genuinely interested in leisure in general and water-based sports in particular. The energy crisis and the anxiety over resources in general should affect the demand for sailing and boating less than many pursuits as the consumption of fuel for boating is minute compared with other activities. However, some governments have restricted or banned certain leisure pursuits during crises (the Swiss banned pleasure boating in November, 1973 during an oil shortage). A fuel crisis may cause access to marinas by public transport to assume greater importance but, if air travel and other activities demanding a high consumption of energy are reduced, this would have little effect on the increase in leisure harbours whose catchment has never relied on a substantial proportion of long-distance travellers.

---

13. Marina management: Check list

New marina management

Determine overall aims and objectives
Convene board and consider company formation
Assemble project team and provide outline development brief
Prepare outline programme of procedure with team
Inputs and key dates
See the Management Diary, page 321

Book-keeping and records

Prepare list of essential records
Allocate personnel responsible for their upkeep
Establish regular audit

Providing the service

Prepare planned programme of provision for services and facilities linked to total budget
   estimates
Establish preferences from patrons and examine in light of profitability
Consider
   profit-making facilities
   loss leaders
Establish programme under
   facilities built during construction
   those added afterwards
Devise payment system for services
Monitor profitability of individual services

Staff	Establish initial and future staffing and prepare chart showing overall structure and responsibilities
	Draft the policy for recruitment
	salary structure
	training and promotion
	Settle proposals for PAYE
	insurance
	sick pay and pensions
	bonuses and holidays
	provision of clothing
	accommodation
Security	Devise (with police or specialists) total security framework including fire precautions
	anti-vandal measures
	general safety, etc.
	Integrate the above with insurance arrangements
	Consider electronic central control system
	Allocate staff responsibilities and, if necessary, consider training scheme
	Ensure regular review and overhaul of security system including lighting
	patrolling
	security fences
	alarms
	Determine prosecution policy
Fire prevention	Consult local fire service
	Decide on mains system for grounds and moorings
	Review provision of sprinklers etc.
	Consult local bye-laws on means of escape
	Train staff and allocate responsibility
	Acquaint owners of dangers
	initiate emergency drill
	Discuss fuel station with petroleum company
	Ensure swift reliable communication to emergency services
	Draft maintenance programme for all fire equipment
	Link fire prevention with overall marina security
Marina regulations	Formulate regulations as extension of management policy
	See list on page 333
	Review regulations at periodic intervals
	Consider information booklet

13. Management: Breakdown of fields of marina management responsibility

The following list brings together under 3 main headings some of the matters with which a marina management should expect to deal over the years. Many subjects will, of course, merge and cross from one heading to another. The order of presentation does not imply relative importance nor any chronological order.

1 Financial	Capitalisation, shares
	General financial management
	Economic environment and forecasting
	Accounts/record keeping
	Fund raising/short term/long term
	Development loans/grants
	Pricing policy/reviews
	Return on capital/cash flow analysis
	Salary administration/rating
	Taxation
	Insurance management
	Depreciation/amortisation and maintenance policy
	Auditing/external and internal

2 Technical

Liaison with:
    development team
    manufacturers/suppliers/agencies
    competitors
    local authority and utility companies
    boat owners
New techniques/equipment/buildings/materials
New trends/demands
Suggestions/complaints
Improvements/expansion
Safety/security/noise/cleanliness
Maintenance

3 Managerial

Basic management concepts/aims/objectives/targets
Corporate planning/organisation/mergers/acquisitions
    long-range planning/market research/business forccasting
Business logistics/warehousing/materials/handling
    inventory of supply and control
General level of service/efficiency/information for owners
Advertising/publicity/public relations
Personnel administration/labour relations/staff training and development
Legal problems
Office administration

---

13. Marina management: Bibliography

Bates, James, *The Financing of Small Business*, Sweet & Maxwell Ltd., London 1963. Second edition 1971.

Beazley, Elizabeth, *Designed for Recreation. A Practical Handbook for all concerned with providing Leisure Facilities in the Countryside*, Faber & Faber Ltd., London 1969.

Bolton Committee, The, *Report of the Committee of Inquiry on Small Firms*, Cmd. 4811. H.M.S.O., London 1971.

Bolton Committee, The, 'A Postal Questionnaire Survey of Small Firms: Non-Financial Data, Tables, Definitions and Notes', *Bolton Committee Research Report 17*, H.M.S.O., London 1971.

Brech, E. F. L. (ed.), *Principles and Practice of Management*, Longman Ltd., London 1963.

Brough, R., 'Business Failures in Britain', *Business Ratios*, Summer 1970.

Bruno, H. A. and Associates Inc., *Marina Operations and Service*, National Association of Engine and Boat Manufacturers Inc., New York 1967.

Bryans, J. R., 'Requirements of the Yachtsman'. Paper presented at the *Symposium on Marinas and Small Craft Harbours*, Department of Civil Engineering, University of Southampton, April 1972.

Bûgg, D. E., *Burglary Protection and Insurance Surveys* (Second Edition), Stone and Cox Ltd., London 1966.

Confederation of British Industry, *Britain's Small Firms, their Vital Role in the Economy*, C.B.I., London 1969.

Confederation of British Industry, *Problems of Small Firms*, C.B.I. Evidence to the Bolton Committee, C.B.I., London 1970.

Conover, H. S., *Grounds maintenance handbook* (Second Edition), F. W. Dodge Corporation, New York 1958.

Department of Education and Science, *Planning for Leisure*, H.M.S.O., London 1969.

Economists Advisory Group, *Financial Facilities for Small Firms*, Committee of Inquiry on Small Firms, Research Report 4, H.M.S.O., London 1971.

Hackett, B., 'Maintenance Costs and Landscape Design', *Municipal Journal*, 6 March 1953.

Heyel, Carl (ed.), *The Encyclopedia of Management*, Reingold Book Corporation, New York 1968.

Inselberg, Henry S., *Accounting Manual for Marina and Boatyards* (Fourth Edition), National Association of Engine and Boat Manufacturers Inc., New York 1963.

Jones, Glyn P., *A New Approach to the Standard Form of Building Contract*, Medical and Technical Publishing Company Ltd., Oxford and Lancaster 1972.

Laird, Guy, 'Managing for Better Profits—Records and what they Mean to You', *International Boat Industry Magazine*, August 1971.

Langdon-Thomas, G. J., *Fire Safety in Building: Principles and Practice*, A. & C. Black Ltd., London 1972.

Marina Merchandising: The Marina Operator's Magazine (Bi-monthly), Trend Publications Inc., Pamph, Florida.

Matthews, Tony and Mayers, Colin, *Developing a Small Firm*, British Broadcasting Corporation, London 1968.

Molyneau, D. D., *Working for Recreation*, Town Planning Institute Journal, April 1968.

National Association of Engine and Boat Manufacturers, *Some Boat-Owner Impressions of Marina Services*, The Association, New York, September 1967.

Parfitt, R., *Management Information Part I: Regular Statements of Profit and Loss; Part II: Budgetary Control*, Information Circular No. 19 Parts 1 and 2, *Available in the Department of the Environmental Library*.

Pritt, D. N., *Employers, Workers and Trade Unions*, Laurence & Wishart Ltd., London 1970.

Sports Council, *Planning for Sport—a Report*, The Council, London 1968.

Urban Land Institute, *Planning and Developing Waterfront Property*, Technical Bulletin No. 49, The Institute, Washington D.C. 1967.

Ward, Colin (ed.), *Vandalism*, The Architectural Press Ltd., London 1973; Van Nostrand Reinhold Co., New York 1973.

14 Legal considerations

1. Moore, Stuart, *History of the Foreshore* (Third edition).

An accurate legal interpretation of the word foreshore is "that part of the shore between the high and low water mark of ordinary tides".[1] Seashore on the other hand means "the bed and shore of the sea and of every channel creek bay or estuary, every cliff, bank, barrier, dune, beach, flat or other land adjacent to the shore and every river as far up that river as the tide flows".[2]

2. Roddis, Roland J., *The Law of Coast Protection*, Shaw & Sons Ltd., London 1950.

Despite these fairly precise definitions the words foreshore and seashore are often interchanged although the latter usually implies an area greater than the tidal limits. Generally speaking the ownership of the foreshore is vested in the Crown but quite often local authorities own areas or have stretches leased to them by the Crown.

The shore can be privately owned with the right to exclude the general public except for navigation and fishing purposes—these being common law rights, and the owner cannot make any claim which interferes with the public's enjoyment of them. Sometimes the local authority will have owned the land for centuries but more often they will have acquired it under Section 164 of the Public Health Act 1875 or by private acts.

Acquisition and dedication

Land purchased under the 1875 Act must be dedicated for public use but bye-laws can be made regarding recreational uses such as bathing, sailing and so on. In other words legislation to allow bathing to the exclusion of water ski-ing or vice versa. In more recent times extensions of foreshore use for recreations are often made by access orders under the National Parks and Access to the Countryside Act 1949.

As far as acquisition is concerned there seems no reason why land should not be acquired and regulated under Section 164 of the 1875 Act, but in practice authorities seem to prefer to do this under a special act of their own or a Parliamentary Bill.

The freehold purchase of land from the previous owner sometimes only involves the area above high-water level and a separate agreement with a different party may be necessary for a long-term lease on both the foreshore and perhaps the bed below low-water mark, extending to the off-shore limits of the proposed development.

With peripheral or boundary watercourses the coastal owner's holding usually extends to the *further edge* of the ditch beyond, unless the title deeds specifically state otherwise.

The law about acquisition, bye-laws and the constraints that may be put upon the public are by no means plain sailing and *any authority or individual who is not absolutely sure of the subject would be wise to take experienced legal advice.* The public can, understandably, be very jealous of its rights and any local authority will need to be most careful about what it is acquiring (and *for* whom) and what it is selling (and *to* whom). It is easy by acting in good faith to improve the situation for one section of the community only to deprive another section of its rights.

Marina development often involves a complicated mixture of interests, both social and financial, and where private development is concerned the local authority (and the private developer) must be quite sure and clear about the freeholds, leases and concessions that they grant or receive.

Local authority powers

3. See also the *National Parks and Access to the Countryside Act* 1949, of which the 1968 Act is an extension and which it partly repeals.

In the past local authorities have had insufficient powers to provide the quays, slipways and hardstanding necessary for sailing and boating. Even when the council is itself a harbour authority it is not always easy to regulate its bye-laws to cover recreational uses. Section 8 of the *Countryside Act* 1968[3] helps them to overcome some of these problems because, in simple terms it:

1 Allows the local authority to provide facilities and services for sailing boating, bathing and fishing where water already exists in a country park

2 Allows local authorities to do this where the park adjoins the sea or a waterway

3 Allows local authorities to erect buildings and jetties and do other work outside the park boundary on land adjoining the sea or other waters

4 Anything local authorities propose doing under 1, 2 and 3 must be with the consent of any river or other authority having a function relating to land or water within the area

5 Having done the work the local authority may then make the necessary bye-laws but must first consult the Countryside Commission

6 Nothing in 1–5 above must contravene Section 34 of the *Coast Protection Act* 1949 (Works Detrimental to Navigation) or Section 9 of the *Harbours Act* 1964 (Control of Harbour Development)

Whilst the above gives useful *enabling* powers an equally good law of *restraint* is Section 82 of the *Public Health Acts Amendment Act* 1907 which allows the local authority to make bye-laws for the prevention of danger, obstruction or annoyance of persons using the seashore where an order of the Home Secretary has been made.

Main planning legislation and directives

In addition to the Acts already mentioned the following is a brief guide to the principle planning legislation most likely to be relevant to marinas and associated developments:

Acts
National Parks and Access to the Countryside Act 1949
The purpose of this Act is to preserve the countryside whilst making it available to townspeople (only Part III applies to Scotland):
Part I Provides for setting up the National Parks Commission
Part II Gives the Commission powers to designate National Parks
Part III Provides for establishing nature reserves and the Nature Conservancy
Part IV Deals with rights of way, footpaths, bridle ways (other aspects are largely superseded by the *Highways Act* 1959)

Part V Deals with access to cliffs, foreshore, banks and dunes
Part VI Deals with areas of outstanding natural beauty. Improvement to derelict and neglected land

Local Authorities (Land) Act 1963
Extends powers to acquire land by agreement and improve and use derelict and neglected land

Town and Country Planning Act 1962 (not Scotland or Northern Ireland)
Section 4 Deals with surveys of planning areas
Section 12(b) Indicates that depositing waste material generally denotes development or agricultural tenants
Section 16 Deals with the notification of owners and agricultural tenants of planning applications
Section 23 Deals with appeals against planning decisions
Section 29 Deals with Tree Preservation Orders
Section 155 Extinguishment of public rights of way
Section 199 Powers relating to Crown lands.

Relevant ministry circulars

27/'53   *Tree preservation orders*
36/'56   *Trees* Encourages councils to seize redevelopment opportunities to introduce more trees
39/'61   *Rivers (Prevention of Pollution) Act* 1961 Deals with discharge of effluent into rivers. Section 9 applies to tidal waters
20/'61   *Pollution of water by Tipped Refuse*
52/'62   *Liaison Between Planning Authorities and River Boards* Indicates that development must take drainage and flooding problems into consideration
10/'63   *Rivers (Prevention of Pollution) Act* 1961 Discharging sewage or trade effluent without the River Board's consent is a punishable offence (see S.I. 1963 No. 320 and 322)
39/'63   *Treatment of Derelict Land* See handbook *New Life for Dead Lands*
45/'63   *Rights of Way: Effect of Development* Indicates that planning approval does not entitle a developer to obstruct rights of way
49/'63   *Disposal of Surplus Government Land*
56/'63   *Coastal Preservation and Development* Indicates that the special amenity of the coast must not be spoilt by recreation
55/'64   *Derelict Land* Indicates that local authorities must obtain information for the Minister and keep it up to date
1/'66   *Tree Preservation Orders*
7/'66   *The Coast* Deals with long-term policies for safeguards and promotion of enjoyment of the coast by the public. Coastal preservation policy maps are to be prepared
28/'66   *Elevational Control*
57/'66   *Surplus Land* Indicates that surplus land owned by Government should be offered to former owners, then local authorities to ensure its use in the public interest
17/'67   *Rehabilitation of Derelict, Neglected or Unsightly Land*
59/'66   *Derelict Land* Extends 55/'64.

14.1, 2, 3 Three design phases of Thamesmead Marina, London, to illustrate the care taken to integrate a marina into a high-density urban framework

14.1 This plan, one of a series at different levels, shows the proposed basin in its town-centre context. The central area is envisaged as a Y shape, the arms of which embrace the harbour

Key

1 Car park
2 Boatyard
3 College of further education
4 Dance hall
5 Theatre
6 Boat showroom
7 Boat chandler
8 Information
9 Hotel
10 Evangelical project
11 Boat hire
12 Sports centre
13 Main health centre
14 Ecumenical project
15 First school
16 Mixed upper school
17 Light industry
18 GPO Sorting office
19 Garage
20 Polytechnic
21 Theatre

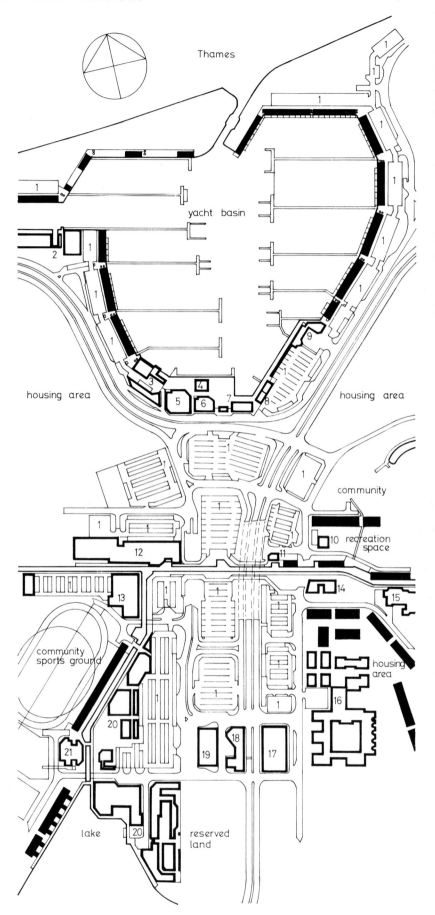

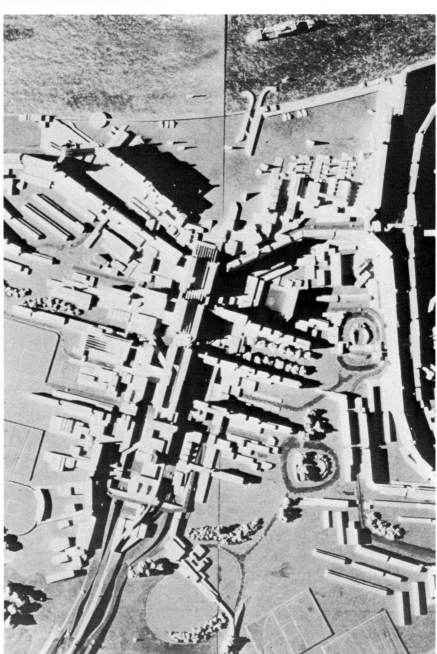

14.2 This vertical view shows an earlier sketch model using the sculpting technique not as an end product but as a working tool in the design process

14.3 Another polystyrene model with residential blocks curving around the harbour. This model and the plan were colour-coded to denote functions and uses

Development plans
Report of the Planning Advisory Group (The 'PAG' Report) 1965. This report inter alia advocates more purposeful planning of rural and recreational areas

Regional studies (economic)
Department of Economic Affairs, *The North West*, HMSO, London, 1965

Department of Economic Affairs, *A strategy for the South East*, HMSO, London, 1967

Department of Economic Affairs, *The West Midlands*, HMSO, London, 1965

East Midlands Economic Planning Council, *The East Midlands Study*, HMSO, London, 1966

Ministry of Housing and Local Government, *The South East Study 1961–1981*, HMSO, London, 1964

The North East: a Programme for Regional Development and Growth, Cmnd. 2206, HMSO, London, 1963

Northern Economic Planning Council, *Challenge of the Changing North*, HMSO, London, 1966

South East England, Cmnd. 2308, HMSO, London, 1964

West Midlands Economic Planning Council, *The West Midlands: Patterns of Growth*, HMSO, London, 1967

Yorkshire and Humberside Economic Planning Council, *A Review of Yorkshire and Humberside*, HMSO, London, 1966

Development applications
Generally speaking all new development requires planning permission. This includes engineering, mining, dredging, tipping and new buildings, extensions and structural alterations. Planning application on official forms need to be submitted to the local planning authority—usually a county borough or district council in whose area the proposal is situated. The district council will pass the application to the county office with its comments and recommendations, except where it is able to deal with it under delegated powers. This would not apply to most marina developments for they are unlikely to be designated as such within the Statutory Development Plan and, as a departure from it, would be handled by the county council.

Whilst it is useful for any developer to know about planning legislation it is much more important for him to approach the planning authorities at local and county levels and establish a working relationship as early as possible. Whilst council officers cannot anticipate their committees' decisions, if the developer's case is straightforward, he will soon be told if it is *likely* to be approved or not, where the difficulties lie, what the objections are and the need to alter and improve etc.

If sufficient progress is made the planning authority will usually advertise the proposal and invite public comment both from individuals and relevant groups and organisations. The application may be for outline or full permission. The former will require draft plans of the area involved and will, if accepted, receive approval *in principle*. Full approval is given when details of density, building form and location, materials and planning are agreed.

Relevant authorities
The local authority can help the potential developer by taking responsibility

for informing and inviting comment from relevant bodies which may include the following:

1   The Department of the Environment

The Secretary of State has overall responsibility for development. His inspectors hear any public inquiry

2   District Council

Will probably invite comment from the rural district or parish councils

3   Harbour Authority

Gives a local view of navigational harbour development and conservancy matters

4   Department of Trade and Industry

Comments upon navigation and coastal protection

5   Crown Estate Commissioners

Concerned with ownership and protection of the sea bed on estuaries and coasts

6   River Authorities

Concerned with matters affecting the regime, drainage, flood prevention and pollution of banks, rivers and estuaries

7   Countryside Commission

Determines policy for rural protection and designation of heritage coasts

Local consultations, often within the county, may involve the highways, public health, police and fire departments. Further important bodies, particularly for inland sites, include local water boards and reservoir authorities, the Home Office which has overall responsibility for water safety, the British Waterways Board whose Divisional Office controls canals, locks and towpaths and the Board of Trade which are responsible for all maritime matters and the licensing of craft carrying 12 or more passengers for hire or reward.

Inquiries and conditions

If difficulties arise between authorities concerning all or parts of the application or considerable public opinion is mounted against the proposal, it is generally 'called in' by the Secretary of State for the Environment. He decides either to leave the decision to the local authority or to hold a Public Local Inquiry.

Whether an Inquiry is held or not, an approval may contain conditions to which the developer is obliged to agree before work can proceed. Such conditions may be fairly trivial but can sometimes be onerous enough to put into question the viability of the project. Aspects most likely to cause concern in this way are:

1   Restrictions upon area or numbers of berths
2   Building height or bulk
3   Limits upon future expansion
4   Car parking requirements
5   Statutory services
6   Public health demands
7   Questions of coastal or waterside protection

Adequate road access is another common pre-requisite particularly with remote sites and negotiations may be necessary with the local highway authority. Some local authorities have designated certain stretches of coast or waterside as boating areas within which marina development is considered favourably. Heritage coastal areas figure in Chapter 2 and county planning officers would respect the Countryside Commission's delineation.

Evolving a boating policy

Local authorities at all levels should initiate a monitoring system which will enable them to review the situation and its policy implications continuously not only within its own area but on a regional basis. In this way information upon leisure in general and boating in particular would be brought together with that from neighbouring towns or counties, making it possible to form a comprehensive picture for the whole region.

The size and distribution of moorings governing the number of participants in boating and allied activities is a fundamental factor in planning the future of any region. Within this framework of anticipated growth in moorings the plan should be twofold—to provide the greatest capacity commensurate with the plan as a whole and to direct this effort to the areas of greatest need and highest potential. Because of deference to protected waterside and the prior development of the more obvious or desirable sites, there are in Britain already difficulties in obtaining approval for marina development.

The answer may lie in the overall upgrading of those waterside areas presenting a wasteful muddle of poor employment or subsistence farming or fishing: what Reyner Banham calls "the limbo-land of agro-industry".

It is in these areas that public and private finance and expertise could most usefully combine to create the improvements so necessary if the leisure explosion is to be contained and guided to give the maximum benefit in areas of permitted development and ensure protection to areas offering the greatest landscape value. There is no doubt that the financial involvement of a local authority will add confidence to any investment company which is considering funding a development in a 'speciality' field.

---

14. Legal considerations: Check list

Legal advice
Contact and retain experienced, specialist, professional legal adviser

Basic information
Record names, addresses etc., of persons and organisations concerned in the project including
  client
  agent
  vendor
  tenant
  solicitors
  authorities
  development team and advisers
Record site address and map reference

Authorities
Establish which authorities and bodies have a consultative interest in the development
  local authority engineer, surveyor, planning officer
  public health inspector
  county planning department
  Department of Trade and Industry (local office)
  Catchment Area Board
  river, harbour, coastal, canal, water authorities
  National Trust
  Countryside Commission
  Crown Estate Commissioners
  Nature Conservancy

Statutory undertakings

Contact
    gas
    electricity
    water boards
    GPO
    police
    fire services
Consider sewage disposal system, type of connection etc.
Examine retention and use of existing service runs

Background information

Establish the necessary rights of access for all who need entry
Determine client's liabilities as regards site
                        buildings
                        effects
Insure site, buildings etc., from time of purchase
Give notification of intention to develop to
    tenant
    adjoining owners etc.
Check for hidden charges, liabilities etc.
Obtain from previous owner or solicitor existing documents and drawings pertaining to
    site or its surroundings, e.g.
        buildings
        drainage
        surveys
Investigate
    all town planning permissions
    zoning
    land use constraints
Establish the local authority's boating policy

Legal inquiries

Determine any general factors which might affect development
Determine lease
    period
    responsibilities
    reversion
    ground rent position
    rates
Determine sublease position as above
Investigate any restrictive covenants e.g.
    restrictions on use type
    form of buildings
    rent of buildings
    tenure
    sale
Contact Land Registry
    registration in new ownership should be completed
Investigate bye-laws in addition to Building Regulations especially
    waterside changes
    coastal clauses
    riparian clauses
    public access to waterside
    safeguards for fishing, navigation
Discuss rights of way with local authority e.g.
    termination
    realisation
    relocation
Are any existing approvals already granted?
Consider way leaves: public utilities may have established powers for
    cables
    pipes
    lines
    rights of entry
Establish rights of light and air
Establish boundaries of site

Planning permission	Establish current position
	Investigate local and national waterside and coastal preservation policies
	Is any change of use envisaged?
	Determine information necessary for outline permission
	Determine information necessary for full planning consent
	Check on
	building lines
	improvement lines
	restrictions on access
	planning requirements: density
	plot ratio
	building height—restrictions
	parking standards
	light angles
	Investigate possible orders e.g.
	historic schedules
	conservation orders
	tree preservation orders
	Will an Industrial Development Certificate be required?
	Will an Office Development permit be needed?
	Confirm submission dates for planning and building regulations
Grants	Explore possibilities of grant aid for
	industrial development
	historic buildings (maintenance)
	derelict land restoration
	recreation in the countryside
	leisure provision
	flooding and sea defences
	housing improvement
Further site information	Examine
Sewers	position
	depth
	fall
	age
	material
	connections to existing sewers
	If none, will authority accept cesspits?
	If so what
	type
	size
	collection arrangement
	Establish whether any trade effluent is likely
Road adoption	Have access roads been adopted?
	If not, establish
	what standard is needed for adoption
	what cost
	what proportion will be borne by authority/developer
General	Check on mining subsidence
	tunnels
	water table
	flooding
	pollution
Fences and walls	Establish
	condition
	ownership rights of support
	Record condition
Party Wall Notices	Give notice to adjoining owners of intention to carry out work
Adjoining buildings	Should schedules of dilapidations be prepared and photographs taken?

Smokeless zones	Remember these can influence fuel type and methods of heating
	Establish all boundaries and ownerships of
	seashore
	foreshore
	off-shore tidal limits
Specialist surveys	Consider
	soil investigation and analysis
	climatic survey
	noise report
	market and amenity survey and report
Public inquiries	Is an inquiry unavoidable?
	If an inquiry is unavoidable, retain and brief counsel
	Establish scope and nature of inquiry
	Become acquainted with inquiry procedure for all stages
	Obtain opposition evidence as early as possible
	Verify date of hearing
	Prepare programme of work
	Assemble maps, drawings, models, etc.
	Call coordination meetings of inquiry team and witnesses
	Estimate and allow for costs involved
	Find out likely date of inspector's report
	Determine a course of action if
	development is approved
	development is rejected
	if conditions are attached
	If conditions are attached, prepare feasibility report on their effect upon the development
Construction	Examine credentials of all tenderers
	Establish any general or special requirements for foundations
	structure
	waterproofing
	insulation
	Establish any design requirements on
	height
	form
	projections beyond waterline
	fenestration
	materials
	Agree responsibilities regarding the organisation and management of the contract between
	architect
	quantity surveyor
	contractor
	client etc.
	Supervise tender procedure and examine all contract documents
	Consider arrangements for overall job control and site supervision
	Check completion procedure and maintenance arrangements
Management	Advise on responsibilities for safety of
	persons
	craft
	equipment
	Check insurance position as regards the marina
	equipment
	accidents to staff, owners and visitors
	Review the marina handbook on rules and regulations

14. Legal considerations: Bibliography[1]

1. For main planning legislation and directives see pages 340–341.

Bigham, D. A. *The Law and Administration Relating to Protection of the Environment*, Oyez Publishing Ltd., London 1973.

Campbell, Ian, *Law of Footpaths: A Practical Guide*, The Commons, Open Spaces and Footpaths Preservation Society, London 1972.

Clarke, John J., *The Gist of Planning Law. A Guide to the Town and Country Planning Acts 1962 and 1963 and Cognate Legislation* (Second edition), Macmillan Ltd., London 1968: St. Martin's Press Inc., New York 1968.

Heap, Desmond, *Outline of Planning Law* (Fifth edition), Sweet & Maxwell Ltd., London 1969.

Keating, Donald, *Law and Practice of Building Contracts including the Law Relating to Architects and Surveyors*, Sweet & Maxwell Ltd., London 1969.

Ministry of Housing and Local Government, *Planning Appeals: A Guide to Procedure*, H.M.S.O., London 1969.

Moore, Stuart, *History of the Foreshore* (Third edition),

Noise Abatement Society, *The Law on Noise*, The Society, London 1969.

Pritt, D. N., *Employers, Workers and Trade Unions*, Lawrence & Wishart Ltd., London 1970.

Roddis, Roland J., *The Law of Coast Protection*, Shaw & Sons Ltd., London 1950.

Roddis, Roland J., *The Law of Parks and Recreation Grounds* (Third edition), Shaw & Sons Ltd., London 1970.

Whittaker, Chris, Brown, Peter and Monahan, Jane, *The Handbook of Environmental Powers*, The Architectural Press Ltd., London 1976.

Wisconsin Department of Resource Development and State Recreation Committee, *Conservation Easements and Open Space Conference*, Madison 1961.

Wisdom, A. S., *The Law on the Pollution of Waters*, Shaw & Sons Ltd., London 1966.

Index

Figures in italics refer to pages in which illustrations occur and to their captions. Illustrations and diagrams have been indexed selectively. The contents of check lists and bibliographies have not been indexed.